Challenges in Physics Education

Series Editor

Marisa Michelini, Dipartimento di Fisica, University of Udine, Udine, Italy

This book series covers various aspects of physics teaching and learning across all educational levels and learning environments. The individual volumes cover a wide array of topics, including innovative approaches and pedagogical strategies for physics education, the development of effective methods to integrate multimedia into physics teaching and learning, pioneering lab experiments, and the utilization of web-based interactive activities. Selected research-based contributions will be prominently featured throughout.

The GIREP, the International Research Group on Physics Teaching, collaborates on this publication series, incorporating selected papers from globally recognized experts and monographs. Book proposals from other sources are also encouraged.

"Challenges in Physics Education" is targeted towards professionals, teachers, researchers, instructors, and curriculum developers, aiming to enhance physics teaching and learning and, consequently, the overall perception of physics within society.

Proposals for publication should be submitted to Marina Forlizzi (marina.forlizzi@springer.com).

Zuzana Ješková · Jozef Hanč
Editors

Advancing Physics Education

Innovations from Early Learning to Higher Education

Springer

Editors
Zuzana Ješková
Faculty of Science
Pavol Jozef Šafárik University
Košice, Slovakia

Jozef Hanč
Faculty of Science
Pavol Jozef Šafárik University
Košice, Slovakia

ISSN 2662-8422 ISSN 2662-8430 (electronic)
Challenges in Physics Education
ISBN 978-3-031-94206-8 ISBN 978-3-031-94207-5 (eBook)
https://doi.org/10.1007/978-3-031-94207-5

This Springer imprint is published by the registered company Springer Nature Switzerland AG
The registered company address is: Gewerbestrasse 11, 6330 Cham, Switzerland

Preface

Physics education plays a pivotal role in shaping the scientific literacy of individuals and fostering critical thinking, problem-solving skills, while deepening an appreciation for the natural world. As the world undergoes rapid scientific and technological advancements, the methods used to teach physics must evolve accordingly. Traditional instructional approaches, while foundational, often fall short in engaging students and fostering deep conceptual understanding. The integration of new technologies, interdisciplinary methodologies, and inquiry-based learning techniques has the potential to revolutionize physics education, making it more accessible, interactive, and relevant to students' lives.

The **GIREP-EPEC 2023 conference** provided a unique forum for educators, researchers, and practitioners to explore innovative strategies, share best practices, and discuss the challenges and opportunities in teaching and learning physics at all educational levels.

This book, *Advancing Physics Education: Innovations from Early Learning to Higher Education*, presents a carefully curated selection of contributions from the conference that reflect these innovative efforts, showcasing how educators are leveraging modern tools and pedagogical frameworks to enhance student engagement and learning outcomes. The 17 papers included in this volume were selected through a rigorous double-blind review process with each paper evaluated by at least two reviewers from among the 65 papers submitted to the GIREP-EPEC conference on Physics Education, held in Kosice, Slovakia, from July 3 to 7, 2023.

The volume is organized into four main parts, each highlighting a distinct facet of physics education:

1. *Keynote Contributions*—Featuring insights from leading experts in physics education, this part discusses contemporary challenges and opportunities in the field, including the role of artificial intelligence, the integration of current physics research into curricula, and strategies for addressing the global shortage of physics teachers.

2. *Pre-school and Primary Education*—Early exposure to physics concepts lays the foundation for future scientific inquiry. This part presents research on innovative ways to introduce young learners to physics through storytelling, drama, and hands-on exploration of fundamental concepts like magnetism and nanoscience.
3. *Secondary Education*—At this stage, students encounter more complex physics topics that require engaging and effective teaching strategies. Papers in this part focus on the teaching of measurement uncertainties, spectroscopy, quantum mechanics, and radioactivity, highlighting the use of digital tools, experimental activities, and immersive learning experiences.
4. *Pre-service Teacher Education and University Education*—The final part highlights the critical role of teacher education and university-level physics instruction. Topics include professional development programs for physics teachers, interactive engagement in university courses, the redesign of laboratory courses, smartphone-based experiments, alternative approaches to teaching quantum entanglement, and the use of extended reality technologies in physics education.

Through these diverse contributions, this volume underscores the evolving nature of physics education and the necessity of evidence-based, innovative practices to ensure effective teaching and learning. By bridging research and practice, these studies offer valuable insights for educators, researchers, and policymakers striving to advance physics education from early learning through higher education.

Košice, Slovakia

Zuzana Ješková
Jozef Hanč

Contents

Keynote Contributions

Challenges and Selected Solutions in Physics Education Generated by Artificial Intelligence

Ľubomír Antoni and Gabriel Semanišin

Abstract Artificial intelligence (briefly AI) is an area of computer science that aims to create intelligent machines that can perform tasks that typically require human intelligence, such as decision-making, visual perception, speech recognition, and natural language generation. We present the basics of AI and discuss how AI can help bridge the gap between the current state of knowledge and the graduate skills, for example, in STEM education. We also present a case study on the use of AI in innovative physics education.

Keywords Artificial Intelligence (AI) · Physics education · Machine learning · Generative AI · Neural networks

1 Introduction and Foundations of Artificial Intelligence

Artificial intelligence (for short also AI) is one of the fastest growing fields in today's world. It is concerned with understanding and building intelligent machines that can compute how to act effectively and safely in a diversity of novel challenges. The 21st century is characterised as a knowledge-based and information-centric society. In this modern world, knowledge, information, and data have become essential resources that play a critical role in the development of society (Russel and Norwig 2021).

In the early modern period, Gutenberg's printing press played a key role as a new medium; the book made it possible to disseminate information and knowledge to the public. Different ideas about knowledge began to emerge. One side argued that knowledge and information should only be accessible to the upper echelons of society. The other side argued that, on the contrary, the lower classes deserved equal access. Learning was a sphere that dealt with so-called academic knowledge. After

Ľ. Antoni (✉) · G. Semanišin
Institute of Computer Science, Faculty of Science, Pavol Jozef Šafárik University, Košice, Slovakia
e-mail: lubomir.antoni@upjs.sk

G. Semanišin
e-mail: gabriel.semanisin@upjs.sk

Z. Ješková and J. Hanč (eds.), *Advancing Physics Education*, Challenges in Physics Education, https://doi.org/10.1007/978-3-031-94207-5_1

the invention of the printing press, the library not only took on new dimensions but also became a rival to the university lecture hall (Burke 2009).

Aristotle was the first to create a precise set of laws governing the logical component of the mind. In the 17th century, Thomas Hobbes proposed the idea of a thinking machine. George Boole developed the mathematical ideas of formal logic in the 19th century, and the theory of probability can be seen as the generalisation of logic to uncertain circumstances, which is of great importance for AI. The Turing test was proposed in 1950 as a thought experiment to answer the question of whether a machine can think.

In the summer of 1956 at Dartmouth, John McCarthy, Marvin Minski and others first used the official term artificial intelligence and made the conjecture that any aspect of learning could be described in such a way that a machine could simulate it. After the initial excitement and high expectations, the so-called winter of artificial intelligence arrived. Advances in high-performance computing, powerful hardware, and the invention of the World Wide Web led to the development of algorithms for large data sets (Russel and Norwig 2021).

In the 21st century, (deep) neural networks contributed to victories over top human Go players in the AlphaGo programme, or to visual object and speech recognition. This led to interest in AI among businesses, universities, students and the general public. In 2014, generative adversarial networks were introduced, and generative AI applications were able to create authentic images, videos, text and audio of real people. Today, generative AI is being used in a wide range of applications, from creating art to supporting education and improving healthcare. AI can be useful in helping students explain topics in all kinds of subjects. From solving mathematical equations to explaining the principles of physics (Goodfellow et al. 2016; Sejnowski 2018).

In this paper, we present the challenges and selected solutions in physics education generated by AI. In Sect. 2, we describe the different types of machine learning tasks, the selected classification methods and the evaluation of their performance. In Sect. 3, we present possible applications of artificial intelligence technologies in physics education. In particular, we provide examples of image classification of physics concepts, analysis of text documents, and exploration of hidden patterns in tabular data.

2 Types of Machine Learning Tasks

In the past, computers were mainly used to process numerical (quantitative) data. More recently, they have also been used to represent relational data, which allows us to explore the relationships between instances of objects. In addition, the information and knowledge extracted from the data helps to solve many problems and challenges related to science or society.

Machine learning can be seen as a subset of artificial intelligence that uses scientific methods, processes, algorithms and systems to gain knowledge and insights from data in various forms, both structured and unstructured. Machine learning uses techniques and practices from mathematics, computer science, statistics and other related fields (Alpaydin 2010; Dhar 2013; Han and Kamber 2022; Neil and Schutt 2014).

In machine learning, we distinguish several basic types of tasks (related to data analysis):

- supervised learning,
- unsupervised learning,
- reinforcement learning,
- semi-supervised learning,
- self-supervised learning,
- transfer learning (Alpaydin 2010; Dhar 2013; Han and Kamber 2022; Neil and Schutt 2014).

Supervised learning can be seen as a type of machine learning where the model (set of parameters) is trained on the so-called labelled data. This means that for each input we have a corresponding output (correct answer, ground truth, target attribute) and the algorithm learns to map inputs to the correct outputs based on this training data. It's commonly used for tasks such as classification and regression, where the goal may be to predict outcomes based on historical examples.

Unsupervised learning can be thought of as a type of machine learning where the model is created with unlabelled data, meaning there are no predefined outcomes (no target attribute). The algorithm explores the data to find hidden patterns or groupings, such as clustering or association. It's particularly useful for discovering insights in datasets where the structure is unknown, such as segmenting customer behaviour, generating association rules or detecting anomalies.

Reinforcement learning can be thought of as a type of machine learning where the model is trained through trial and error. The model interacts with an environment, making decisions and receiving feedback about rewards or punishments. After several iterations, it learns to maximise cumulative rewards by adjusting its strategies. This approach is widely used in robotics, gaming, and real-time decision making scenarios.

Semi-supervised learning can be thought of as a combination of supervised and unsupervised learning, using a small amount of labelled data along with a large amount of unlabelled data. The model is trained using the labelled data to guide its learning process, but also explores the unlabelled data to identify structures or patterns. This approach is useful when it is costly or time-consuming to obtain labelled data, but a large amount of unlabelled data is available, which is typical of tasks such as image recognition or natural language processing.

Self-supervised learning can be seen as a type of machine learning in which the model generates its own labels from raw, unlabelled data by using the inherent structure of the data. It uses one part of the data to predict other parts. This approach allows models to learn data representations without the need for human labelled data. This approach is useful in natural language processing or computer vision, where large datasets are abundant but labelling can be labour intensive. For example, models such as Chat-GPT and BERT use self-monitoring techniques to predict missing words in sentences (Alpaydin 2010; Dhar 2013; Han and Kamber 2022; Neil and Schutt 2014).

Transfer learning can be thought of as a type of machine learning where a model trained on one task is reused for a different but related task. Rather than starting the learning process from scratch, the model uses knowledge from a previous task to improve performance on a new task, often with limited data. This method is particularly effective in domains where labelled data is scarce, such as medical imaging or natural language processing. For example, a model trained to recognise objects in generic images can be fine-tuned to identify specific types of objects with minimal additional training (Alpaydin 2010; Dhar 2013; Han and Kamber 2022; Neil and Schutt 2014).

2.1 *Supervised Learning*

In supervised learning, we distinguish two main tasks: data classification and data prediction (regression).

Data classification involves a two-step process: the learning phase, where a classification model is built, and the classification phase, where the model is used to predict class labels for new data. In the first phase, a model is generated to represent a set of data classes. This model is built by analysing input objects, also called examples, instances, data points, or samples, which are characterised by specific attributes. During this learning phase, a classification algorithm constructs the classifier by "learning" from a training set consisting of database records and their corresponding class labels. A tuple, X, is represented by an n-dimensional attribute vector, X = (x1, x2, …, xn), representing n measurements made on the tuple from n database attributes, respectively, A1, A2, …, An. Each tuple, X, is assumed to belong to a predefined class as determined by another database attribute called the class label attribute.

Since the class label is provided for each training tuple, this process is known as supervised learning, meaning that the classifier's learning is "supervised" because it knows which class each training tuple belongs to. This contrasts with unsupervised learning (or clustering), where class labels are not provided, and the number or types of classes may be unknown in advance. The learned model can be expressed as classification rules, decision trees, or a mathematical formula. For instance, in a customer credit database, classification rules can be used to identify customers with excellent or fair credit ratings. The second step is to apply the model to classify new objects.

Data prediction involves building a model to estimate the class of an unlabelled sample or predict the likely value of a target attribute for a given object. In this sense, classification and regression represent two main types of prediction task: classification is used to predict discrete or nominal values, while regression predicts continuous or ordered values. From another perspective, data prediction differs from classification in that the class label variable can take continuous values.

Classification and prediction have a wide range of applications, including credit approval, performance forecasting, medical diagnosis, targeted marketing, predicting water consumption in specific regions, estimating user ratings of movies, and recommendation systems.

2.2 *Selected Classification Methods*

In this subsection, we present several well-known classification methods and algorithms that form the basis of many analytical techniques. These algorithms offer unique advantages and are widely used across various applications in data science and artificial intelligence:

- decision trees and random forests,
- artificial neural networks,
- SVM (Support Vector Machine),
- regression analysis,
- Bayesian networks,
- hidden Markov models,
- genetic algorithms and other algorithms (Dhar 2013; Alpaydin 2010; Han and Kamber 2022; Neil and Schutt 2014).

The core concept behind decision trees is the recursive splitting of a training dataset into two parts, forming a binary tree structure. At each step, a single explanatory variable is used, and a dividing point is chosen to segment the data. This point is chosen to maximise the homogeneity or purity of the resulting subsets. Various statistical measures, such as the Gini index, information gain, entropy, and variance reduction, are used to assess homogeneity. The tree-building process stops when all observations in the terminal nodes (leaves) belong to the same class or when a predefined condition, like the minimum number of observations in a leaf or the maximum tree depth, is met. Each leaf node, representing a series of logical conditions, is assigned the class that occurs most frequently in the corresponding subset of the training data.

Random forest algorithms, on the other hand, generate multiple decision trees independently, often in parallel. Each tree is built using a different random sample of observations from the training set, and randomness is also introduced by selecting a random subset of variables at each split, rather than considering all variables. The final prediction is made by averaging the predictions of all trees (in regression) or taking the most frequent class (in classification) (Alpaydin 2010; Dhar 2013; Han and Kamber 2022; Neil and Schutt 2014).

Neural networks are popular methods of artificial intelligence for text or image analysis. The mathematical model of a neuron, the basic element of the neural network, contains a tuple of inputs (real numbers) with modifiable weights to classify input examples into two or more categories. The main idea of neural learning is based on the principle that if the neuron output to a given input example is correct, we do not modify the weights. If the answer is incorrect, then we need to change the weights with respect to the value of the error, the learning rate, and the values of inputs. Intuitively, if we expected the output to be 1, but the correct answer is 0, we increase the weights. Conversely, if we expect the output of the neuron to be 0, but the correct answer is 1, we decrease the weights. The most commonly used architectures of neural networks are:

- feed-forward neural networks,
- convolutional neural networks,
- recurrent and recursive nets,
- deep generative models (Goodfellow et al. 2016).

2.3 Performance of Classification Methods

When evaluating the effectiveness of classification methods, it is essential to measure how accurately artificial intelligence models classify input data represented as objects. Four basic measures are used to evaluate accuracy (see Fig. 1):

- TP (True Positives): Examples that have been correctly classified as positive.
- FP (False Positives): Examples that have been incorrectly classified as positive.
- FN (False Negatives): Examples that have been incorrectly classified as negative.
- TN (True Negatives): Examples that have been correctly classified as negative.

By using TP, FP, FN, and TN, we can derive other measures of classification performance:

- Recall: TP/(TP + FN), i.e., the number of correctly classified positive examples from all the positive examples in the training or testing set.
- Specificity: TN/(TN + FP), i.e., the number of correctly classified negative examples out of all negative examples in the training or testing set.
- Precision: TP/(TP + FP), i.e., the number of correctly classified positive examples out of all positively classified examples in the training or testing set.

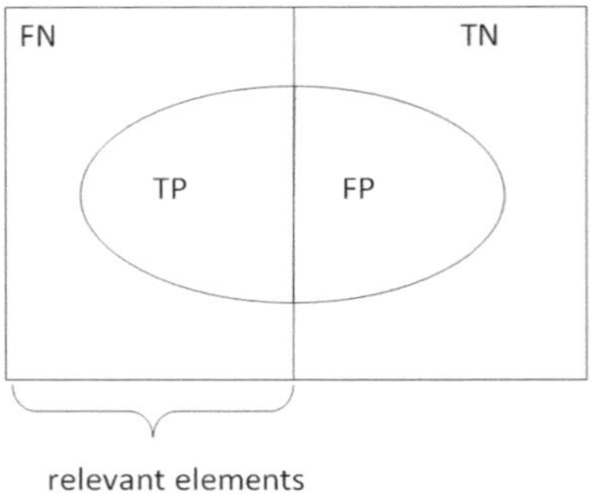

Fig. 1 Evaluation of classification performance

- Accuracy: (TP + TN)/(TP + TN + FP + FN), i.e., the number of correctly classified examples out of all examples.
- F1-measure (F1-score): 2TP/(2TP + FP + FN), i.e., the harmonic mean of precision and recall.
- ROC curve.
- Kappa coefficient, which takes into account the number of observations in each class based on probability, where a value of 1 means full accuracy.

2.4 *Data Science*

Data science can be seen as a combination of different, partially overlapping, (sub) disciplines (see Fig. 2). The boundaries between disciplines are not precisely defined and may change over time.

As mentioned in the introduction, the origins of data science and artificial intelligence are closely related to statistics. Statistics can be divided into descriptive statistics, which summarise data from the whole dataset using measures such as mean, standard deviation, and frequency, and inferential statistics, which draw conclusions about the whole population based on information derived from samples.

In the context of data science, we often use terms such as big data, data mining, and knowledge discovery. The key characteristics of big data include volume (quantity), velocity (speed of generation), variety (diversity), veracity (uncertainty), and value (van der Aalst 2016; Chollet 2018).

The profession of data scientist emerged in 2008 on social networks such as LinkedIn and Facebook, where American mathematicians and computer scientists

Fig. 2 Data science elements (modified from van der Aalst 2016; Chollet 2018)

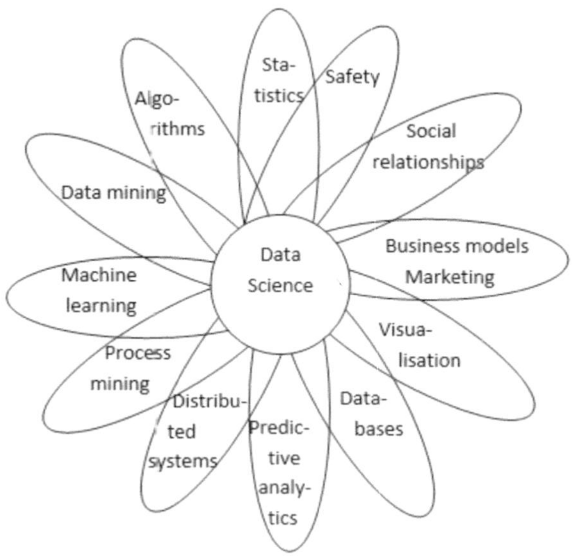

Dhanurjay Patil and Jeff Hammerbacher first used the term to describe their roles. They outlined the responsibilities of data scientists and highlighted their growing importance and scarcity. The skill set of a data scientist is illustrated in Fig. 3.

Data scientists have an important role in transforming data into fruitful outputs, addressing steps such as:

- What happened? (Reporting)
- Why did it happen? (Diagnosis)
- What is likely to happen? (Prediction)
- What should be done? (Recommendation).

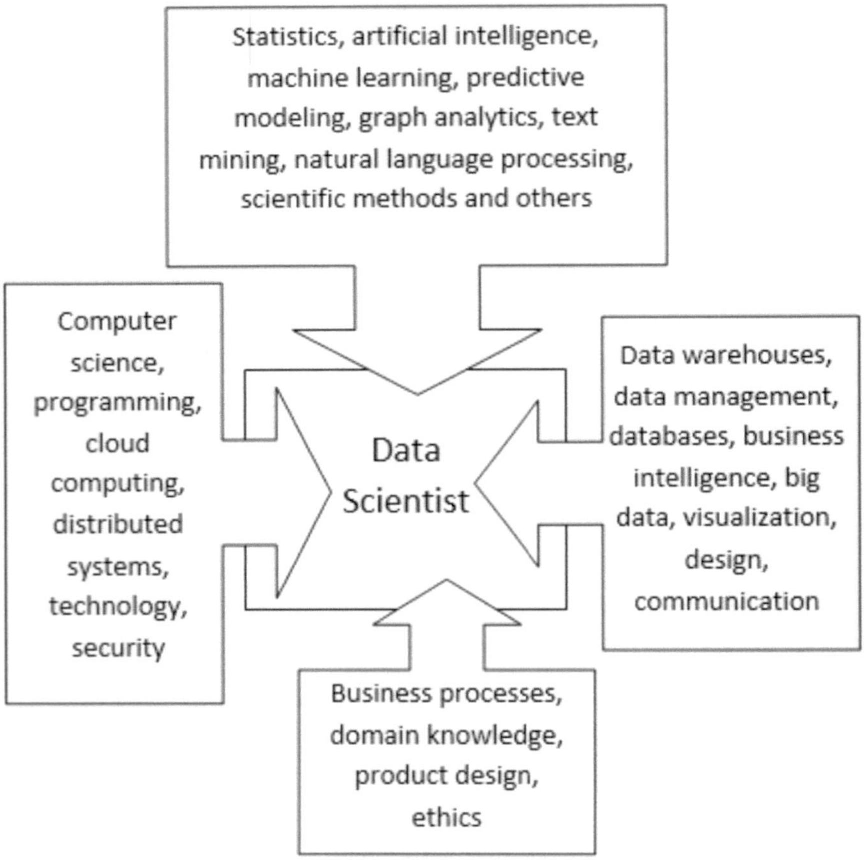

Fig. 3 Key skills required for a data scientist (modified from van der Aalst 2016; Chollet 2018)

3 Applications of Artificial Intelligence Technologies in Education

Generative artificial intelligence can help bridge the gap between the current state of knowledge and the knowledge of graduates. Throughout the history of physics, the emergence of major paradigm shifts like the development of quantum physics has stimulated extensive and novel research, leading to a surge in publications and the establishment of novel subfields, ranging from nuclear to condensed-matter physics (Sun and Latora 2020; Sinatra et al. 2015; Tria et al. 2014).

Generative AI has the potential to bridge the gap between the current state of knowledge and the knowledge of graduates by generating educational content that is tailored to the needs and interests of individual learners. Generative AI can create personalised learning paths for students based on their learning history and preferences, can generate simulations and scenarios that simulate real-world situations, or can help to analyse large amount of data and identify patterns and hidden information that are difficult to detect by humans (Goodfellow et al. 2016; Sejnowski 2018).

Automated AI assessment, machine learning, or learning analytics can improve the learning and instruction quality in STEM education. Several types of AI applications in STEM education have been studied, such as educational programming or social robots, intelligent tutoring systems (personalised learning path recommendation, resource recommendation, instructional content delivery), student behavior detection, learning prediction (performance prediction, at-risk student prediction), automation (automated question generation, automated assessment), or others (artificial intelligence textbooks). Computer systems, robots, experimental instruments, wen open sources, mobile phones, or e-books play the role of educational media (Xu and Ouyang 2022).

The application of AI in innovative physics education can enhance the experience for both students and teachers. For example, visualization and evaluation techniques can help to find hidden patterns in the students' solutions which allow to prevent student mistakes or misconceptions. In the following subsections, we present examples of the application of AI in innovative physics education with the help of several machine learning methods and algorithms.

3.1 *Classification of Images in Physics Education by Artificial Intelligence Tools*

The integration of AI in educational contexts has the potential to enhance students' engagement with images in physics. Microsoft Lobe, a user-friendly tool for training image classification models, can be used to develop an educational project that classifies and categorizes various physics concepts based on their visual representations. By training the model with a diverse set of images, such as force diagrams, energy transformations, wave patterns, or chest X-rays (see Fig. 4), students can interactively

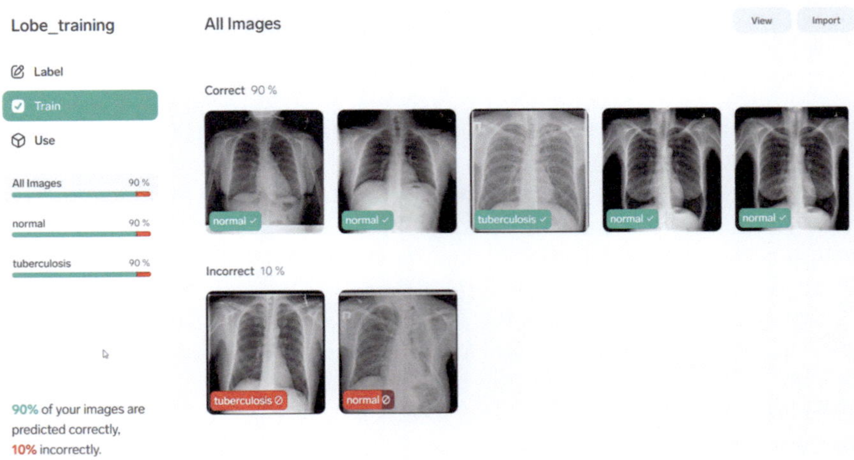

Fig. 4 Example of tuberculosis chest X-ray classification in Microsoft Lobe

explore these topics. The project aims to train an artificial intelligence model where teachers can upload images physics-related and receive immediate feedback on the model's classification performance, thereby enhancing the learning experience and fostering a deeper understanding of the physical principles.

The implementation of image classification tools (e.g., Microsoft Lobe, TensorFlow) in physics education will not only facilitate a more dynamic learning environment but also encourage students to interact creativily with physics content. Through hands-on experimentation, students will be able to submit their own images or diagrams for classification, promoting active learning and critical thinking. In addition, the proposed project can explore the effectiveness of visual learning aids in improving student's understanding and retention of physics concepts. By combining technology with educational practices, this initiative aims to enable both teachers and students to make physics more accesible and enjoyable for all.

3.2 Analysis of Text Data in Physics Education Using Artificial Intelligence Tools

The analysis of textual data by means of AI is closely related to concepts such as natural language processing, text corpora, or Word2Vec (Mikolov et al. 2013). Word2Vec is a neural network model that generates vector representations, known as word embeddings, for each word in a given language. This model is employed in applications such as Google Translate.

The applications of text data analysis include:

- Text classification (spam, sentiment, text language, author);

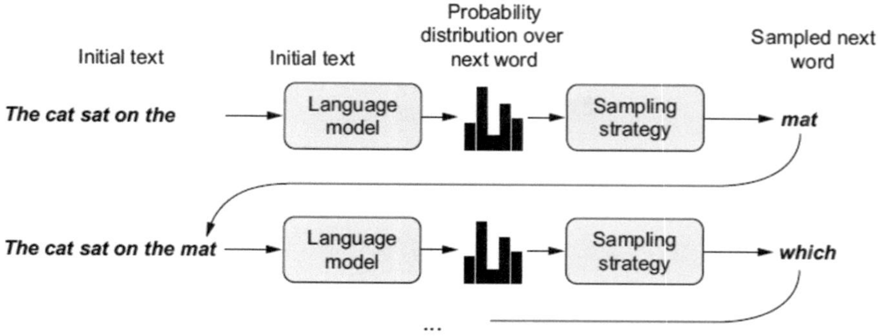

Fig. 5 Process of word-by-word text generation in ChatGPT (Chollet 2018)

- Text generation (e.g., in ChatGPT);
- Summarization, translations;
- Keyword extraction (e.g., in PoolParty tool);
- Sign language processing;
- Detection and correction of grammatical errors and others.

ChatGPT-3 is a 175 billion parameter text generation model trained by the startup OpenAI on a large text corpus, including most digitally available books, Wikipedia, and a large fraction of a crawl of the entire Internet. The process of word-by-word text generation is shown in Fig. 5.

3.3 Analysis of Tabular Data in Physics Education by Artificial Intelligence Tools

Tabular data can be effectively extracted from a variety of sources, including documents, images, or signals. Tabular data can illustrate the relationship between students' task solutions and their corresponding components or steps, allowing educators to identify common patterns in problem-solving approaches. In addition, a structured analysis can be conducted by mapping educational tasks to their respective educational objectives, which helps to assess the alignment of instructional activities with desired learning outcomes. By systematically organizing these relationships in tabular format, educators and researchers can derive insights that help to improve educational strategies and enhance students' learning experiences.

Weka, Minitab, or RapidMiner Studio are software tools designed for data analysis, machine learning, and artificial intelligence. They empower a wide range of users, including students, teachers, researchers, and data analysts, to perform intelligent data analyses, create predictive models using machine learning and artificial intelligence, and evaluate the performance of the generated machine learning and

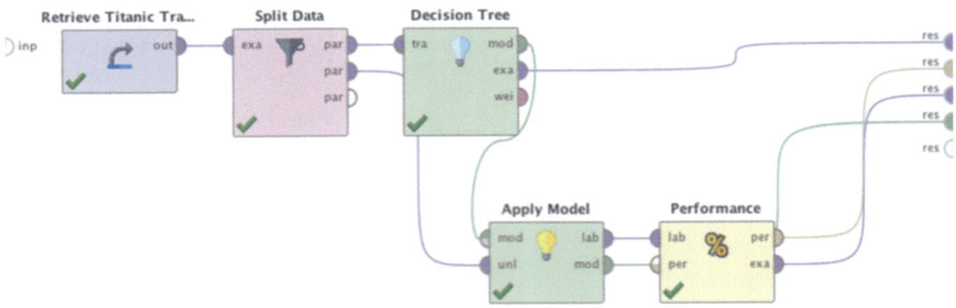

Fig. 6 Process of tabular data analysis in RapidMiner studio

artificial intelligence models. The data analysis process in these tools includes the following phases (see Fig. 6):

- data pre-processing (missing values, transformations, integrity),
- data splitting (training set and testing set),
- modeling (e.g., training model by method of decision trees described in Sect. 2.2),
- evaluation of model (described in Sect. 2.3).

Moreover, in the context of educational processes, tabular data plays a critical role in analysing and understanding student performance and learning outcomes. This data can be organized into meaningful comparisons and relationships, such as the correlation between students and their assessments, where each student's performance is assessed against pre-defined criteria.

4 Conclusion

Artificial intelligence can help solve many problems in society and bridge the gap between the current state of knowledge and the knowledge of graduates. In this paper, we presented the challenges and selected tasks in physics education generated by artificial intelligence. We described the different types of machine learning tasks, the selected classification methods and the evaluation of their performance by precision, recall, F1 score and other measures. We also presented the potential applications of artificial intelligence technologies in physics education.

The implementation of artificial intelligence tools (e.g. Microsoft Lobe, ChatGPT, TensorFlow) in physics education will not only facilitate a more dynamic learning environment, but also encourage students to interact creatively with physics content. The application of AI in innovative physics education can improve the experience for both students and teachers.

Acknowledgements This research is partially supported by the Erasmus+ project no. 2023-1-PL01-KA220-HED-000166765—Developing Talents in Artificial Intelligence to Solve Disruptive Environmental Problems. This research is also supported by the the European Commission's grant HORIZON-CL4-2023-HUMAN-01-CNECT under GA No 101135737 AI-BOOST—Artificial Intelligence for better opportunities and scientific progress towards a trustworthy and human-centric digital environment.

References

van der Aalst W (2016) Data science in action. In: Process mining. Springer, Berlin, Heidelberg, pp 3–23. https://doi.org/10.1007/978-3-662-49851-4_1
Alpaydin E (2010) Introduction to machine learning, vol 2 MIT Press, Cambridge, MA
Burke P (2009) Society and science. Karolinum press, From Gutenberg to Diderot
Chollet F (2018) Deep learning with python. Manning Publications Co, Shelter Island, NY
Dhar V (2013) Data science and prediction. Commun ACM 56(12):64–73. https://doi.org/10.1145/2500499
Goodfellow I, Bengio Y, Courville A (2016) Deep learning. MIT Press, Cambridge, MA
Han J, Kamber M (2022) Data mining: concepts and techniques, 4th edn. Morgan Kaufmann Publishers, Cambridge, MA
Mikolov T, Sutskever I, Chen K, Corrado GS, Dean J (2013) Distributed representations of words and phrases and their compositionality. Adv Neural Inf Process Syst 26:1–9
Neil CO, Schutt R (2014) Doing data science: straight talk from the frontline. O'Reilly Media, Sebastopol
Russell S, Norwig P (2021) Artificial intelligence, a modern approach. Pearson Education Limited, Harlow, UK
Sejnowski TJ (2018) The deep learning revolution. MIT Press, Cambridge, MA
Sinatra R, Deville P, Szell M, Wang D, Barabási AL (2015) A century of physics. Nat Phys 11:791–796. https://doi.org/10.1038/nphys3494
Sun Y, Latora V (2020) The evolution of knowledge within and across fields in modern physics. Nat Sci Rep 10:12097. https://doi.org/10.1038/s41598-020-68774-w
Tria F, Loreto V, Servedio VDP, Strogatz SH (2014) The dynamics of correlated novelties. Nat Sci Rep 4:5890. https://doi.org/10.1038/srep05890
Xu W, Ouyang F (2022) The application of AI technologies in STEM education: a systematic review from 2011 to 2021. Int J STEM Educ 9(1):1–20

Fundamental Research in Physics and Its Role in School

Mojca Čepič

Abstract The paper considers different aspects of introducing contemporary research results in physics to pre-university physics education. For introduction to the classroom adapted units have to be developed, which allow for experimenting and investigation of new phenomena. To develop them a strong collaboration between researchers, educators and teachers is needed. Next, new topics in physics, which are not publicly well known could be used for identification of gifted students, which come from families with lower economical and intellectual background. Such students often do not develop good reading, writing, drawing and math skills, as the work for school is not appreciated and supported at home. Finally, to train teachers in units related to contemporary research results, and to use the adapted units as a part of an instrument for identification of students with deprivileged home background, the regular collaboration between educators and teachers has to exist, and the most advisable form are professional learning communities of teachers. These communities, how they are supported, and how they function are discussed in the last part of the paper.

Keywords Active learning · Inquiry-based learning · Contemporary research in teaching · Gifted students · Teacher-researcher collaboration · Professional learning communities

1 Introduction

Physics in school is considered a difficult subject, which only highly makes students understand, and regular students do the best, if they try to figure out the way to avoid learning it (Osborn et al. 2003). Why is that so? Partly, because some teachers remember how difficult the study was, and want to stress that they have succeeded

---

M. Čepič (✉)
Faculty of Education, University of Ljubljana, Ljubljana, Slovenia
e-mail: mojca.cepic@pef.uni-lj.si

Jožef Stefan Institute, Ljubljana, Slovenia

Z. Ješková and J. Hanč (eds.), *Advancing Physics Education*, Challenges in Physics Education, https://doi.org/10.1007/978-3-031-94207-5_2

by discussing difficult tasks only a few chosen ones understand. Partly, because the language of physics or even science is almost a foreign language to students. Expressions that are used regularly in everyday life have different meanings in physics. Sentences, constructed from everyday words, are often not understandable in the physics context, and finally, physicists love math and its language that allows for measurable predictions or extracting physical quantities from measurements. The language of mathematics is abstract, how mathematics describes physics phenomena requires higher cognitive levels because of transformation between different scientific representations and has to be learned and trained. Finally, because physics wants to bring mathematical language closer to students' knowledge of mathematics, uses simplifications, neglects many everyday circumstances, and therefore very often examples, on which the student should train, are in contradiction with everyday experience. The most prominent example are Newton's laws and forces. Associating everyday experience, for example, pushing a brick along the surface, requires force, but the concept of force in physics is something entirely different than the feeling of push or pull in the muscles.

There is another important aspect, why students consider physics boring and irrelevant. Physics topics are old. Even modern physics is more than a hundred years old, an eternity for students. Real research from laboratories rarely finds the way to the school. School visits to research institutions are usually filled with colorful presentations, loads of new concepts and data, visits in labs show complicated technologies, which usually communicate with computers and the studied phenomena are not directly observable (Čepič 2017).

However, physicists passionately love physics, yearn to investigate properties of new materials, development of space through time, behavior of very small particles with high energies etc. On the other hand, physicists are also normal people, who play basketball with friends, hike with their families, and listen to music. This part usually remains hidden, and students remain under the impression of weirdos, who spend their life in laboratories communicating with machines instead of people.

I am a theoretical physicist by training, and my passion is to model the behavior of relatively big molecules, which form liquid crystals. In physics, everything is a model. As Feynman stressed in his lectures (Feynman and Sands 1963–65), we observe, investigate and measure the phenomena in nature, and the only thing we can do, is, we reason, we hope to find cause-consequence relations, we make predictions based on our understanding, experiment and measure, and hope, the measurements will turn out consistent with our understanding, that is, we construct models, which are more or less consistent with natural phenomena. But we are never able to prove our model is correct, on the other hand, a single experiment, which is inconsistent with predictions of the model, disproves it. Scientists love their babies, that is, their models, and they often try to refute disproval. But usually, this persistence does not go further than carefully repeating experiments, which disproof the model. I was lucky in my research, and a few times we have been able to predict a new structure, a new phenomenon or a new model, which explained observations. However, the thirst to find out, what can cause such and such behavior of some liquid crystals was always going around my head, and crucial ideas occurred in very strange situations, a few of them during long cycling routes.

To understand the scientific passion, the fundamental research community should design active units applicable during teaching physics, which include relatively simple hands-on experiments providing experience to students, on which they can build the understanding. When those experience are gained, they should investigate further, with some hints of supervisors and find new "discoveries" by themselves. This is the essence of the active learning methods like IBL (Erasmus + project 3DIPhE Vol I 2020a), ISLE (Etkina 2019) and similar. However, those methods focus on curricular topics and, sometimes, on physics of everyday life, but usually do not lead to the contents of contemporary research. Section 2 considers prerequisites for the introduction of contemporary topics to education, some experience and some suggestions for lecturers, who would like to try it.

Contemporary fundamental research topics cannot be found in regular curricula. They are usually not known to teachers besides attending lectures by scientists here and there. However, some contemporary topics, prepared in advance, may serve for identification of more able students that come from economically or intellectually disadvantaged backgrounds, immigrants, who are not fluent in language of schooling, and other students with problems, for example twice exceptional students, who are able to reach mediocre standard by high abilities obscured by learning difficulties such as dyslexia, ADHD and similar. How to use units with experiments on contemporary topics to identify students, who are gifted but lack intellectual skills, which are fluent reading, writing, math and/or drawing, is the content of the Sect. 3.

Without the support of teachers, we in academia could only dream about various ideas to be implemented in schools. On the other hand, after a few years of experience schoolteachers usually yearn to meet colleagues, to exchange experience and problems, and solutions for them, ideas for work in general. Besides, they always want to learn new experiments, which are simple enough to be assembled and carried out in their practice. The answer for both problems, the access of academia lecturers to schools and vice versa, an exchange of academia teaching knowledge with teachers on their demand and expectations are professional learning communities of teachers (Erasmus + project 3DIPhE Vol I, Vol II 2020a, b, Erasmus + project RISE 2023). How to establish such communities and how to facilitate activities in them is discussed in the Sect. 4.

As being trained as a schoolteacher initially and later as a theoretical physicist of soft matter during my PhD studies, where I am still slightly active, and finally working for decennia, to be clear, more than 30 years as an educator of future physics teachers, the three sections mentioned above describe my evolution in physics education. I hope the reading will be interesting and will give the reader some ideas on various aspects of the lecturer's potential influence.

2 Physics in School and Physics in Contemporary Laboratories

As my passion is theoretical physics of soft matter, more focused, liquid crystals, even more detailed, polar liquid crystals. The antipolar variant, which is liquid crystals behaving like antiferroelectrics were discovered two years before I started my PhD. I have to admit, it is a very unique situation when something new and promising is discovered, when one can read all the published papers on the topic and follow reports on every new experiment. In those times, "schools", which offer one or another model to explain experimental observations, and when experiments, which disprove some ideas are shown, "schools" sometimes fight for their models unscrupulously. In my theoretical life I was a part of such discussion several times (Takezoe et al. 2010; Jukić and Čepič 2019, Szydłowska et al. 2023). But everything was very similar to a classroom with a set of good and very motivating experiments. So, the idea started to pester me, can we introduce such topics to students? Can we find experiments which allow for investigation and disproval of models? Can we introduce contemporary topics to students in a way that they can pick up a little of researchers' enthusiasm?

At the University of Ljubljana, Faculty of education, in the period from 2008 to 2015, we were lucky to have a group of young PhD students who systematically developed experiments for teaching about liquid crystals. For such work a strong connection with a motivated researcher, me in this case, is necessary, because the fundamental knowledge of the topic under development and adaptation for school is essential. A newcomer in the field can easily make scientific mistakes. On the other hand, the researcher has to be familiar with teaching methods, not only having lectures for undergraduate students, but introducing new concepts to students at the preuniversity level and even lower, that is, to students without a preliminary knowledge of undergraduate/graduate students. My colleagues, fundamental researchers on liquid crystals, directly commented on this idea, "we do not have time for such thing, it is much time consuming and does not result in highly cited papers", which is all true. So, again, we were left with me in the role of researcher.

The work of years resulted in a model for introduction of contemporary physics to the classroom depicted in Fig. 1 (Čepič 2017). The process is illustrated in four levels, which influence each other forwards and backwards. If the reader is willing to risk the endeavour needed for introduction of a contemporary research topic to teaching, he/she should answer the "Why" question first. One might find several motives. Motives are always internal, and they vary, from loving the topic so much to feel that everybody should know something about it, as happened in my case, to simply struggling for a topic worth to investigate for a PhD in physics education research. Although the second example seems rather bizarre, nevertheless, when one works on something serious enough, the affection usually appears and becomes a passion. One might on the other hand want to introduce something timely to students, who are bored by old physics. But the motivation has to be strong enough for all the work hidden in other levels.

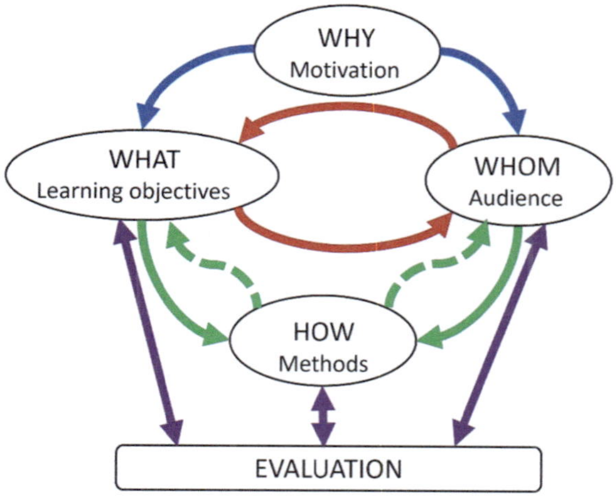

Fig. 1 The model for introduction of contemporary physics to pre-university class

The question "What" seems simple, but when one searches for already studied or even prepared units of examples of contemporary physics, written resources are often available, but they usually leave students unimpressed. Experimental work has to be included in such lectures, not as demonstration only, but as a set of activities, which allow students for preliminary experience related to the topic, and to observe important phenomena by themselves. And preparing experimental work for students to be active usually creates problems. One should search for alternative experiments, models and similar. For example, for optical anisotropy typical for liquid crystals one can use activities with adhesive tapes and polarizers, which illustrate phenomena perfectly and are easily accessible (Čepič 2014). When students obtain necessary preliminary experience, then the lecturer can start discussing front end science.

The question on the middle right "Whom" is another difficulty. If one wants to discuss newly discovered phenomena in introductory lectures for graduate students, who will continue the research, one can use the whole arsenal of mathematical tools and very often leaves the construction of understanding to students. But at the pre university level, the preliminary knowledge is limited. So, the audience also greatly affects the answer to the "How" question, the methodology of teaching. The constructor of the teaching unit has to take into account the lack of preliminary knowledge, devote some part of the unit to introduce experience and knowledge which graduate students already have or are able to learn without the lecturer's help, and the most important, to the appropriate teaching methodology. For lower secondary students, observations are much more important than precise measurements of dependencies. Upper secondary students, who already mastered some measuring techniques, are able to do some investigations more thoroughly. Anyway, our experience has shown that front end science is the most appreciated, when it is not accompanied by thorough theoretical explanation based on math.

Finally, and importantly, after the unit is constructed, after it is carried out with students, partially or in full, the thorough testing gives the lecturer the feedback of its success. The evaluation of impact of the unit has to be systematically planned and focused on various aspects of the teaching unit.

For these topics, the comparison of the knowledge before and after the unit is not very important. Our studies (Pavlin et al. 2013; Fazio et al. 2021) have shown that knowledge about contemporary topics practically does not exist. Between 20 and 60% of students have heard words related to contemporary research topic, and this share is strongly topic dependent, however, less than 20% associate these words with some devices or the places, where they heard about them. About 1% declared they know something about the topic. So, any new information will seriously increase the level of knowledge about the topic, which is reasonable but not a tremendous success. However, the evaluation of the knowledge before and after the unit and if circumstances allow, also a few months later, has to be carried out as an important indicator of the gained students' knowledge. One does not want to invest the precious teaching time on the topic using the unsuccessful methodology, which yields no results, of course. But for the evaluation, more important is, how structured is the new knowledge, do students remember experiments and what have they learned when they have carried them out? Also motivating factors are important. Did they like the activities? Do they have an impression they have learned something interesting? Be careful, this question is about an opinion, not the fact. Do they want more lectures/units on other contemporary topics? The evaluation is also a rich resource for improvement of the unit's structure. For example, during the experimental work with liquid crystals, students observed colours of anisotropic materials between crossed polarizers. Only the evaluation has shown that they mistakenly thought these colours are a source of display colours, which is not true, as they are obtained by colour filters. In the following implementation we changed the sequence of activities in the unit, which corrected the issue.

To conclude this Section, Contemporary research has difficulties to find its way to regular teaching, mainly because the motivation of researchers is focused on transfer of new knowledge to potential new researchers, PhD students at the graduate level. On the other hand, a few units during the upper secondary school may make a difference toward the attitude of students to physics, because they may realize, how interesting it may be. An efficient implementation of such units needs direct support and help of developers of units, which means a good collaboration between fundamental researchers, educators and pre-university level teachers. To my best knowledge, well-structured units on front end science topics are very difficult to find. A thorough study of the literature yielded only a few, and even they were mostly examples of interesting experiments and not systematics guides to the teacher with the theoretical background, suggestions for methodology of teaching, set ups of experiments and such. As there are not many units available, also systematic studies of the influence on motivation for learning or even study physics later are not available, which would also be worth to investigate in future. Contemporary science topics and their introduction to pre university level allows for other applications, one of them is discussed in the next Section.

3 Gifted Students with Deprivileged Backgrounds and Contemporary Research Topics

This Section sums up some of our activities related to the development of an instrument for identification of students having deprivileged background, that is, economical and/or social difficulties at home, intellectually weak background, ignorance or absence of parents or alternatively, poor knowledge of learning language typical for refugees and immigrants in general, or students being twice exceptional, that is, having learning difficulties like ADHD or dyslexia, but being gifted masks their learning problems and makes them poorly successful.

Why is this group of students important? First, research has shown that gifts are equally dispersed among different groups, which vary in place, color, religion etc. However, tools for identification regularly used, are not understandable to them in an equal way (Borland and Wright 1994; Borland 2004). The study of Borland is extremely interesting. They tested students in one of the problematic schools with criminal, drugs, teachers changing every year or even every few months etc. in one of the districts in New York. For the students, who had highest marks, IQ was measured around 70. Then, the few of those best students were moved to a regular, supportive school. In two years, they were among the best students in the new school and new tests revealed IQs around 140. Are standard identification tests for giftedness really equal for everybody?

The problem of identification is especially critical in societies, where giftedness is strongly associated with academic success measured in school marks (Cankar et al. 2016). Students with a deprivileged background very often come from homes, where school, learning, and knowledge are not respected, even more, sometimes parents even force students to work manually instead of learning or even humiliate their learning efforts. Under an umbrella of intellectual skills, we consider automatized reading, writing, simple calculus, estimations of quantities, drawing, but also articulating and arguing ideas. At homes without books, with parents not really talking to each other or to children, and if they do, they use a very limited vocabulary, children usually develop reading and writing skills later than their peers or they do not automatize reading at all. This results in a lack of understanding of written instructions in school, and consequently low marks in knowledge tests. Next, as teachers often assess student's abilities by their academic success i.e. marks (Endelpohls-Ulpe and Ruf 2005), these students are considered as students, who are lazy, not motivated, without talents etc.

Our idea was to develop an instrument which relies on students' ability to observe, draw conclusions and apply new information gained through these specially designed observations in new situations. To effectively apply new knowledge in new situations is one of the important characteristics of giftedness. But the cycle "observe, draw conclusions and apply new information in a new situation" is also characteristic for inquiry-based learning, however the method has to be adapted to contemporary research. Why not have any topic? Because when working in groups, students are usually led by academically successful students, who are the most probable generators of ideas and conclusions. Besides, those students usually have good intellectual support at home, which also includes visits to science centres and other science popularization activities, reading of popular science books, doing experiments suggested in them, etc. and they are often already familiar with the topic taught in an inquiry-based approach. The same does not hold for contemporary research (Čepič 2018).

Contemporary research is new for everybody, that is, all students have equal negligible level of knowledge. The main idea for the approach is to develop relatively simple experiments related to contemporary topics. For example, experimenting with anisotropic materials like adhesive tapes, which provide experience that can be used for understanding of properties of liquid crystals. These experiments are accompanied by a very short text, if any, carrying out the experiment(s) is not demanding, and conclusions are new to students. For example, by observing the magnified computer screen, they learn that only three colours are present, nevertheless their combinations give the impression of all colours we meet in our surrounding by changing the "intensity" of their components. Next, they learn the additive colour math through observation. Application of the new knowledge comes through difficult questions, which we call "indicative questions", because the answers indicate students' potentials. If a student is asked to suggest how to change the three colours on the screen to change the pure red to pink, the answer, to add equal "amounts" of blue and green, is in high contrast with experience from art classes, where such mixture results in an ugly grey brown. So, we believe if a student provides the correct answer, she/he was able to combine the new experience from studying colour math to a new, even counterintuitive situation, which strongly indicates the ability (Fig. 2).

We tested the unit on screen colours with recognized gifted students who achieved high results in physics and math competitions, and with students in a regular lower secondary school. Indeed, while students successful in physics/math competitions were mostly successful in answering indicative questions, the discrepancy between academic achievements expressed in school marks and answers to indicative questions were much more pronounced in lower secondary school.

To conclude, the idea of identification of gifted students who for whatever reasons lack intellectual skills, that is, are weak in reading, writing, simple math and drawing, by experimenting with new generally not familiar phenomena and testing the application of the new experience in new circumstances, seems promising. But there is a lot to be done. To develop new units, for which collaboration among educators and researchers is needed, to test this units for applicability in the classroom and for identification potential, to motivate and train the teachers, and to persuade authorities to recognize this type of identification. Mission impossible, most likely, but still, a very motivating, interesting and difficult research problem.

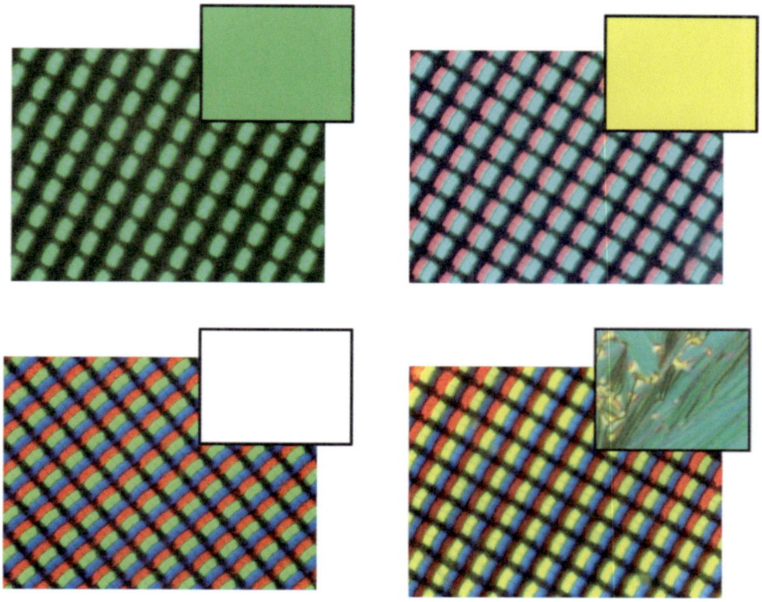

Fig. 2 Small rectangles show what is seen on the screen. Corresponding bigger rectangles show what is seen by USB microscope. All single component (RGB) and double component (YCM) colours are checked experimentally by students, three examples of general colours are also a part of the introduction of the screen colours. The experience is the basis for two indicative questions, how to make red pink, and how to make red dark red

4 Professional Learning Communities of Teachers

In the last six years University of Ljubljana, Faculty of Education was lucky to collaborate with partners from Ireland, Poland and Belgium in three Erasmus + projects. The first one, based on the idea of Wim Peeters: "Let us introduce inquiry-based learning approach to different levels in education, to use the same approach to experimental work for students, to practitioner inquiry of teachers, and to practitioner inquiry of coaches" (Peeters). We stubbornly neglected rejections and reapplied two more times after the first application, and the third try in 2017 was successful. The Erasmus + project Three Dimensions of Inquiry in Physics Education (3DIPhE) (2020) established professional learning communities of teachers (PLCT), in which teachers from schools, from elementary level to schools, which prepare for universities, directly collaborated with lecturers from partners of the project, mostly universities. Regular meetings were held once a month, partners prepared interesting, mostly new experiments to use in the classroom (Fig. 3 left), and an important part of the meeting were discussions of problems in classes, sometimes teachers shared their ideas and units, and finally, teachers discussed their practitioner inquiries, mostly on topics they learned in the group, or about inquiry-based approach they adapted to something they were teaching at the time. When COVID hit, we carried out activities

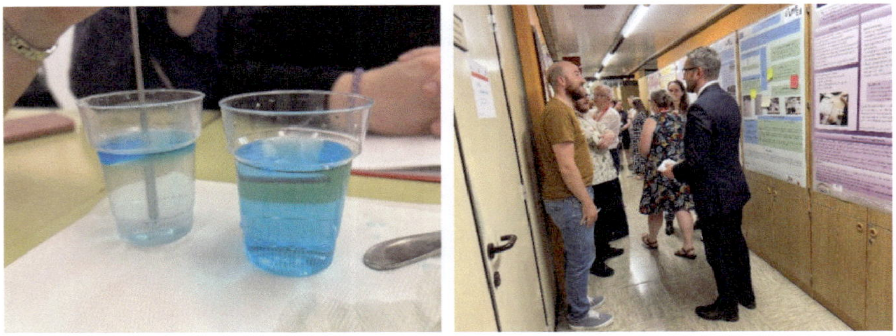

Fig. 3 Workshop during the PLCT meeting. Melting of ice in tap and salted water. The left one is the tap water, the right one is salted, and ice takes longer to melt in the salted water. Ice was coloured for better visibility (left). Teachers presenting posters as an addition to their oral presentation in parallel session of the final STAMPEd and RISE conference. This allowed to all teachers to get familiar with all presentations and to discuss them (right)

remotely and in the two new projects, which started in 2020 (Supporting Transitions in Math and Physics Education—STAMPEd and Remote Inquiry in Science Education—RISE) were also remote during the repeating lockdowns. We developed many units applicable also remotely, teachers were happy to meet remotely, especially when they were searching for teaching ideas under remote conditions.

After COVID, the projects were already in the final stage, but nevertheless, we were able to organize the final conference in person. Teachers came to the conference, which was held in June 2024 in Ljubljana and presented their work as oral contributions and posters, joined workshops and discussion groups. More than 50 presentations were given by more than 50 teachers and partners (Erasmus + project RISE 2023). Finally, the combined STAMPEd and RISE conferences allowed merge finances, which enabled more teachers to come in person. The result of these two conferences was an international community of teachers, plans for future collaborations etc. As also the final conference of the 3DIPhE project was international although remote, two teachers coming from the part of Slovenia where they have learned German as a foreign language, were stimulated to learn English and they presented their work as an oral presentation and have communicated with teachers from other countries actively (Fig. 3 right).

Active professional learning communities of teachers in collaboration with educators and researchers from universities and laboratories are crucial for both topics discussed before, for introduction of contemporary fundamental research results to teaching and learning, and for using those results for identification of gifted individuals coming from socially and/or economically deprivileged home situation resulting in low intellectual skills. The main problem to continue working in such groups is formal support. Headmasters very often do not support their teachers in visiting meetings without a formal framework of a project or similar. The formal initiative for the meeting usually remained on lecturers, because they provide premises easier than teachers, they are used to organization, they offer ideas for a part of the program at least. This all takes time, and not to forget, coffee and cookies, which make the whole meeting something entirely different allowing teachers some time for discussion among them, which also requires some financial resources, and here comes the urge for the formal support. Formal support could come from the institution, that is, the faculty, but with tight financial situations this is usually difficult, and it is not directly valued as it does not lead to publishable research results directly. It could come from new projects. But we all know that there are years when no projects applications are successful, and the opposite, when one gets more projects than the manpower to work on. In years of "starved cows", the Slovenian proverb, which notifies years without scarce finances and failed applications, it is nearly impossible to support and maintain the community. The teachers have life of their own and when a new opportunity arises, the community is already gone and has to be established anew. So, formal support for professional learning communities could make a huge difference in teachers' development.

5 Conclusion

The introduction of contemporary research to teaching at the pre-university level is a difficult problem as discussed in this paper. First, motivation to introduce it is quite indirect and is partly related to motivating students for physics and potentially also to study physics later. To properly introduce contemporary research into education, a strong collaboration between the researcher active in fundamental research, a curious educator and the motivated teacher is needed. This combination is in pressure of grants, demanding physics curriculum and teaching obligations of educators very difficult to establish. However, if it is established, it may lead to interesting and motivating teaching units, but also to various physics education research studies. For example, research on how pre university students approach new research ideas and results, how many new concepts are able to comprehend and to what level? Is introduction of contemporary research results motivating students or not? The research with these respects is non-existent to my best knowledge.

Next, contemporary science introduced to the school can be used as a support for identification of gifted students who come from deprivileged background and have consequently weak intellectual skills. These topics allow for conditions equal to all students, where the preliminary knowledge is practically negligible and cannot result in differences in comprehension of the learned topic. If experiments do not need long instructions, are easy to perform and conclusions drawn from observations are straightforward and firm, students can apply newly gained knowledge in new circumstances immediately, which is an important indicator of abilities despite not being realized in academic success. The latter is quite common for students with weak intellectual skills.

Finally, for testing ideas, academia and teachers have to collaborate as equals. Great support with this respect is professional learning communities where teachers and academics regularly meet and collaborate to search for solutions of problems together. Unfortunately, learning communities need support. If there is a project, a common goal of teachers and lecturers, everything is fine. But when this project focus is gone, the reason for further collaboration has to be established anew and new goals have to be found. Teachers have always many ideas but without a formal frame it often happens that authorities like headmasters in schools do not support expenses which inevitably occur through travel expenses or modest meeting's organization costs. We strongly believe and evidence from our Erasmus + projects strongly support the need of formal support of active collaborations of teachers and academia within professional learning communities.

References

Borland JH (2004) Issues and practices in the identification and education of gifted students under-represented groups. University of Connecticut, The National Research Center on the Gifted and Talented, Storrs

Borland JH, Wright L (1994) Identifying young potentially gifted economically disadvantaged students. Gift Child Q 38(4):164–171. https://doi.org/10.1177/001698629403800402

Cankar G, Bren M, Zupanc D (2016) For better equity of Slovenian school system, research report. RIC, Ljubljana

Čepič M (2014) Liquid crystals through experiments. Morgan & Claypool Publishers, San Rafael

Čepič M (2018) Inquiry based learning of contemporary physics topics and gifted students. In: Sokolovska D, Michelini M (eds) The role of laboratory work in improving physics teaching and learning. Springer, Cham, pp 203–215

Čepič M (2017) Introduction of current scientific results to education: experiences from the case of liquid crystals. In: Greczylo T, Debowska E (eds) Key competences in physics teaching and learning: selected contributions from the international conference GIREP EPEC 2015, Wrocław Poland, pp 6–10 July 2015, In: Greczylo T, Debowska E (ed) (Cham: Springer) vol 190, pp 41–54

Endelpohls-Ulpe M, Ruf H (2005) Primary school teachers' criteria for identification of gifted pupils. High Abil Stud 16(2):219–228. https://doi.org/10.1080/13598130600618140

Erasmus+ project RISE and publications on the pages (2023) https://rise.splet.arnes.si/en/

Erasmus+ project 3DIPhE (2020) Čepič M, Sokolowska D (eds) Vol I: inquiry based learning to enhance teaching. https://www.3diphe.si/files/2021/12/3D_volume1_v1_MC.pdf. Accessed 6 Dec 2024

Erasmus+ project 3DIPhE (2020) Čepič, M, De Lange J (eds) Vol II: practitioner's inquiry in the context of inquiry based learning. https://www.3diphe.si/files/2021/12/3D_volume2_v1_MC.pdf. Accessed 6 Dec 2024

Etkina E, Planinšič G, Van Heuvelen A (2019) A college physics: explore and apply. Pearson, New York

Fazio C et al (2021) Inquiry based learning and responsible research and innovation: examples of interdisciplinary approaches at different schooling levels. In: Levrini O et al (eds) Engaging with contemporary challenges through science education research, contributions from science education research, vol 9. Springer, Cham, pp 31–44

Feynman RP, Sands ML (1963–65) The Feynman lectures on physics. Addison-Wesley Pub Co, Reading, MA

Jukić D, Čepič M (2019) Chiroclinic effect and the phase diagram of achiral polar molecules in the antiferroelectric smectic B2. Phys Rev E 99:062705. https://doi.org/10.1103/PhysRevE.99.062705

Osborne J, Simon S, Collins S (2003) Attitudes towards science: a review of the literature and its implications. Int J Sci Educ 25(9):1049–1079. https://doi.org/10.1080/0950069032000032199

Pavlin J, Vaupotič N and Čepič M (2013) Liquid crystals: a new topic in physics for undergraduates. Eur J of Phys 34(3):745–761. https://doi.org/10.1088/0143-0807/34/3/745

Peeters W (n.d.) Private communication

Szydłowska J, Čepič M, Vaupotič N et al (2023) Ferroelectric nematic-isotropic liquid critical end point. Phys Rev Lett 130:216802-1–216802-5. https://doi.org/10.1103/PhysRevLett.130.216802

Takezoe H, Gorecka E, Čepič M (2010) Antiferroelectric liquid crystals: interplay of simplicity and complexity. Rev Mod Phys 82:897–937. https://doi.org/10.1103/RevModPhys.82.897

Remaking Physics Teacher Professional Learning: Designing a National Upskilling Programme for Physics Teachers

Eilish McLoughlin

Abstract One of the major challenges facing physics education internationally is the shortage of qualified physics teachers at the second level. This study discusses the design and implementation of a professional learning programme to upskill in-service second-level teachers to gain qualification as physics teachers. The programme design is informed by international research and practices and incorporates principles that are fundamental to effective physics teacher education, i.e., strong physics knowledge is essential for good teaching, and teachers carrying out a practitioner inquiry on their own practice as part of a professional learning community will enhance their content knowledge for teaching physics. An overview of 54 participating teachers' prior teaching experiences, classroom and reflective practices, is presented. This profile provides invaluable insights into teachers' beliefs and practices when they start the programme and what are their current needs. Further studies will examine the influence of this upskilling programme on teachers' confidence and competence for teaching physics as well as the influence on their broader beliefs and practices.

Keywords Physics teacher education · Teacher professional learning · Pedagogical content knowledge · Professional learning communities · Inquiry-based learning · Practitioner inquiry

1 Introduction

Teacher quality is widely recognized as the primary influence on student learning (Darling-Hammond 2000; Hattie 2009). Unfortunately, meeting the demand for highly qualified science teachers can be challenging and in many countries across the world, science teachers may have to teach outside of their subject specialism–a practice commonly referred to as out-of-field teaching (Mizzi 2021; Luft et al., 2020).

E. McLoughlin (✉)
School of Physical Sciences and CASTeL, Dublin City University, Dublin, Ireland
e-mail: eilish.mcloughlin@dcu.ie

Z. Ješková and J. Hanč (eds.), *Advancing Physics Education*, Challenges in Physics Education, https://doi.org/10.1007/978-3-031-94207-5_3

The impact of such practices is that teachers may lack relevant pedagogical content knowledge in areas that they did not study at advanced or degree level (Kind 2009) and this has been shown to negatively influence instruction and constrain teachers' professional development–particularly in the case of newly qualified teachers (Nixon et al. 2017)

Worryingly, many physics teachers report feeling isolated in their schools and do not have another physics teacher to discuss classroom practices and problems (Krakehl et al. 2020; Sheppard et al. 2020). This sense of isolation can lead to teachers feeling frustrated and if not supported teachers are likely to burn out or leave the profession. In the USA, the shortage of physics teachers has been highlighted as critical in an annual survey of school districts (AAEE 2020). As a result, most high school physics classes are taught by a teacher with little to no formal preparation in physics (Hill and Gruber 2011). PhysTEC's National Report Card on Physics Teacher Preparation examines the need for physics teachers in each US state and concludes that teacher shortage is due to the low teacher preparation rates at US colleges and universities and stresses the need for support to expand physics teacher preparation programmes. As education systems prepare learners to live and work in a complex and connected society, it is becoming increasingly apparent that teacher preparation and continued education also needs to evolve (OECD 2020). The challenge of designing and implementing effective teacher professional learning programmes that influence both teachers (teachers' individual characteristics and their teaching skills) and student learning outcomes is widely debated (McLoughlin and Sokolowska 2023).

The aim of this study is to provide insights into the design and features of a professional learning programme to upskill in-service second-level teachers to become qualified as physics teachers. The Professional Diploma in Teaching Physics (PDITP) was funded by the state government, as part of a response to a lack of qualified physics teachers and high levels of out-of-field physics teaching in Ireland. Two research questions are considered in this study:

1. What are the key features of a national upskilling programme for teaching physics?
2. What are the background and prior experiences of teachers participating in an upskilling programme for teaching physics?

2 Background and Context

Second level education in Ireland has experienced a period of sustained growth with pupil enrolments increasing by more than 30,000 over a five-year period and projected to peak in 2024 (Department of Education 2020). This significant increase in enrolments in second-level schools has created challenges in both infrastructure and the availability of qualified second-level teachers. The shortage of qualified physics teachers has been highlighted in statistics provided by the Teaching Council of Ireland (2023). In 2021, 1115 teachers were registered as qualified to teach physics compared to 2097 qualified to teach chemistry and 3544 qualified to teach biology. On average 46 new teachers registered as qualified physics teachers annually, over the period 2019–2022. This compares to an average of 103 teachers entering as chemistry teachers and 253 entering as biology teachers over the same four-year period.

In 2018, the Department of Education and Skills in Ireland established a Steering Group on Teacher Supply to facilitate multi-agency engagement and lead on a coordinated programme of actions to address emerging concerns. This Group identified an urgent need for a national upskilling programme in three targeted subject areas—Physics, Mathematics, and Spanish. In response to this finding, the Department of Education and Skills (DES) funded a national programme in 2020—the Professional Diploma in Teaching Physics (PDITP)—to develop in-service teachers' content and pedagogical content knowledge for teaching physics. This programme is available to all second-level teachers, with priority given to in-service teachers teaching physics out-of-field. PDITP is a level 8, two-year, part-time 75 ECTS (European Credit Transfer and Accumulation System) programme designed to upskill in-service, second level teachers to meet the requirements stipulated by the Teaching Council of Ireland for teaching physics to Leaving Certificate (upper secondary level). The programme is funded by the Department of Education in Ireland and delivered through collaboration between three Irish universities: Dublin City University, University of Galway and University of Limerick. Four cohorts (154 teachers) have participated in the PDITP programme since it began in January 2021.

The objectives of the PDITP programme are to ensure that teachers:

- Acquire the theoretical and experimental knowledge of physics as well as pedagogical content knowledge that is necessary for effective physics teaching at second level.
- Demonstrate an ability to connect physics content modules and the school physics curriculum.
- Develop a high standard of practical competence in physics teaching as reflective practitioners.

3 Theoretical Basis

In 1986, Shulman articulated and connected three types of content knowledge required by teachers (Shulman 1986, p. 26): subject matter knowledge (comprehension of a subject relevant to a content specialist), pedagogical knowledge (understanding of how a subject is comprehended or misconstrued) and curricular knowledge (consideration of how knowledge is organised for instruction). Pedagogical Content knowledge (PCK) was later proposed by Shulman as one of seven categories of teachers' knowledge: general pedagogical knowledge, knowledge of learners and their characteristics, knowledge of educational contexts, knowledge of educational ends, purposes, values, and their philosophical and historical grounds, content knowledge, and curriculum knowledge (Shulman 1987, p. 8). Developing a teacher's PCK provides them with the required knowledge and skills to transfer discipline content knowledge into specific teaching and learning activities and actions in the classroom and forms the basis for in-service teacher education programmes worldwide (de Winter and Airey 2020).

3.1 Strategies for Teacher Professional Development

Guskey proposes that changes in teachers' attitudes and beliefs depend on teachers' observations of a positive influence on student learning outcomes from changes in their classroom practices (Guskey 1986). Teacher professional development is considered an essential mechanism for deepening teachers' knowledge and developing their teaching practices (Desimone et al. 2002). Teacher engagement in professional learning has been shown to have a positive influence on learners' academic achievement in science and this is increased when it is extended over a long period; involves the engagement of external experts; and provides opportunities to take part in a professional community of practice (Timperley et al. 2007).

Darling-Hammond (2017) examined models of teacher professional development from around the world and highlighted some leading practices in relation to pre-service, induction and in-service teacher professional learning across countries such as, Australia, Canada, Finland, and Singapore. In particular, she identified that supporting teachers to learn from one another and profession-wide capacity building that encourages sharing of research and good practices are two important aspects of successful teacher professional learning (Darling-Hammond 2017). Several strategies and models for teacher collaboration have been reported in recent years, including lesson study and learning study (Cheung and Wong 2014); professional learning communities and communities of practice (Vangrieken et al. 2017), practitioner inquiry (Cochran-Smith and Lytle 1993), co-design (Penuel et al. 2007) and research practice partnerships (Penuel and Gallagher 2017).

Many reviews of science teacher professional development programmes emphasize the need for science content and teachers' active participation in following or experiencing model lessons, planning lessons, enacting and reflecting on lessons, as well as sufficient duration and support from experts. In the context of inquiry-based science education, effective teacher professional development should include: sufficient time for learning, support from experts and peers, authentic experience, similar to that which they will later enact in their classroom, coherency with the curriculum, design of lessons, working through learning content as modelled inquiry, time and structure for reflection, transference of the professional development materials to the classroom and developing content knowledge (Capps et al. 2012).

Recent research highlights the need for professional development to explicitly consider teacher identity, and that identity formation is an important consideration for out-of-field teachers (Kenny et al. 2020). They emphasise that alignment between the specific professional development needs of out-of-field teachers and the planned learning experiences is essential to maximise sustained learning and applicability to teachers' classrooms. Sancar et al. (2021) suggest several issues that should be considered in the professional development process: the formative assessment process; large-scale research processes; diversification of the scope of research processes, especially across individual, institutional, and governmental levels; dissemination for applying the PD process; long-term process-oriented research over short-term research; collaboration between individuals, institutions, and organizations; context-oriented adjustments; long-term support and supervision.

3.2 Strategies for Physics Teacher Education

Physics teachers require a complex set of knowledge and skills to effectively engage and facilitate their students' learning. This includes professional knowledge, such as physics content knowledge (Meltzer et al. 2014), as well as the skills to plan a physics lesson and reflect on physics instruction (Vogelsang et al. 2022). Leinonen et al. (2020) developed a physics knowledge for teaching (PKT) framework, based on Shulman's PCK model, to analyse physics teacher education programmes across different domains of teacher knowledge. In their study with graduate physics teachers, they reported that domains related to content knowledge (referring to pure physics content e.g. methods, theories, laws terminology) were appropriately addressed while deficiencies were evident in teacher's PCK across various sub-domains (knowledge of content and students, knowledge of content and teaching and knowledge of content and curriculum) (Leinonen et al. 2020).

Etkina (2010) highlights the importance of building both teacher's physics knowledge and physics PCK in her physics teacher preparation programme and provides evidence of teachers learning physics content, learning physics processes and the ability to engage students in active learning of physics. Etkina et al. (2018) conceptualized Tasks of Teaching as activities that permeate every aspect of a physics teacher's professional life related to student learning: anticipating student thinking around science ideas; designing, selecting, and sequencing learning experiences and activities; monitoring, interpreting, and acting on student thinking; scaffolding meaningful engagement in a science learning community; explaining and using examples, models, representations, and arguments to support students' scientific understanding; and using experiments to construct, test, and apply concepts. Vokos et al. (2022) propose that teachers need to create an inclusive classroom in which every student grows intellectually and emotionally and develops a robust physics identity. They surmise that the foundation of such teaching lies in teacher dispositions and strategic enactment of content knowledge for teaching physics. They further warn that these are not enough and only when teachers develop productive habits can they enact faithfully the everyday Tasks of Teaching, which lead to all students being engaged in "doing" physics and feeling capable of doing so.

3.3 Inquiry Approaches in Physics Education

Understanding physics phenomena is one of the most necessary endeavours for today's learners and society and requires teachers to design and adopt inquiry approaches for learning and teaching physics that embody the true nature of this discipline (McLoughlin and Sokolowska 2022). The use of Inquiry Based Learning (IBL) has been widely promoted in European Science Education since 2007, because of funding provided by the European Union Seventh Framework programme to increase the use of IBL in primary and second level schools (Finlayson and McLoughlin 2017). Practitioner inquiry (PI) has been promoted as an effective strategy for teacher professional learning defined as the systematic, intentional study of one's own professional practice (Cochran-Smith and Lytle 1993). PI empowers teachers to make evidence informed professional judgements and changes in classroom practice that influence their student learning (Cochran-Smith et al. 2009). PI involves teachers identifying problems, constructing inquiry questions, gathering, and analysing data to make evidence-based conclusions and recommendations with respect to their chosen problem. Teachers engage in systematic reflection and act for change by asking questions or "inquiries", gathering data to explore their inquiry, analysing the data, making changes in practice based on knowledge constructed, and sharing learning with others as part of professional learning communities (Dana and Yendol-Hoppey 2015). The design and content of the pedagogical approach adopted in this programme was strongly informed by the author's collaboration in the Erasmus + project Three Dimensions of Inquiry in Physics Education and (3DIPhE) and key learnings from this project which have been prepared as eBooks for educators on IBL, PI and PLCs (de Lange 2020; Lovatt et al. 2020; Peeters 2020; Sokolowska 2020).

4 PDITP Programme Design

4.1 Physics Content Modules

The programme design reflects specific demands outlined by the Department of Education in Ireland which could not be varied. While completing the PDITP programme teachers would remain employed in their schools and undertake the programme via a blended learning approach. Graduates of the programme would meet the Teaching Council of Ireland (https://www.teachingcouncil.ie/) requirements for fully qualified secondary physics teachers. The programme design also includes principles that are fundamental to effective physics teacher education. These include strong physics knowledge is essential for good teaching, and teachers carrying out a practitioner inquiry on their own practice as part of a professional learning community, which will enhance their content knowledge for teaching physics.

The PDITP programme (shown in Table 1) consists of 15 modules (75 ETCS) delivered on a part-time basis over two years and delivered through online and face-to-face lectures, tutorials, laboratories and workshops. The ten physics modules (50 ECTS) are delivered online using both synchronous and asynchronous modes. For each module, teachers are required to watch 24 h of pre-recorded lectures and complete weekly tutorials assigned on associated virtual learning platforms (e.g. Moodle, Mastering Physics or Webworks). Teachers attend 2-h online tutorial sessions on Mondays or Wednesday evenings from 6–8 pm, held weekly during the term (using Zoom). The assessment element of these modules is based on the number of completed weekly tutorials (40%), one mid-term online examination (30%) and one final online examination (30%).

Physics laboratory modules (10 ECTS) are held in person on the campus of the three partner universities. Laboratories are facilitated in the University physics laboratory on six Saturdays each semester. Teachers have the option of choosing which venue is most convenient for them and must attend for 18 h of laboratory work. The focus of laboratories each semester is aligned with the content of modules delivered during that period. Teachers collaborate in pairs to complete experimental activities that promote the development of understanding and skills using an inquiry-based approach in the physics laboratory developed at Dublin City University (Wemyss

Table 1 Overview of PDITP programme and modules

Year	Period	Modules	
1	January–May	Mechanics (5 ECTS) Waves, light and modern physics (5 ECTS) Thermal physics (5 ECTS) Physics laboratory (2.5 ECTS)	
1	June–August	Pedagogy and research in physics education (5 ECTS) Practitioner inquiry in physics education 1 (5 ECTS)	
1	September–December	Electricity (5 ECTS) Modern physics (5 ECTS) Physics laboratory (2.5 ECTS)	
2	January–May	Environmental physics (5 ECTS) Semiconductor devices (5 ECTS) Life, the universe and everything (5 ECTS) Physics laboratory 2 (2.5 ECTS)	Practitioner inquiry in physics education 2 (5 ECTS)
2	September–December	Electricity and magnetism (5 ECTS) Optics (5 ECTS) Physics lab 3 (2.5 ECTS)	

2010). The assessment element of these modules is based on submission of laboratory reports for the first three semesters. In the final semester teachers are required to design and carry out their own investigations on a physics topic of their choosing.

4.2 Physics Pedagogy Modules

The three physics pedagogy modules (15 ECTS) are facilitated online, through summer schools and evening workshops. The first pedagogy module, Pedagogy and Research in Physics Education, is facilitated as a four-day (20 h) online summer school in June of year one. The module introduces teachers to physics pedagogy, practices and physics education research. Teachers are introduced to the Irish Senior Cycle (Leaving Certificate) Physics curriculum and assessment requirements. Teachers consider the importance of developing physics content knowledge as well as physics pedagogical content knowledge to being effective teachers of physics. An emphasis on the development of student key skills, and on incorporating literacy and numeracy, within the physics classroom is also promoted. Nine 1.5-h workshops are facilitated over the four days by university teacher educators as well as invited workshops from experienced second level physics teachers. Topics include Introduction to what research says about teaching physics; Equity and inclusion in physics learning and teaching; The Physics Teacher's Toolbox; Understanding mental models in physics education; Outdoor learning and STEM Education; Physics Hooks; Social Scientific Issues and Argumentation in Physics; Interplay between mathematics and physics; and designing a Unit of Learning in Physics. In addition, teachers are facilitated to collaborate in pairs to select and design a unit of learning on a topic from the Leaving Certificate Physics syllabus. Teachers are required to refer to literature that informs the design and implementation of their learning unit. The assessment element of this module is based on teachers presenting their unit of learning and associated lesson plans that incorporate appropriate pedagogical strategies, learning approaches and a variety of modes of assessment to support diverse learners in their physics classroom. The design and content of the second and third pedagogy modules were adapted from the approach developed by Lovatt et al. (2020) as part of the Erasmus+ project Three Dimensions of Inquiry in Physics Education (3DIPhE). The three authors of this report collaborated to design two linked pedagogy modules to support teachers in adopting inquiry approaches in physics education, based on their co-reflections and evaluation of learnings from the 3DIPhE project, and consideration of the needs of PDITP teachers.

The second pedagogy module, Practitioner Inquiry in Physics Education 1, is facilitated as a five-day (24 h) online summer school in June of year one. The module is delivered as a series of workshops that introduces teachers to Inquiry Based Learning (IBL) and Practitioner inquiry (PI) and fosters a professional learning community (PLC) among teachers and academics, as outlined in Table 2 (adapted from Lovatt et al. 2020). The teachers are facilitated to carry out six hours of experimental activities over the first 3 days with the focus of each activity mapped to stage(s) of an

IBL cycle (Sokolowska 2020, p. 12). Teachers are required to collect and prepare the necessary resources so that they can complete these experiments in-person with remote facilitation by teacher educators. Five hours of PI activities are facilitated over the first four days mapped to stage(s) of a PI cycle (de Lange 2020). Three hours of workshops are facilitated towards building and fostering a PLC among participants and this approach is adopted throughout all module activities (Peeters 2020). Time is allocated during each day to support teacher collaboration and reflection. The assessment element of this module is based on teachers' submission of two written assignments. The first assignment is an individual reflection on their current understanding of IBL, PI and PLCs and draws on evidence from module activities and their own classroom practice. The second assignment involves collaborating in a PLC group (comprising of two or three individuals) to prepare an Inquiry Brief–made up of a PI question and detailed plan of how they will conduct their PI.

Table 2 Overview of practitioner inquiry in physics education 1 module

Day		Focus of activities			
		IBL	PI	PLC	
Monday	• Generating ideas	*			Time allocated for teacher collaboration and reflection
	• Forming IBL questions	*			
	• Introduction to PI		*		
	• Building a PLC			*	
Tuesday	• Different types of inquiry	*			
	• Forming PI questions		*		
	• PI data types and collection		*		
	• Understanding each other's contexts			*	
Wednesday	• Planning and carrying out inquiries	*			
	• Drawing conclusions	*			
	• Sharing findings	*			
	• Refining PI questions		*		
	• Ethical considerations		*		
	• Planning a PI		*		
Thursday	• Zones of comfort			*	
	• PI data analysis and interpretation		*		
Friday	• Sharing of inquiry brief	*	*	*	
	• Providing peer feedback	*	*	*	
	• Refining inquiry brief	*	*	*	

The third pedagogy module, *Practitioner Inquiry in Physics Education 2*, is designed to support teachers as they conduct their own PI. The module is delivered online from September of year two – September of year three). Each PLC group is assigned a tutor who provides guidance and supports the group during the second summer school and throughout the academic year. During September to December check in meetings, both individual and whole group, are facilitated to support teachers in preparing their PI, collecting data and making initial observations on their data. The assessment element of this module is based on teachers completing three assignments. Teachers present a brief update on their PI plan and initial findings in January and give a poster presentation on their PI question, plan and key findings in May. The focus of all meetings is to provide opportunities for teachers to share their learning at different stages of the PI cycle and receive feedback from their peers and tutors. The third assignment is the submission of a written report and reflection on their PI due in September of year three.

5 Profile of PDITP Teachers

The following section presents a profile of the 54 (30 female, 24 male) teachers that had self-selected to enrol and complete the PDITP programme during 2022–2024. All participating teachers were practising in-service teachers and were registered with the Teaching Council in Ireland as qualified teachers—mainly qualified to teach mathematics, chemistry and biology. They ranged in teaching experience from 1 to 20 + years. An online questionnaire was shared with teachers at the end of semester one in June 2022. This questionnaire was co-designed by the researchers who designed the PI pedagogy modules and adapted from the approach developed by these researchers as part of the 3DIPhE project (Lovatt et al. 2020). At the time of completing this questionnaire the teachers had completed assessments and received grades for their first three physics modules but had not yet commenced the summer school pedagogy modules (see Table 1). Researchers provided an overview of the questionnaire, including its purpose, content and use after which teachers were required to complete an informed consent form in accordance with ethical approval received for this study from the researchers University Ethics Committee.

Teachers were asked to provide details on their years of prior teaching experiences, understanding and use of inquiry-based learning, classroom and reflective practices. Collecting data and examining the profile of teachers on the PDITP programme provides invaluable insight into teacher's experiences and practices in advance of the first two pedagogy modules and data was used for forming working groups of teachers to engage with workshop activities. Responses from the PDITP teachers are presented for the 30 female and 24 male teachers in the following tables (Table 3 and Figs. 1, 2, 3, 4 and 5). Most of the teachers (72%) reported they had been teaching at second level for over 5 years and 15 of the teachers (28%) reported they were newly qualified and were teaching less than 4 years (Table 3).

Table 3 Overview of PDITP teachers' years of teaching

Years of teaching	Total (N = 54) (%)	Female (N = 30) (%)	Male (N = 24) (%)
1–4 years	26	27	29
5–10 years	43	43	38
10–15 years	18	20	17
15–20 years	6	7	4
20 + years	7	3	12

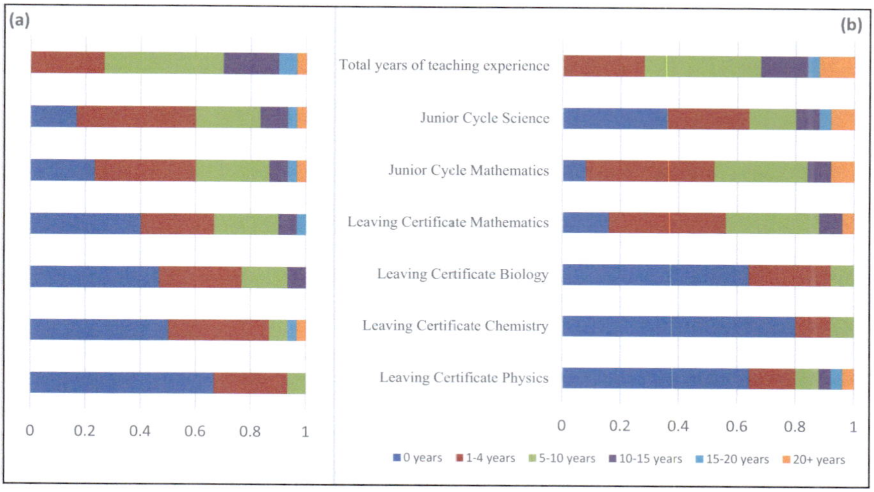

Fig. 1 Overview of teachers teaching experiences **a** female teachers (N = 30) and **b** male teachers (N = 24)

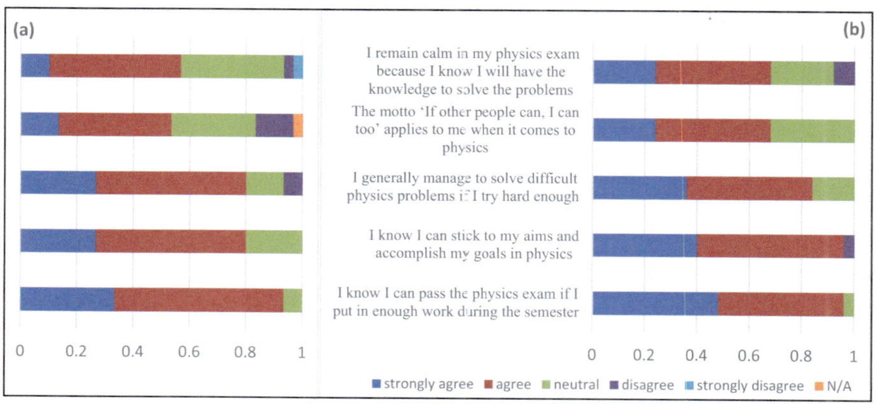

Fig. 2 Teachers' self-efficacy **a** female teachers (N = 30) and **b** male teachers (N = 24)

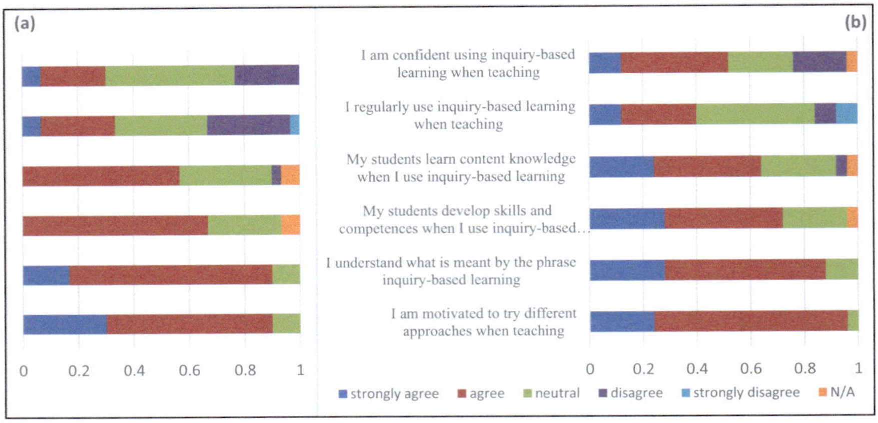

Fig. 3 Teachers experience and use of using inquiry based learning **a** female teachers (N = 30) and **b** male teachers (N = 24)

Teachers were asked to provide details on their experiences of teaching STEM subjects at Junior Cycle level (lower secondary with students aged 12–15 years) and Leaving Certificate level (upper secondary with students ages 15–19 years) (Fig. 1). One-third of teachers (33%) reported they were already teaching Leaving Certificate physics in their schools and are designated as out-of-field physics teachers. Most of the male teachers had experience in teaching mathematics, 19 (83%) at leaving certificate level and 22 (92%) at junior cycle level, compared to the female teachers, 18 (60%) at Leaving Certificate level and 23 (77%) at Junior Cycle level. Female teachers had more experience of teaching Biology and Chemistry at Leaving Certificate level compared to male teachers. Only 9 (38%) male teachers reported they were qualified to teach Leaving Certificate Biology, compared to 16 (53%) female teachers. Only 5 (21%) male teachers reported they were qualified to teach Leaving Certificate Chemistry compared to 15 (50%) female teachers.

5.1 Teachers' Self-Efficacy

To measure participants' self-efficacy for learning physics, Lindstrøm's and Sharma's (2011) validated, self-efficacy questionnaire was used. Teachers were asked to respond to five statements using a five-point Likert scale (strongly disagree, disagree, neutral, agree and strongly agree) and responses for female (N = 30) and male (N = 24) teachers are shown in Fig. 2. The calculation of teacher's self-efficacy scores shows that the overall self-efficacy for learning physics is higher for male teachers compared to female teachers. Most of the teachers (over 95% for all teachers agreed/ strongly agreed) reported very high confidence that they "can pass physics exams

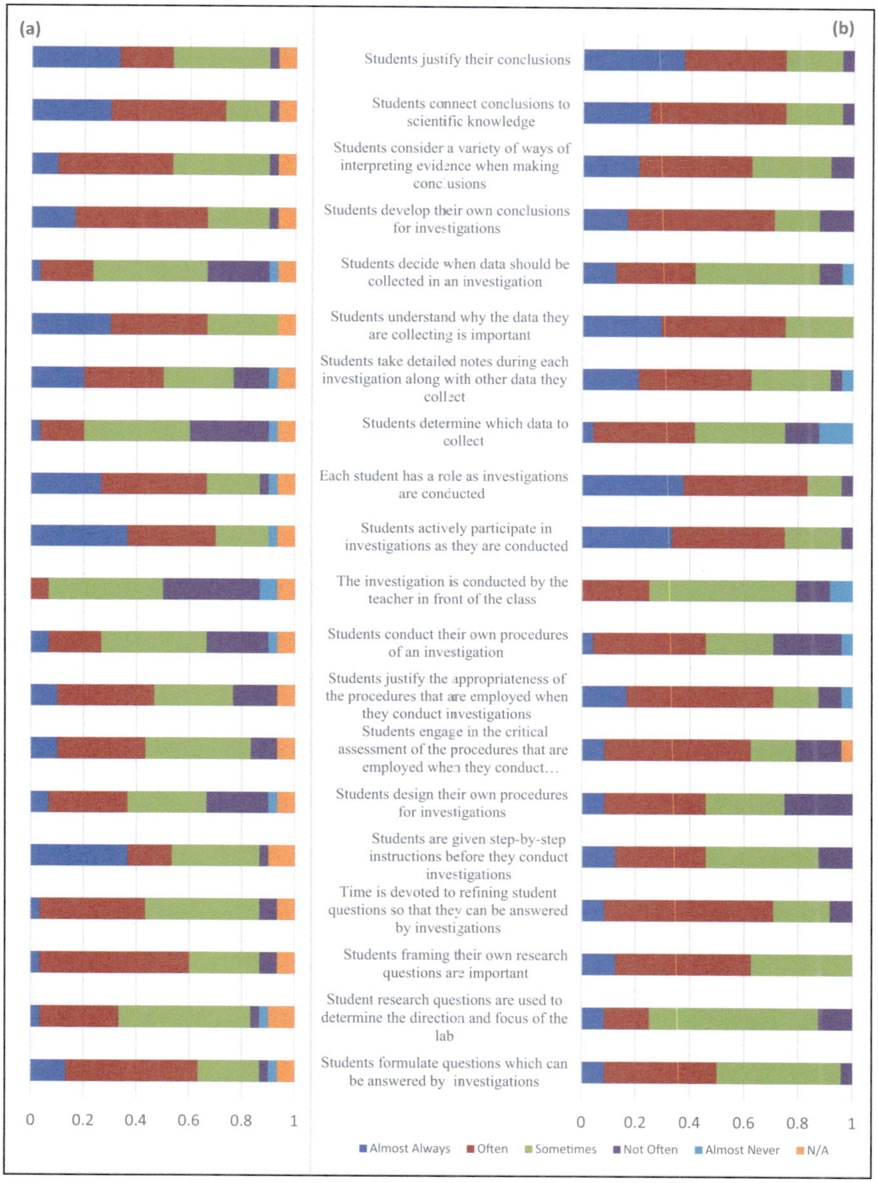

Fig. 4 Teachers classroom practices **a** female teachers (N = 30) and **b** male teachers (N = 24)

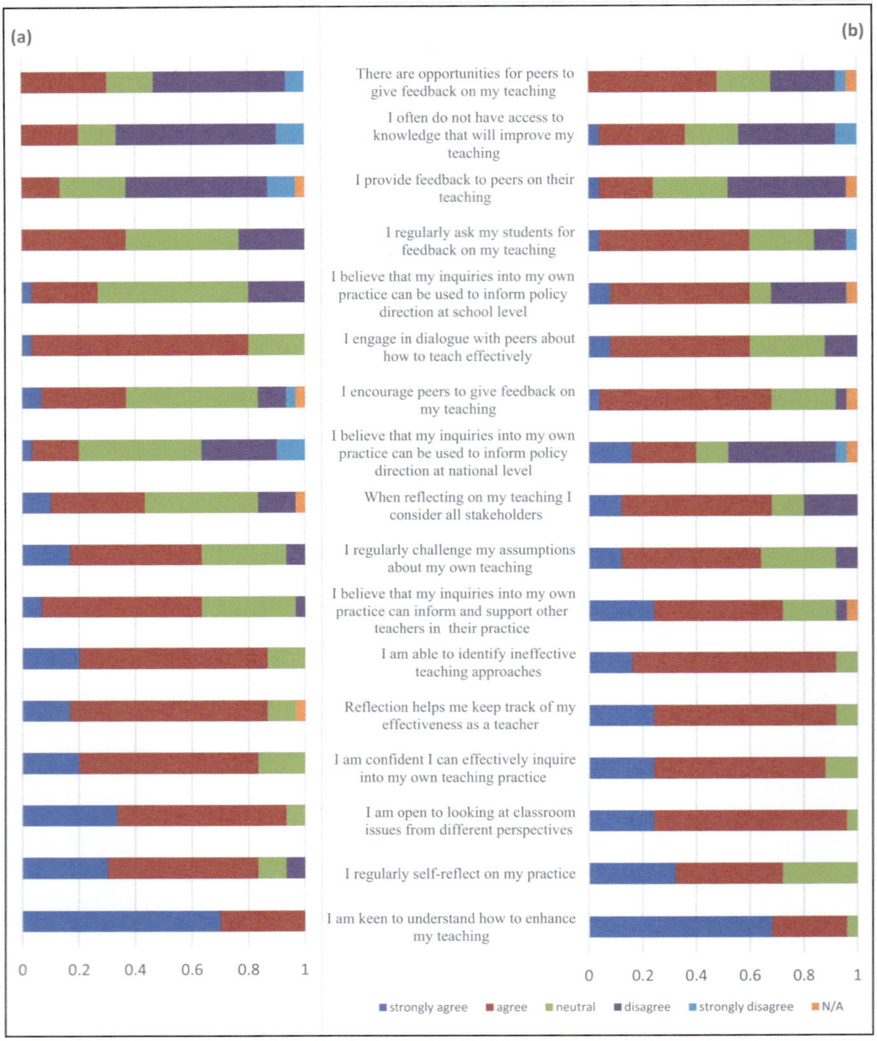

Fig. 5 Teachers' reflective practices **a** female teachers (N = 30) and **b** male teachers (N = 24)

if I put in enough work during the semester". Teachers demonstrated less confidence in "having the knowledge to solve the problems" (68% of male teachers and 57% of female teachers agreed/strongly agreed). The self-efficacy of this cohort of teachers has been shown to be higher than that of undergraduate students and gender differences have been shown to be not significant (Gammell et al. 2024).

5.2 *Teachers' Use and Understanding of Inquiry-Based Learning*

Teacher responses to six statements that examined their use and understanding of inquiry-based learning (IBL) are presented in Fig. 3. This questionnaire was developed by European physics teacher educators (including the author) during the 3DIPhE project (De Lange 2020). For each statement, participants were presented with a five-point Likert scale (strongly disagree, disagree, neutral, agree and strongly agree). Teacher reported a strong understanding of IBL (over 92% agreed/strongly agreed) and motivation for trying different approaches in their classroom practices (over 90% agreed/strongly agreed). Male teachers indicated strong belief that students could develop both content knowledge (67% agree/strongly agree) and skills (75% agreed/strongly agreed) through IBL, compared to female teachers (57% agreed for content knowledge and 67% agreed for skill development). However, less than half the teachers (42% of male and 33% of female teachers) agreed/strongly agreed that they regularly use IBL in their teaching, with male teachers expressing stronger confidence in using IBL (54% of male and 30% of female teachers agreed/strongly agreed).

5.3 *Teachers' Classroom Practices*

To measure participants' perspectives on the extent to which students are engaged in scientific inquiry during their classroom practices, Campbell et al.'s (2010) validated questionnaire was used. For each of the twenty items on the questionnaire, participants were presented with a five-point Likert scale (almost never, not often, sometimes, often, almost always). As shown in Fig. 4, female and male teachers had comparable responses to practices involving students' active participation in investigations, collecting and analysing data, and drawing conclusions. For none of the twenty items, was the proportion of female responses of often/almost always greater than that of the male teachers. Male teachers reported greater use of student-led approaches in facilitating students to design and justify their decisions for planning and carrying out investigations compared to female teachers. For example, most of the male teachers (71% response often/almost always) indicated that "Time is devoted to refining student questions so that they can be answered by investigations" compared to less than half of the female teachers (43% response often/almost always). Three-quarters of the male teachers (75% response often/almost always) indicated that "Students justify their conclusions" compared to just over half of the female teachers (53% response often/almost always).

5.4 Teachers' Reflective Practices

Teacher responses to statements that examined their reflective practices are presented in Fig. 5. This questionnaire was developed by European physics teacher educators (including the author) during the 3DIPhE project (De Lange 2020). For each statement, participants were presented with a five-point Likert scale (strongly disagree, disagree, neutral, agree and strongly agree). Female and male teachers had comparable responses on their confidence in reflecting and the importance of reflection. Male teachers reported stronger engagement in giving and receiving feedback, including feedback from peers and their students, than female teachers. For example, 50% of males agreed with the statement "There are opportunities for peers to give feedback on my teaching", compared to 30% of female teachers. Female teachers reported engagement in self-reflection and dialogue with peers, compared to the male teachers. Most female teachers (80%) agreed/strongly agreed with the statement "I engage in dialogue with peers about how to teach effectively compared to a smaller proportion of male teachers (60% agreed/strongly agreed)". Most female teachers indicated "I regularly self-reflect on my practice (83% agreed/strongly agreed) compared to a smaller proportion of male teachers (71% agreed/strongly agreed)".

5.5 Teachers' Challenges in Teaching Physics

Teachers were asked to complete an open response question on what challenges they encounter/perceive in teaching physics. A total of 85 individual responses were received (from the 54 respondents) and these were coded into the themes presented in Fig. 6. The percentages indicated for each theme represents the proportion of responses coded to each theme. Over half of the responses were related to gaps in teachers own knowledge and skills, e.g. 19% of responses related to teachers' lack physics knowledge. A small number of teachers commented on lacking relevant PCK (9% of responses) or lacking confidence in teaching a new subject (8% of responses). Over a quarter of responses highlight concerns about lack of resources and support, e.g. 21% of responses mentioned lack of equipment. Less than one-fifth of the responses considered what was the background and needs of students studying physics, e.g. 2% of responses mentioned diversity of student's needs. A small number of responses (2%) raised concerns about the influence of state examinations at the end of Junior Cycle Science (Lower Secondary) and Leaving Certificate Physics (Upper Secondary) on pedagogical approaches and assessment strategies.

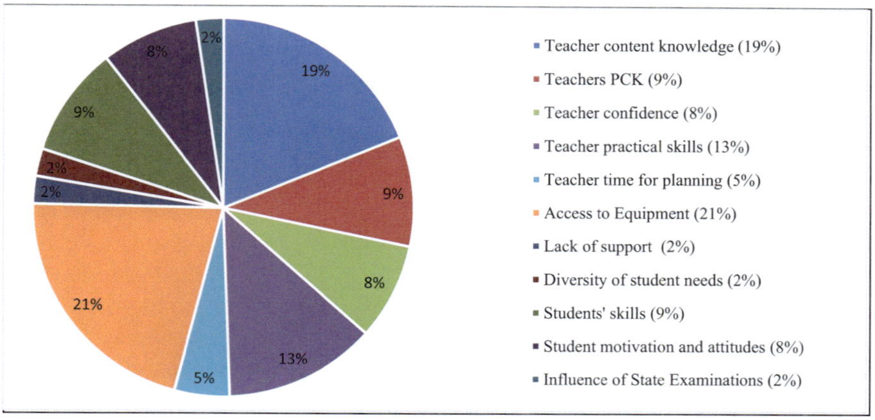

Fig. 6 Overview of PDITP teachers challenges in teaching physics (number of statements = 85)

6 Discussion

This paper discusses the design of a national blended learning programme for 'out-of-field' physics teacher professional learning in Ireland. The design PDITP programme was developed to meet the anticipated needs of in-service teachers who self-selected to enrol in this national upskilling programme and to obtain a qualification to teach physics. The programme design was shaped by restrictions set by the Department of Education in Ireland (2020), such as '*Participation on the programmes should not impact on class contact time for currently serving teachers. Flexible provision should enable delivery on a national scale with a regional aspect as appropriate*'. These restrictions dictated that all synchronous activities must take place outside of school hours, such as evenings, weekends and school breaks.

Collecting data and examining the profile of teachers enrolled in the PDITP programme provided educators with invaluable insights into the differences between individual teachers' experiences and practices and supported the formation of groups and support activities throughout the two years. Figure 6 provides useful insights into participating teachers' confidence and concerns about teaching and learning physics. Less than one-fifth of teachers expressed concern about their lack of physics content knowledge or concern about their ability to support the diverse needs of their students. Most of the teachers (over 90%) did not express concern about lacking pedagogical content knowledge (PCK) or lacking confidence for teaching physics (Fig. 6) – this may be attributed to the fact that 39 of the teachers have already been teaching for over 5 years, whilst the other 15 teachers have been teaching 1–4 years. Schiering et al. (2023) report that pre-service physics teachers' PCK development depends on different factors, such as content knowledge, teaching experiences and beliefs, and transitions into higher levels of PCK additionally require teaching experience and adequate beliefs about teaching and learning. The three PDITP pedagogy modules were designed to support teachers in developing their competence and confidence for teaching physics, drawing on these different factors. Facilitating teachers to collaborate as professional learning communities (PLCs) supported experienced and inexperienced teachers to openly share their knowledge, experiences and classroom practices.

A core consideration in the design of the PDITP programme is related to the role of mathematics in the teaching and learning of physics. De Winter and Airey (2022) reported on pre-service physics teachers' perceptions of the role mathematics plays in the learning of physics at university and school levels. They identified three concerns expressed by pre-service teachers: The necessity of mathematics to be successful at physics; Success in mathematics can often mask a lack of conceptual understanding in physics; There can be a disconnect between the physics and mathematical aspects of courses (De Winter and Airey 2022). One of the eligibility requirements for enrolment in the PDITP programme was that teachers must present some evidence of proficiency in mathematics at second or University level. The data presented in Fig. 1 shows that most of the teachers that participated in the PDITP programme in January 2022 had a strong background in mathematics and most teachers (45 of the 54 teachers) were qualified to teach mathematics at second level—23 female teachers and 22 male teachers were currently teaching Junior Cycle mathematics compared to 18 female teachers and 20 male teachers currently teaching Leaving Certificate mathematics. The remaining 9 teachers had completed mathematics modules at university level as part of the undergraduate education and were currently teaching one or more of the Science subjects at upper secondary—Biology (8 teachers), Chemistry (5 teachers) or Physics (3 teachers). Each of the ten physics modules provided additional support materials on the mathematical concepts used. This targeted support provided invaluable support for teachers who had not recently studied mathematics. Notably, teachers did not identify a lack of mathematical knowledge or skills as a challenge in teaching physics (Fig. 6).

Another objective of the PDITP programme was to provide teachers with hands-on experience of experimental investigations and to develop teachers' confidence for planning and facilitating practical work with their own students. Many of the teachers (40 of the 54 teachers) were currently involved in teaching Science and had experience of planning and carrying out investigations (Fig. 1) –40 teachers teaching Junior Cycle science, 24 teaching Leaving Certificate Biology and 20 teaching Leaving Certificate Chemistry. 14 of the teachers (all mathematics teachers) had no laboratory experience, and they found laboratory work very daunting initially. Despite most of the teachers having experience completing Biology and Chemistry laboratories, nearly half of the teachers expressed concerns about a lack of relevant knowledge, skills and support for physics experimental activities (Fig. 6). To address this challenge, teachers were facilitated to carry out laboratory activities in pairs and were facilitated to select their laboratory partner. A guided-inquiry approach was used in the laboratory to support teachers in sharing and discussing their ideas towards developing both conceptual understanding and practical skills. The physics laboratory sessions provided a unique opportunity for the teachers to meet in person, and teachers engaged in rich discussions and collaboration at all stages of the investigation—setting up experiments, recording measurements, analysing data and drawing appropriate conclusions.

The design of this and other national up-skilling programmes for second-level teachers has been informed by the design of national professional learning programme for out-of-field teachers of mathematics, the Professional Diploma in Mathematics for Teaching (PDMT). Goos et al. (2020) discuss and reflect on the evolution of the design principles for this programme over an 8-year period (2012–2019). They identified new practices and innovations that were introduced over this period, drawing on the capacity of blended learning to improve the learning environment and programme delivery. The authors' retrospective analysis of the evolution of the programme implementation were based on the concept of boundary crossing (Akkerman and Bakker 2011). Their analysis of the PDMT drew on three perspectives of boundary crossing: the perspective of the out-of-field teacher moving between different disciplinary fields; the perspective of those who designed and taught in the programme and negotiated discontinuities between the practices of mathematics and mathematics education; and the blended learning perspective that coordinates face-to-face and computer mediated teaching and learning (Goos et al. 2020). The authors observed that the PDMT programme evolved in an ill-defined educational environment that was challenging for operationalising a blended learning design and reflected that the practice-based knowledge they had gained was of benefit to the wider research community.

The design of the PDITP programme evolved since it was first considered in 2020. The first roll-out of the programme in January 2021 coincided with school closures in Ireland due to the Covid-19 pandemic. Consequently, the first cohort of PDITP teachers were facilitated to complete virtual and remote physics experiments during the first year of the programme and all physics pedagogy modules were re-designed to be facilitated as online workshops. For the second cohort of PDITP teachers that began in January 2022 and subsequent cohorts, the pedagogy modules have remained as online workshops, in response to teachers requests and other national considerations (e.g. accommodation shortages, increased transport costs). All physics modules have remained as fully online modules. All modules incorporated a range of synchronous and asynchronous activities (Heilporn 2021). The importance of a clear, unified course structure and ensuring a sustained pace throughout the duration of the course has been highlighted by Heilporn (2021) as the core consideration for choosing between synchronous and asynchronous activities.

A blended approach, using both synchronous and asynchronous activities, was adopted in the PDITP to provide teachers with a healthy balance between activities that they can carry out at time of their own choosing and scheduled sessions. Asynchronous activities, such as video recorded lectures and homework assignments, were provided for all physics modules so teachers could complete these activities during the week prior to an online synchronous tutorial facilitated by the educator. Synchronous activities, real-time interactions between PDITP teachers and educators, were scheduled, to allow teachers to receive immediate feedback from peers and educators, and consolidate learning (Farrell et al. 2019; Owston et al.

2020). Synchronous activities were scheduled to be aligned with teacher's availability and included online physics tutorials two evenings per week during semester (Mondays and Wednesdays 6–8 pm), on-campus physics laboratories three Saturdays per term, and online workshops for pedagogy modules as summer schools and evening workshops.

Use of technology is inherent to physics/science education. Teacher education that focuses on digital technologies and digital learning within physics/science contexts is a current focus of research, policy and practice (Nipyrakis and Stavrou 2022). The PDITP programme design and implementation (since 2020) has been informed by increased focus in literature around understanding and usage of terms for modern learning contexts which include a range of in-person and technology-mediated instruction types as well as a range of methods for teaching and learning in these contexts (Margulieux 2016). Learning contexts may be somewhere on a continuum from face-to-face (F2F) instruction, where students gather with the instructor in person, to wholly technology-based learning, as in online and distance learning (Margulieux 2016). Mixed instruction types use a combination of F2F and instruction via technology, and include hybrid, blended and combination modalities (Margulieux 2016). When discussing learning contexts, they can be differentiated by the delivery medium (how instruction is given) and instruction type (what instruction is given) (Margulieux 2016).

7 Conclusion

The key feature of the PDITP programme is that it provides opportunities for teachers to deepen their knowledge and understanding of physics concepts, alongside developing the pedagogical content knowledge that is necessary for effective physics teaching at the second level. A core objective of the PDITP programme was to develop a high standard of practical competence in physics teaching as reflective practitioners. A key challenge for the PDITP programme was to devise a programme structure and timetable for synchronous and asynchronous activities that incorporate both online and in-person learning environments, at convenient times for in-service teachers, who were generally in full-time employment. This paper provides insights into the key considerations for the design of the PDITP programme that addresses the needs of the participating teachers and scaffolds their learning over a two-year period. Details of the programme structure, design principles and considerations are shared to provide insights and inspiration to educators from other disciplines and contexts.

The profile of participating teachers presented in this study provides invaluable insights into the diversity of experiences and practices of in-service teachers completing the PDITP programme. In general, PDITP participating teachers are qualified and actively teaching STEM subjects, mathematics or chemistry, or biology at leaving certificate level and mathematics or science at junior cycle level in Ireland. Most of the teachers (76%) are already teaching physics at the lower second level as part of junior cycle science. Many of the teachers (36%) are already teaching Leaving Certificate Physics as 'out-of-field' teachers. Most of the teachers (72%) had over 5 years of teaching experience, and had different levels of experience in their classroom and reflective practices. Teachers' backgrounds and needs were anticipated in the initial design of the programme and changes to design, and implementation have been informed by analysing this profile data and obtaining regular feedback from teachers and educators. Additional statistical analysis of the data collected from teachers on their self-efficacy for learning and teaching physics, gender, experiences and practices have been carried out (Gammell et al. 2024; McLoughlin et al. 2024). This analysis reported that self-efficacy for learning physics is high amongst PDITP participants (compared to undergraduate physics students), that there is no evidence of a difference in self-efficacy based on gender, and that prior teaching of some subjects correlates with increased self-efficacy for learning physics.

Additional studies will be carried out to examine the influence of the PDITP programme on teachers' beliefs and practices within one year of completing the programme. Such studies will support teacher reflection on the influence of the programme on their teaching and learning of physics and other STEM subjects. Teacher reflection on their own experiences is expected to deepen their awareness and anticipation of student thinking around physics/STEM concepts and influences their planning and facilitation of learning. Teachers will be asked to reflect on the influence of this programme on their own professional practices of teacher collaboration, reflection and classroom-based research. Teacher collaboration as part of a Professional Learning Community (PLC), sharing and learning from each other's experiences, was a core principle of the design and implementation of the pedagogy modules. This study will examine if teachers continue to engage in PLCs, conduct practitioner inquiries that focus on student learning outcomes (knowledge, skills, attitudes and values) or engage in self- or peer- reflection practices towards improving the quality of physics teaching and learning in secondary schools.

Acknowledgements The Professional Diploma in Teaching Physics (PDITP) acknowledges funding support from the Department of Education and Skills in Ireland that has enabled four cohorts of teachers to participate in the programme since January 2021. This programme has been implemented through a collaboration between academic and researchers from across three Irish Universities, Dublin City University, University of Galway, and University of Limerick. Paul Grimes and James Lovatt have collaborated to co-design the two PI pedagogy modules and co-design and facilitate workshops with teachers that support online and blended learning. Paul Grimes and James Lovatt have collaborated to co-design the teacher questionnaire described in this study. Academics from three Irish Universities have collaborated to design and facilitate physics lecture and laboratory modules for synchronous and asynchronous delivery. In particular Liz Coleman, Pierre Cazade and Stephen Gammell have made important contributions to the success of the PDITP programme. We gratefully acknowledge the commitment and dedication of the in-service teachers that have participated in the PDITP programme and given consent for their data to be published in research articles.

References

Akkerman S, Bakker A (2011) Boundary crossing and boundary objects. Rev Educ Res 81:132–169. https://doi.org/10.3102/003465431140443

American Association of Employment in Education (AAEE) (2020) 2019–2020 Educator supply and demand report. https://specialedshortages.org/wp-content/uploads/2021/07/2020-2021-AAEE-Educator-Supply-and-Demand-Report.pdf. Last accessed 4 Dec 2024

Capps D, Crawford B, Constas M (2012) A review of empirical literature on inquiry professional development: alignment with best practices and a critique of the findings. J Sci Teach Educ 23(3):291–318. https://doi.org/10.1007/s10972-012-9275-2

Cheung WM, Wong WY (2014) Does lesson study work? A systematic review on the effects of lesson study and learning study on teachers and students. Inte J Lesson Learn Stud 3(2):137–149. https://doi.org/10.1108/IJLLS-05-2013-0024

Cochran-Smith M, Lytle SL (1993) Learning from teacher research: a working typology. In: Inside/outside: teacher research and knowledge. Teachers College Press

Cochran-Smith M, Barnatt J, Friedman A, Pine P (2009) Inquiry on inquiry: practitioner research and student learning. Action Teach Educ 31(2):17–32. https://doi.org/10.1080/01626620.2009.10463515

Dana NF, Yendol-Hoppey D (2015) The PLC book. Corwin Press

Darling-Hammond L (2017) Teacher education around the world: what can we learn from international practice? Eur J Teach Educ 40(3):291–309. https://doi.org/10.1080/02619768.2017.1315399

Darling-Hammond L (2000) Teacher quality and student achievement. Educ Pol Anal Arch 8:1. https://doi.org/10.14507/epaa.v8n1.2000

Department of Education (2020) Call for upskilling programmes. https://hea.ie/assets/uploads/2019/12/Call-forproposals_upskilling-programmes.pdf Last accessed 4 Dec 2024

Desimone LM, Porter AC, Garet MS, Yoon KS, Birman BF (2002) Effects of professional development on teachers' instruction: results from a three-year longitudinal study. Educ Eval Pol Anal 24(2):81–112. https://doi.org/10.3102/01623737024002081

De Lange J (2020) Practitioner inquiry in the context of inquiry based learning, 3DIPhE. In: De Lange J, Čepič M (eds) University of Ljubljana, Faculty of Education, vol 2 (e-book). https://www.3diphe.si/e-book/ Last accessed 4 Dec 2024

De Winter J, Airey J (2020) What makes a good physics teacher? Views from the English stakeholder community. Phys Educ 55:015017. https://doi.org/10.1088/1361-6552/ab5215

De Winter J, Airey J (2022) Pre-service physics teachers' developing views on the role of mathematics in the teaching and learning of physics. Phys Educ 57:065007 https://doi.org/10.1088/1361-6552/ac8138

Etkina E (2010) Pedagogical content knowledge and preparation of high school physics teachers. Phys Rev Spec Top Phys Educ Res 6(2):020110. https://doi.org/10.1103/PhysRevSTPER.6.020110

Etkina E, Gitomer D, Iconangelo C, Phelps G, Seeley L, Vokos S (2018) Design of an assessment to probe teachers' content knowledge for teaching: an example from energy in HS physics. Phys Rev Phys Educ Res 14:010127. https://doi.org/10.1103/PhysRevPhysEducRes.14.010127

Farrell O, Brunton J, Costello E, Donlon E, Trevaskis S, Eccles S, Ní Shé C (2019) An investigation of effective online teaching: a needs analysis of online educators and online students. Dublin: #Openteach Project. https://doras.dcu.ie/23888/

Finlayson O, McLoughlin E (2017) Building teacher confidence in inquiry and assessment: experiences from a Pan-European collaboration. In: A companion to research in teacher education. Springer Nature Singapore Pte Ltd., Springer Singapore

Gammell S, McLoughlin E, Grimes P, Lovatt J (2024) In-service teachers' self-efficacy for learning physics. Proc SMEC 2024 (Under review)

Goos M, O'Donoghue J,Ní Ríordáin M, Faulkner F, Hall T, O'Meara N (2020) Designing a national blended learning program for "Out-of-Field" mathematics teacher professional development, ZDM. Int J Math Educ 52(4) https://doi.org/10.1007/s11858-020-01136-y

Guskey TR (1986) Staff development and the process of teacher change. Educ Res 15(5):5–12. https://doi.org/10.3102/0013189X015005005

Hattie JAC (2009) Visible learning: a synthesis of 800+ meta-analyses on achievement. Routledge

Heilporn G, Lakhal S, Bélisle M (2021) An examination of teachers' strategies to foster student engagement in blended learning in higher education. Int J Educ Technol Higher Educ 18:1–25. https://doi.org/10.1186/s41239-021-00260-3

Hill JG, Gruber, KJ (2011) Education and certification qualifications of departmentalized public high school-level teachers of core subjects: evidence from the 2007–08 schools and staffing survey, statistical analysis report, national center for education statistics, U.S. Department of Education, Washington, D.C. https://nces.ed.gov/pubs2011/2011317.pdf. Last accessed 4 Dec 2024

Hobbs L, Porsch R (2021) Teaching out-of-field: challenges for teacher education. Eur J Teacher Educ 44(5):601–610. https://doi.org/10.1080/02619768.2021.1985280

Kenny J, Hobbs L, Whannell R (2020) Designing professional development for teachers teaching out-of-field. Professional Dev Educ 46(3):500–515. https://doi.org/10.1080/19415257.2019.1613257

Kind V (2009) A conflict in your head: an exploration of trainee science teachers subject matter knowledge development and its impact on teacher self-confidence. Int J Sci Educ 31:1529–1562. https://doi.org/10.1080/09500690802226062

Krakehl R, Kelly AM, Sheppard K, Palermo M (2020) Physics teacher isolation, contextual characteristics, and student performance. Phys Rev Phys Educ Res 16:020117. https://doi.org/10.1103/PhysRevPhysEducRes.16.0201

Leinonen R, Haaranen M, Kesonen M, Koponen M, Hirvonen PE, Asikainen MA (2020) Finnish graduated physics teachers' views about their teacher education program—the disparity between the needs and delivery. J Technol Sci Educ 10(1). https://doi.org/10.3926/jotse.820

Lindstrøm C, Sharma MD (2011) Self-efficacy of first year university physics students: do gender and prior formal instruction in physics matter? Int J Innovat Sci Math Educ 19(2)

Lovatt J, Grimes P, McLoughlin E (2020) Educational design research for teacher professional learning, 3DIPhE. In: Lovatt J, Gimes P, McLoughlin E, Čepič M (eds) University of Ljubljana, vol 4 (e-book), Faculty of Education. https://www.3diphe.si/e-book/ Last accessed 4 Dec 2024

Luft JA, Hanuscin D, Hobbs L, Törner G (2020) Out-of-field teaching in science: an overlooked problem. J Sci Teac Educ 31(7):719–724. https://doi.org/10.1080/1046560X.2020.1814052

Margulieux LE, McCracken WM, Catrambone R (2016) A taxonomy to define courses that mix face-to-face and online learning. Educ Res Rev 19:104–118

McLoughlin E, Sokolowska D (2022) Inquiry approaches in physics education. In: Connecting research in physics education with teacher education, vol 3. Online: International Commission on Physics Education. https://doi.org/10.5281/zenodo.5792967

McLoughlin E, Gammell S, Grimes P, Lovatt J (2024) In-service teachers' self-efficacy for teaching physics. Proc SMEC 2024 (Under review)

McLoughlin E, Sokolowska D (2023) Physics teacher professional learning. In: Taşar MF, Heron PRL (Eds) The international handbook of physics education research: teaching physics, AIP Publishing LLC https://doi.org/10.1063/9780735425712_015

Meltzer DE, Otero VK (2014) Transforming the preparation of physics teachers. Am J Phys 82(7):633–637. https://doi.org/10.1119/1.4868023

Mizzi M (2021) Supporting science teachers teaching outside specialism: teachers' views of a professional development programme. Eur J Teac Educ 44(5):706–725. https://doi.org/10.1080/02619768.2020.1793951

Nipyrakis A, Stavrou D (2022) Integration of ICT in science education laboratories by primary student teachers. In: Papadakis S, Kalogiannakis M (eds) STEM, robotics, mobile apps in

early childhood and primary education. Lecture Notes in Educational Technology. Springer, Singapore. https://doi.org/10.1007/978-981-19-0568-1_4
Nixon RS, Luft JA, Ross RJ (2017) Prevalence and predictors of out-of-field teaching in the first five years. J Res Sci Teach 54(9):1197–1218. https://doi.org/10.1002/tea.21402
OECD (2020) TALIS 2018 results (Volume I): teachers and school leaders as lifelong learners. TALIS, OECD Publishing, Paris. https://doi.org/doi.org/10.1787/1d0bc92a-en
Owston R, York DN, Malhotra T, Sitthiworachart J (2020) Blended learning in STEM and non-STEM courses: how do student performance and perceptions compare? Online Learn 24(3):203–221. https://doi.org/10.24059/olj.v24i3.2151
Peeters W (2020) Building professional learning communities: guide, workbook and tools, 3DIPhE. In: Peeters W, Čepič M (eds) vol 3 (ebook) University of Ljubljana, Faculty of Education. https://www.3diphe.si/e-book/. Last accessed 4 Dec 2024
Penuel WR, Gallagher DJ (2017) Creating research practice partnerships in education. ERIC. Harvard Education Press. https://eric.ed.gov/?id=ED573803
Penuel WR, Roschelle J, Shechtman N (2007) Designing formative assessment software with teachers: an analysis of the co-design process'process. Res Pract Technol Enhanced Learn 2(01):51–74. https://doi.org/10.1142/S1793206807000300.
PhysTEC national report card on physics teacher preparation. https://phystec.org/report-card/ [retrieved: 01.07.2023] Last accessed 4 Dec 2024
Sancar R, Atal D, Deryakulu D (2021) A new framework for teachers' professional development. Teach Teacher Educ 101:103305. https://doi.org/10.1016/j.tate.2021.103305
Schiering D, Sorge S, Keller MM, Neumann K (2023) A proficiency model for pre-service physics teachers' pedagogical content knowledge (PCK)—What constitutes high-level PCK? J Res Sci Teach 60(1):136–163. https://doi.org/10.1002/tea.21793
Shulman LS (1986) 'Those who understand: knowledge growth in teaching' teaching. Educ Res 15(2):4. https://doi.org/10.2307/1175860-14. https://doi.org/10.2307/1175860
Shulman L (1987) Knowledge and teaching: foundations of the new reform'reform. Harv Educ Rev 57(1):1–23. Available at: https://doi.org/10.17763/haer.57.1.j463w79r56455411.
Sheppard KK, Padwa L, Kelly AM, Krakehl R (2020) Out-of-field teaching in chemistry and physics: an empirical census study. J Sci Teac Educ 31(7):746–767. https://doi.org/10.1080/1046560X.2019.1702268
Sokolowska D (2020) Inquiry based learning to enhance teaching: theory, tools and examples, 3DIPhE. In Sokolowska D, Čepič M (eds) University of Ljubljana, Faculty of education, vol 1 (e-book). https://www.3diphe.si/ebook/ Last accessed 4 Dec 2024
Teaching Council of Ireland (2023) Statistical data provided through private communication. https://www.teachingcouncil.ie/ Three Dimensions of Inquiry in Physics Education (3DIPhE) https://www.3diphe.si/. Last accessed 4 Dec 2024
Timperley H, Wilson A, Barrar H, Fung I (2007) Teacher professional learning and development. Best Evid Synthes Iteration [BES], New Zealand Ministry of Education, Wellington, New Zealand
Vangrieken K, Meredith C, Packer T, Kyndt E (2017) Teacher communities as a context for professional development: a systematic review. Teaching and teacher education. Teac Educ 61:47–59. https://doi.org/10.1016/j.tate.2016.10.001
Vogelsang C, Kulgemeyer C, Riese J (2022) Learning to plan by learning to reflect? Exploring relations between professional knowledge, Reflection skills, and planning skills of preservice physics teachers in a one-semester field experience. Educ Sci 12:479. https://doi.org/10.3390/educsci12070479
Vokos S, Seeley L, Etkina E (2022) Research-guided physics teaching: foundations, enactment, and outcomes In:,Guisasola J, Mc Loughlin E (eds) Connecting research in physics education with teacher education 3. https://doi.org/10.5281/zenodo.5792968
Wemyss T (2010) Implementing an inquiry-based approach in first year undergraduate physics

Use of Quantitative Methods in Physics Education Research: Searching for New Perspectives to Inform Theory and Practice

Italo Testa

Abstract In this paper, I will briefly review the main statistical techniques used nowadays in physics education research and call for an increased use of quantitative studies in order to find consensus about the obtained results within the community. To this end, I will also envisage possible new research avenues for physics education research building on the possibilities offered by advanced statistical methods.

Keywords Physics education research · Quantitative studies · Statistical methods · Psychometric scales · Concept inventories

1 Introduction

A quantitative method of analysis is a method that uses a hypothetical deduction approach to find measurable patterns, relationships and associations between the variables of a study through a statistical modeling technique (McMillan et al. 2013). Quantitative methods are essential in educational research for systematically investigating phenomena, testing hypotheses, and making data-driven decisions (James et al. 2012). These methods involve the collection and analysis of numerical data to understand patterns, relationships, and causal effects (Teo 2013). More concisely, "a quantitative method is used when a researcher wishes to quantify observations through some statistical [modeling] techniques in order to better describe, explain, and make inferences about certain events, ideas or actions in physics education" (Ding and Liu 2012). The aims of using quantitative methods in educational research, as physics education research, are manifold: to improve the validity of the studies; to foster replicability of the studies; to help close the credibility gap of educational research; to make predictions through "causal" relationships among variables (Teo et al. 2013). For instance, in educational research, we could be interested in assessing in a reliable way to what extent the students are learning after being exposed to

I. Testa (✉)
Department of Physics "E. Pancini", University Federico II, Naples, Italy
e-mail: italo.testa@unina.it

Z. Ješková and J. Hanč (eds.), *Advancing Physics Education*, Challenges in Physics Education, https://doi.org/10.1007/978-3-031-94207-5_4

mandated curriculum under certain conditions or to an innovative teaching-leaning sequence (TLS) (Psillos and Kariotoglou 2016).

There are four general types or research studies in which quantitative methods can be used: descriptive/survey studies, correlational studies, quasi-experimental studies, experimental studies (Fraenkel et al. 2012). Descriptive/survey studies aim to describe the characteristics of a population or phenomenon. They answer questions about what, where, when, and how something occurs. Common methods include surveys and standardized questionnaires, observations, and case studies. For instance, surveys can gather data on students' attitudes towards online learning, while observations can document classroom interactions. Descriptive research provides a snapshot of the current situation but does not infer causality (Hamaker et al. 2020). Correlational research studies investigate the relationship between two or more variables without manipulating them. This method helps determine whether and to what extent variables are related. For example, researchers might explore the correlation between students' socioeconomic status and academic performance. While correlational studies can identify associations, they cannot establish causation (Lau 2017). Quasi-experimental studies are used to infer causal relationships when random assignment is not feasible. These designs often involve comparing groups that have been exposed to different treatments or conditions. For example, a study could compare the academic outcomes of students in traditional classrooms versus those exposed to an inquiry-based approach. Quasi-experimental designs are valuable in educational settings where practical constraints prevent true experimental manipulation (Kim and Steier 2016). Experimental studies are the gold standard for establishing causality. This research design involves manipulating an independent variable to observe its effect on a dependent variable, with random assignment of participants to control and experimental groups. For instance, an experiment might test the impact of a new teaching method on student achievement. Experimental designs, especially those involving a randomized controlled trial, provide strong evidence of cause-and-effect relationships but can be challenging to implement in educational settings due to ethical and logistical constraints (Handley et al. 2018).

All the above research designs can involve samples at different time frames. Time is an important dimension in educational research since researchers might be interested in both the prevalence and dynamic changes over time regarding a specific theme (Mercer 2008). Longitudinal studies and cross-sectional studies are two distinct research designs used to gather data and analyze trends over time or at a specific point in time (Johnston 1974). Longitudinal studies involve repeated observations of the same variables over a period of time, which can range from several months to many years. The longitudinal design allows researchers to track changes and developments in the subjects being studied. For instance, a longitudinal study might track students' understanding of Newtonian mechanics from middle school through high school. Researchers could administer the same conceptual tests at multiple points in time to observe how students' understanding evolves and identify factors that contribute to their learning progress. Another example could be a study examining the long-term effects of inquiry-based learning versus traditional lecture-based teaching on students' problem-solving skills in physics. By following the same cohort of students over several years, researchers can assess which teaching method has a more lasting impact. Differently, cross-sectional studies collect data from a population, or a representative subset, at a single point in time. This design provides a snapshot of the current state of the variables being studied but does not track changes over time. For example, a cross-sectional study might survey high school students' attitudes towards physics at the end of a school year. This can help identify general trends and differences in attitudes based on factors like gender, socioeconomic status, or prior academic performance. Similarly, researchers might conduct a cross-sectional study to assess the conceptual knowledge of students in different grades (e.g., 9th, 10th, and 11th grades) regarding electromagnetism. This can highlight differences in understanding across grade levels and inform curriculum development. Longitudinal studies can potentially identify causal relationships and long-term trends, while cross-sectional studies can identify associations but not causality. Clearly, both longitudinal and cross-sectional studies are observational in nature, that is researchers measure variables of interest without manipulating them, and hence have their own strengths and limitations. Longitudinal studies are useful to understand how variables change over time and to identify causal relationships. In contrast, cross-sectional studies are useful for providing a quick snapshot of the current state of a population or phenomenon. In physics education research, these methods can be applied to explore various aspects of teaching and learning, from conceptual understanding to attitudes towards the subject.

In the following section, we will briefly describe well known instruments used in physics education research and that can be used in the above-described studies.

2 Overview of Instruments in Physics Education Research to Collect Data

Since early 80's several instruments have been developed in order to collect data in physics education research field (McDermott and Redish 1999; Pfundt and Duit 1994). In this paper, we will distinguish between two main types of data: categorical data and continuous data. Understanding the differences between these types of data is crucial for selecting appropriate statistical methods and accurately interpreting research findings. Categorical data, also known as qualitative data, represents variables that can be divided into distinct groups or categories (Agresti 2013). These categories are typically non-numeric and describe attributes or qualities. Examples are gender (male, female, non-binary), grade level (freshman, sophomore, junior, senior), type of school (public, private). Categorical data can be further divided into nominal data, where the categories have not a specific order, or ordinal data, where the categories have a meaningful order but no consistent difference between them. Therefore, categorical data are often used to classify participants or variables into groups for comparison. For instance, researchers might want to compare the academic performance of students from different types of schools or analyze the distribution of gender in a study sample. Differently, continuous data, also known as quantitative data, represent variables that can take any value within a given range (Teo 2013). These values are typically numeric and can be measured on a continuous scale. Examples include test scores, time spent studying, time reaction to an online prompt. In educational research, most of continuous data (except measured time) cannot be defined by means of a ratio, as it is usual in physics, given the lack of a unit measure. Rather, continuous data are interval data, namely numeric data with meaningful intervals but no true zero point. Note that also this definition is controversial, as Likert type ratings (e.g., 1 = strongly disagree, 2 = disagree, …, 5 = strongly agree) cannot define in principle equal intervals. See (Boone et al. 2006, 2014, 2020) for an in-depth discussion about the issue of interval data. Continuous data are used to measure and analyze variables that can vary in degree or amount. For example, researchers might study the relationship between the number of hours spent studying and test scores, or track changes in students' academic performance over time. From the statistical analysis viewpoint, categorical data can be analyzed using chi-square tests, logistic regression, and other non-parametric methods (Agresti 2013), while continuous data can be analyzed using t-tests, Analysis of Variance (ANOVA), correlation, and regression analysis (Fraenkel 2012).

The instruments developed in physics education research to collect categorical data include (not exhaustive list): Colorado Learning Attitudes about Science Survey (CLASS) (Adams et al. 2006), Physics Metacognition Inventory (PMI) (Taasoobshirazi and Farley 2013), Sources of Self-Efficacy in Science Courses—Physics (SOSESC-P) (Fencl and Scheel 2005), Maryland Physics Expectations Survey (MPEX) (Redish et al. 1998), Attitudes and Approaches to Problem Solving Survey (AAPS) (Mason and Singh 2010), Attitude Toward Physics (ATP) (Testa et al. 2022), Physics Affective Characteristics scale (PACS) (Gungor and De Cock 2021). Table 1 reports the main characteristics of each of these instruments. The instruments developed by physics education researchers to collect continuous data include (not exhaustive list): Force Concept Inventory (FCI) (Hestenes et al. 1992), Force and Motion Conceptual Evaluation (FMCE) (Thornton and Sokoloff 1998), Conceptual Survey of Electricity and Magnetism (CSEM) (Maloney et al. 2001), Brief Electricity and Magnetism Assessment (BEMA) (Ding et al. 2006), Test of Understanding Graphs in Kinematics (TUG-K) (Beichner 1994), Energy and Momentum Conceptual Survey (EMCS) (Singh and Rosengrant 2003), Newtonian Gravity Concept Inventory (NGCI) (Williamson et al. 2013), Quantum Mechanics Concept Assessment (QMCA) (Sadaghiani and Pollock 2014), Quantum Mechanics Evaluation (QME) (Scotti di Uccio et al. 2019). Table 2 reports the main characteristics of each of these instruments.

In the following sections, I will describe how some of these instruments were exploited in quantitative studies.

Table 1 Instruments developed in physics education research to collect categorical data

Instrument	Purpose	Format	Focus
Colorado Learning Attitudes about Science Survey (CLASS) (Adams et al. 2006)	Measures students' beliefs about physics and learning physics	31 Likert-scale statements	Assesses scientific attitude, enjoyment of the discipline, making connections to the real world, and problem-solving strategies
Physics Metacognition Inventory (PMI) (Taasoobshirazi and Farley 2013)	Evaluates the level of metacognition in students when solving physics problems	26 items across six factors	Measures aspects like planning, monitoring, and evaluating one's cognitive processes during problem-solving
Sources of Self-Efficacy in Science Courses—Physics (SOSESC-P) (Fencl and Scheel 2005)	Assesses students' self-efficacy beliefs in physics courses	33 Likert-scale statements	Evaluates beliefs about succeeding in physics, including mastery experiences, vicarious experiences, verbal persuasion, and physiological states
Maryland Physics Expectations Survey (MPEX) (Redish et al. 1998)	Probes students' attitudes, beliefs, and expectations in physics courses	34 Likert-scale items	Measures dimensions such as independence, coherence, concepts, reality link, math link, and effort
Attitudes and Approaches to Problem Solving Survey (AAPS) (Mason and Singh 2010)	Assesses students' attitudes and approaches to problem-solving in physics at the introductory and graduate level	33 Likert-scale items	Evaluates problem-solving strategies, including the use of equations, conceptual understanding, and perseverance
Attitude Toward Physics (ATP) (Testa et al. 2022)	Measures students' attitudes towards physics and physicists	36 Likert-scale items	Assesses factors like interest in physics, perceived utility, self-concept, and motivation
Physics Affective Characteristics Scale (PACS) (Gungor and De Cock 2021)	Measures affective characteristics related to learning physics	51 Likert-scale items	Includes 11 subscales for situational interest, personal interest, relevance to everyday life and future career, course and test anxiety, motivation, self-efficacy, and self-concept

3 Overview of Quantitative Studies in Physics Education Research

Before giving an overview of studies in physics education research that exploit quantitative methods to analyse data, I will briefly describe some of the statistical techniques used in these studies. Some of the most used statistical techniques are: univariate analysis, regression analysis, cluster analysis, logistic regression, factor analysis, network

Table 2 Instruments developed in physics education research to collect continuous data

Instrument	Topic	Number of questions	Sample involved
Force Concept Inventory (FCI) (Hestenes et al. 1992)	Newtonian mechanics	30	Introductory physics students
Force and Motion Conceptual Evaluation (FMCE) (Thornton and Sokoloff 1998)	Force and motion in one dimension	47	Introductory physics students
Conceptual Survey of Electricity and Magnetism (CSEM) (Maloney et al. 2001)	Electricity and magnetism	32	Over 5000 students at 30 institutions
Brief Electricity and Magnetism Assessment (BEMA) (Ding et al. 2006)	Basic concepts in electricity and magnetism	31	Introductory undergraduate students
Test of Understanding Graphs in Kinematics (TUG-K) (Beichner 1994)	Kinematics graphs	21	Over 1000 high school and college students
Energy and Momentum Conceptual Survey (EMCS) (Singh and Rosengrant 2003)	Energy and momentum	25	Over 3000 students in algebra- and calculus-based courses
Newtonian Gravity Concept Inventory (NGCI) (Williamson et al. 2013)	Newtonian gravity	26	Introductory college astronomy students
Quantum Mechanics Concept Assessment (QMCA) (Sadaghiani and Pollock 2014)	Quantum mechanics concepts	31	Upper-division quantum mechanics students at 11 institutions
Quantum Mechanics Evaluation (QME) (Scotti di Uccio et al. 2019)	Basic aspects of quantum mechanics	10 two-tier items	About 450 undergraduate physics students and high school teachers

analysis, structural equation models (SEM) and the Rasch analysis. Univariate analysis involves the examination of a single variable, including descriptive statistics like mean, median, mode, range, variance, and standard deviation. This type of analysis helps summarize and find patterns in data (Ho 2013). Regression analysis is used to examine the relationship between a dependent variable and one or more independent variables. It helps in predicting the value of the dependent variable based on the values of the independent variables (Fahrmeir et al. 2013). Cluster analysis groups a set of objects in such a way that objects in the same group (or cluster) are more similar to each other than to those in other groups. It is used to identify patterns and structures in data (Everitt et al. 2011). Logistic regression is used for modeling the probability of a binary outcome based on one or more predictor variables. It is particularly useful when the dependent variable is categorical (Kleinbaum and Klein 2010).

Factor analysis is a technique used to reduce a large number of variables into fewer numbers of factors. It identifies underlying relationships between variables by grouping them into factors (Yong and Pearce 2013). Network analysis examines the relationships and interactions between entities within a network. It is used to study social structures through the use of networks and graph theory (Carrington et al. 2005). SEM is a multivariate statistical analysis technique used to analyze structural relationships. SEM combines factor analysis and multiple regression analysis and is used to analyze the structural relationship between measured variables and latent constructs (Teo et al. 2013).

Finally, *Rasch analysis* is a form of item response theory used to analyze data from assessments to measure variables such as abilities, attitudes, and personality traits. It provides a way to convert qualitative observations into quantitative data (Boone et al. 2014). In this paper, I will focus on four types of statistical techniques: factor analysis, regression analysis, SEM and Rasch analysis.

Figure 1 reports the distribution of studies published in the journal *Physical Review Physics Education Research* (PRPER) from 2005 to 2023 that used any of these four methods. Figure 1 provides insights into research trends over time within these analytical methods, indicating which methods are gaining or losing popularity among researchers. Results of this, rather limited, analysis suggest that in early years of the journal (2005–2010), there were very few papers (less than 5) that used these four quantitative methods. Basically, from 2005 to 2010 we had only studies using factorial and regression analysis. In the next five years, there was a significant increase in papers across all methods, with regression and SEM having the most at around nine and eight, respectively.

From 2015–2020 there was a significant increase in studies using SEM with over fifteen papers, while other methods also increased but less significantly. Finally, from 2020–2023, there was a decrease in factor analysis and Rasch analysis, while studies using regression analysis remained consistent. SEM studies also slightly decreased but remained high. These trends are likely to reflect the availability of open access statistical software (e.g., *R studio*), which helped to spread the use and the application of these methods.

In the following subsections, I will review studies in physics education research that exploit the above statistical methods, according to the main objectives outlined above.

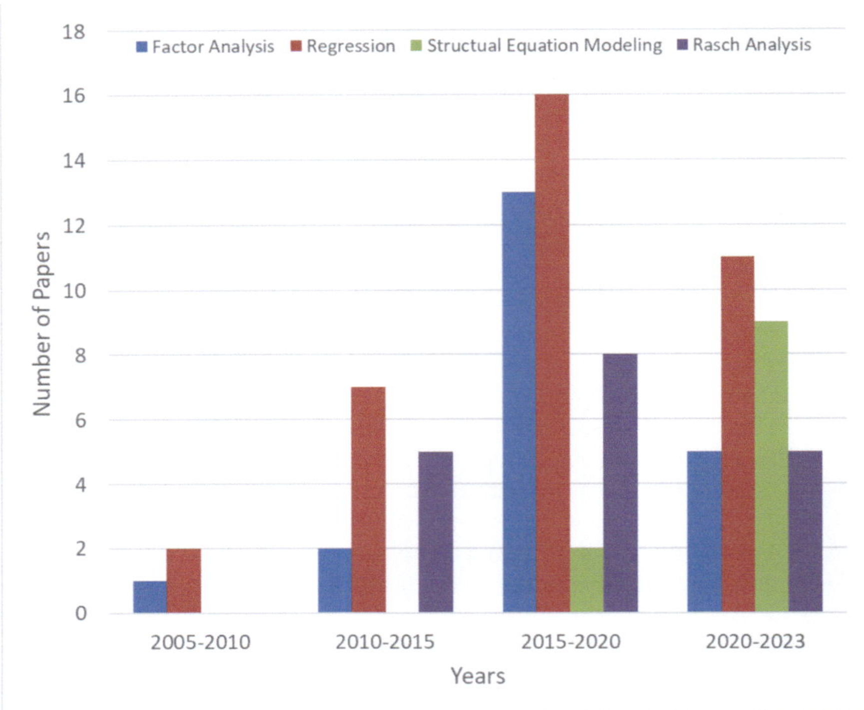

Fig. 1 Distribution of published studies in *Physical Review Physics Education Research* that used factor analysis, regression analysis, SEM and Rasch analysis

3.1 Studies Aimed at Improving the Validity of Instruments: Applications of Exploratory Factor Analysis

One of the first applications of factorial analysis in physics education research was aimed at investigating the factorial structure of the Force Concept Inventory (FCI) (Scott et al. 2012). The reason for which the FCI was first investigated is that it is one of the most productive instruments that we have developed in physics education research and for which we have a large collection of data available online, albeit mainly from the US (Data explorer by Physport). However, the first paper that addressed the question of what the FCI actually measures dates back to 1995 Kleinbaum and Klein (2010). In this study, the authors showed that the proposed dimensions of the FCI had little support from the factor analysis, namely the dimensions had not sufficient psychometric validity given that the items were only loosely related (Huffman and Heller 1995, p. 141). The authors of FCI responded that they did not perform any factorial analysis because they adopted a theoretically driven approach to categorize questions that could be useful for practitioners (Hestenes and Halloun 1995). Moreover, they criticized the use of factor analysis to analyze conceptual tests given that the latent factors should be independent, namely uncorrelated (Hestenes and Halloun 1995, p. 506). Clearly, such critique does not take into account that latent factors can actually be correlated. Moreover, the authors cite in their response a well-known controversial book that criticized the use of statistics, especially factor analysis, in biology and other sciences (Gould 1981). Given the role of FCI in the physics education community, it is no wonder that it took almost 20 years from its first appearance to have a new study that performed a rigorous exploratory factorial analysis (EFA) of the FCI. The EFA reported in ref. (Scott et al. 2012) lead to five factors, which substantially corresponded to the categories of questions put forth initially by the authors of the FCI: identification of forces, Newton's first law with zero force, Newton's second law and kinematics, Newton's first law with canceling forces, Newton's third law. In the following years, the factorial structure of the FCI was confirmed by further studies (Eaton and Willoughby 2018). Factor analysis was later applied also to other instruments as the BEMA and CSEM (Eaton et al. 2019), but results were less in agreement with the original categorizations proposed by the author of these instruments. For the BEMA, five factors emerged from the analysis: Lorentz' force, magnetic field, potential given the electric field, Coulomb's law, Electric field. For the CSME, six factors were extracted: Newton's third law, relationship between electric field and forces, magnetic field from current, Lorentz's force, electric field from the electric potential, electric charge in fields. These results confirm that the two questionnaires address similar but not identical aspects of electromagnetism.

3.2 Studies Aimed at Fostering Replicability of the Results: Applications of Confirmatory Factor Analysis

These studies basically aim at answering the question: how do we know that an instrument is valid across different contexts? We will use CLASS (Adams et al. 2006) as an example. CLASS probes students' beliefs about physics and measures the extent to which students have beliefs that align more or less with professional physicists. CLASS was published in one of the first papers in PRPER and the items are scored with a Likert scale. Example of an item for which agreement indicates expert-like beliefs is: *it is possible to explain physics ideas without mathematical formulas*, while an item for which agreement indicates novice-like beliefs is: *knowledge in physics consists of many disconnected topics*. However, although CLASS was soon adopted in many studies, it was not until 2014 that other authors investigated what were the actual dimensions of CLASS (Douglas et al. 2014). They used a confirmatory factor analysis (CFA), which is a theoretically driven, hypothesis-testing analysis based on the concept of fit (for more details see (Jackson et al. 2009; Byrne 2010). The results of the CFA reported in ref. (Douglas et al. 2014) showed that the actual factors of CLASS were only three: *personal application and relation to real world, problem solving and learning, effort and sense making*. Such results were in stark contrast with the 9 categories proposed by the authors of the CLASS (Adams et al. 2006). The main reason for such discrepancy is that, as shown by a more recent study (Martins and Lindsay 2022), the CLASS substantially mixes items that measure two distinct latent constructs: *the beliefs about physics*; *physics metacognition.* These studies essentially suggest that the CLASS had a poor theoretical basis, as even the authors point out in a reply to the first study using CFA: "*We emphasize that what the CLASS measures was defined by these statements that reflected issues of common concern to students and faculty as expressed in numerous interviews, not by any well-defined theoretical construct that we set out to measure*" (Wieman and Adams 2015, p. 2). An example of a more theoretically driven instrument is the PMI (Taasoobshirazi and Farley 2013; Taasoobshirazi et al. 2015), which was validated through both EFA and CFA and showed also high convergent validity, namely it correlated strongly with another instrument that measures the same construct.

3.3 Studies Aimed at Closing the Credibility Gap of Educational Research: Applications of Rasch Analysis

These studies aim at objectively describing how students' progress through different stages towards scientific views and at measuring the extent to which students' learning is aligned with curriculum or TLS developers. We will focus in this paper on studies that exploit the Rasch analysis. As pointed out above, for a brief introduction to Rasch analysis, the reader can refer to refs (Boone and Scantlebury 2006; Boone et al. 2014; Boone and Staver 2020). One of the first examples to exploit Rasch analysis in physics education to address the above issues is a paper on the learning progression about energy (Neumann et al. 2013). Basically, with the term learning progression we refer to a theoretical framework that allows to describe, interpret, and predict how the students will learn a given "big idea". For more details see our review in ref. (Testa et al. 2023). The results showed that students from grade 6 mostly obtain an understanding of energy forms and energy sources, students of grade 8 demonstrate an understanding of energy transfer and transformation, while only students of grade 10 achieve a deeper understanding of energy conservation. The authors also quantitatively showed that energy dissipation is a difficult concept, as expected. Another example of application of the Rasch analysis is about quantum mechanics. A typical line of research in quantum mechanics concerns what concepts should be taught in high school (Krijtenburg-Lewerissa 2018). However, there is another line of research which investigates what actually the students know about quantum mechanics after being exposed to curriculum instruction at both high school and university level (Testa et al. 2020, 2019; Scotti di Uccio et al. 2020). In one of our previous papers, which described the first validated learning progression in quantum mechanics and the validation of a brief conceptual assessment on key aspects quantum mechanics, Rasch analysis supported the following learning path: *knowledge of atomic models* → *atom stability through Bohr, De Broglie and Schrödinger models* → *knowledge*

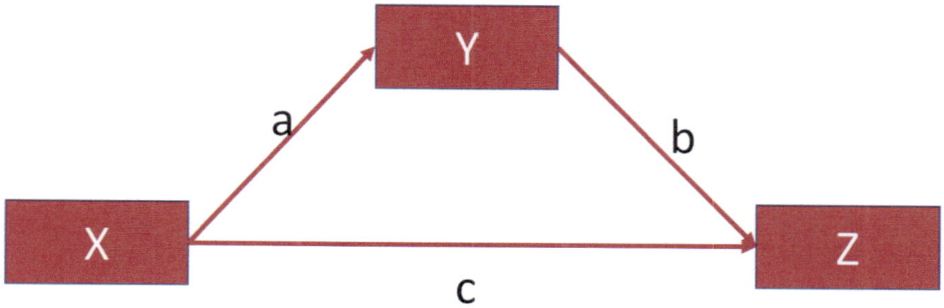

Fig. 2 A simple mediation model

about the wave function. Basically, our study showed that only undergraduate physics had a satisfactory knowledge of the most difficult aspects of quantum mechanics, while most of the students at the end of high school had a scarce knowledge not only of advanced topics of quantum mechanics but also about the atomic description of matter.

3.4 Studies Aimed at Making Predictions Through "Causal" Relationships Among Variables

These studies basically measure the relationship between different variables measured through different instruments. The statistical technique used in these studies is called Structural Equation Modeling (SEM) (Teo et al. 2013). The simplest structural model is shown in Fig. 2, where the variable X has both a direct and indirect effect on the variable Z. The indirect effect is mediated by the variable Y. The regression coefficients a, b and c quantify such relationships.

Here, we will discuss a few studies that measured the structural relationships among affective variables in physics.

In one of our recent studies (Catena and Testa 2024), we used three instruments to measure the relationships between student's attitudes towards physics, students' engagement in physics activities and self-efficacy in physics. We found that the attitudes towards physics had both a direct and indirect effect on self-efficacy, which is a well-known predictor of the performance of students in physics. SEM was also used to validate the well-known structural model of physics identity (Hazari et al. 2010, 2020). Results show that the perception of performance/competence (self-efficacy) and peer recognition have a direct effect on physics identity, while interest has only an indirect effect on physics identity mediated by sense of belonging, peer recognition and self-efficacy. In a series of studies with undergraduate students (Kalender et al. 2019, 2020), Singh and colleagues found that male and female students' self-efficacy and interest dropped after the introductory physics course. However, female students' self-efficacy and interest dropped more than male students. The authors inferred that the university learning environment barely helped students build their physics interest, so that the gender gap in physics was not addressed but actually enlarged. Finally, in a couple of our studies (Colantonio et al. 2021; Galano et al. 2023a, b), we used a specific type of SEM, namely path analysis, to investigate astronomy identity and the effects of gender stereotypes on disciplinary identity in STEM, respectively. In the paper about astronomy identity (Colantonio et al. 2021), we found that interest predicts astronomy identity, and that this effect was also mediated by utility value. Moreover, we found that confidence in one's own knowledge about astronomy mediates the effect of interest on conceptual knowledge. More importantly, we found that the effect of interest on identity is greater for girls than for boys and that the students' interest in astronomy and confidence in their performance decrease with age, as it was confirmed in a later study with a representative sample of Italian students (Galano, Palazzo and Testa 2023b). Concerning the role of gender stereotypes on disciplinary identity (Galano et al. 2023a), we found that the adherence to male role norms and the rejection of hostile sexism have a significant effect on the development of a disciplinary identity in three areas, physics, biology and chemistry and that the effect is fully mediated by self-efficacy and perceived academic control. More importantly, we found that the effects of stereotypes depend on discipline. Specifically, we found that only for physics, girls who endorsed gender stereotypes had higher self-efficacy and thus higher identity.

4 Future Pathways for Quantitative Studies in Physics Education Research

Although the number of quantitative studies in physics education research has steadily increased during the last 20 years, there are still empirical blind spots that we need to address. First, there is a shortage of studies about profiles of students and teachers according to different affective and cognitive indicators. For instance, it would be important to investigate profiles of students that are engaged in out-of-school activities in order to answer questions such as: what is their motivation to participate in such activities? Recently published papers about physics and science Olympiads are addressing this issue (Tschisgale et al. 2024; Steegh et al. 2021), but we need more evidence from other types of out-of-school activities. Similarly, it would be relevant for the field to investigate teachers' profiles according to different indicators such as their practice, self-efficacy or their identity in order to better evaluate their pedagogical content knowledge in physics. We finally need also to investigate in a deeper detail the role of socio-economic status on, for instance, the students' interest or identity in physics.

From the methodological viewpoint, it would be important to have more studies using multiple correspondence analysis (MCA) to analyze categorical data, including open-ended answers. MCA allows to calculate factorial scores related to the frequency with which categories are present in students' responses and that can be used to perform cluster analysis (Testa 2024). To the latter concern, clustering methods have moved towards probability-based assignment of subjects to profiles: latent class and latent profiles analysis are powerful tools in which it is possible to better use categorical and continuous indicators to obtain more meaningful profiles. Examples of latent profile analysis can be found in refs (Galano, Palazzo and Testa 2023b; Tschisgale et al. 2024; Steegh et al. 2021). Finally, there is still a lack of longitudinal studies, which could take advantage of growth curve modelling (Bianconcini et al. 2023) and latent change score approaches (Eriksen et al. 2024). Other examples of person-centered approaches studies in science education can be found in refs (Burns et al. 2023; Kampa et al. 2016).

Finally, I will give some suggestions for studies that can be carried out in a way similar to other fields. First, it would be relevant to investigate whether the gender gap in physics depends on the performance itself of students, that is, if gender differences in physics performance are the same among low, medium and high achievers. This kind of study has been recently carried out for mathematics using quantile regression (Matteucci 2021). Second, it would be important to investigate how physics performance changes through high school using longitudinal data in order to develop suitable interventions to address gaps that emerge from the analysis. Also this kind of study has been carried out for mathematics (Bianconcini et al. 2023) and it could be interesting to compare results obtained for physics, given that, for instance, in Italy the same teacher teaches both math and physics. Third, only a few studies have developed a reliable instrument to assess teachers' pedagogical content knowledge in physics (Kirschner et al. 2016), and we need more evidence about the correspondence between the assessment tool and the theoretical framework. The study about technological pedagogical content knowledge in ref (Schmid et al. 2020) can inform future studies along this direction. Fourth and final, given the increasing relevance of students' drop-out in STEM undergraduate courses, it would be important to assess students' perceptions of the transition between high school and university (Noyens et al. 2019).

Overall, quantitative methods are indispensable in educational research for their ability to provide objective, reliable, and generalizable findings. By employing various research designs and statistical analyses, researchers can uncover important insights into educational processes and outcomes. As the field continues to evolve, the integration of advanced statistical techniques and mixed-methods approaches will further enhance the robustness and applicability of educational research.

References

Adams WK, Perkins KK, Podolefsky NS, Dubson M, Finkelstein ND, Wieman CE (2006) New instrument for measuring student beliefs about physics and learning physics: the Colorado learning attitudes about science survey. Phys Rev ST Phys Educ Res 2:010101. https://doi.org/10.1103/PhysRevSTPER.2.010101

Agresti A (2013) Categorical data analysis. John Wiley & Sons Inc., Hoboken

Beichner R (1994) Testing student interpretation of kinematics graphs. Am J Phys 62:750–762. https://doi.org/10.1119/1.17449
Bianconcini S, Mignani S, Mingozzi J (2023) Assessing maths learning gaps using Italian longitudinal data. Stat Methods Appl 32(3):911–930. https://doi.org/10.1007/s10260-022-00676-9
Boone WJ, Staver JR (2020) Advances in rasch analyses in the human sciences. Springer, New York
Boone WJ, Staver JR, Yale MS (2014) Rasch analysis in the human sciences. Springer, New York
Boone WJ and Scantlebury K (2006) The role of Rasch analysis when conducting science education research utilizing multiple-choice tests. Sci Educ 90:253–269. https://doi.org/10.1002/sce.20106
Burns EC, Martin AJ, Kennett R, Pearson J, Munro-Smith V (2023) High school students' out-of-school science participation: a latent class analysis and unique associations with science aspirations and achievement. J Res Sc Teach 60(3):451–483. https://doi.org/10.1002/tea.21806
Byrne BM (2010) Structural Equation Modeling with AMOS. Routledge, London
Carrington PJ, Scott J, Wasserman S (2005) Models and methods in social network analysis. Cambridge University Press, New York
Catena D, Testa I (2024) Exploring the relationships between undergraduate and high school students' self-efficacy, engagement, and attitudes towards physics: a structural equation model. In: Faletič S, Pavlin J (eds) Teaching and learning physics effectively in challenging times. Springer, Berlin
Colantonio A, Marzoli I, Puddu E, Bardelli S, Fulco MT, Galano S, Terranegra L, Testa I (2021) Describing astronomy identity of upper primary and middle school students through structural equation modeling. Phys Rev Phys Educ Res 17:010139. https://doi.org/10.1103/PhysRevPhysEducRes.17.010139
Data Explorer by PhysPort. https://www.physport.org/DataExplorer
Ding L, Chabay R, Sherwood B, Beichner R (2006) Evaluating an electricity and magnetism assessment tool: brief electricity and magnetism assessment. Phys Rev ST Phys Educ Res 2:010105. https://doi.org/10.1103/PhysRevSTPER.2.010105
Ding L, Liu X (2012) Getting started with quantitative methods in physics education research. In: Henderson C, Harper K (eds) Getting started in PER. American Association of Physics Teachers, College Park, MD, pp 1–42
Douglas K, Yale M, Bennett D, Haugan M, Bryan L (2014) Evaluation of colorado learning attitudes about science survey. Phys Rev Phys Edu Res 10:020128. https://doi.org/10.1103/PhysRevSTPER.10.020128
Eaton P, Willoughby SD (2018) Confirmatory factor analysis applied to the force concept inventory. Phys Rev Phys Educ Res 14:010124. https://doi.org/10.1103/PhysRevPhysEducRes.14.010124
Eaton P, Frank B, Johnson K, Willoughby S (2019) Comparing exploratory factor models of the brief electricity and magnetism assessment and the conceptual survey of electricity and magnetism. Phys Rev Phys Educ Res 15:020133. https://doi.org/10.1103/PhysRevPhysEducRes.15.020133
Eriksen EV, Vestad L, Bru E, Caravita SCS (2024) Social competencies, classroom relationships, and academic engagement: a latent change score modeling approach among lower secondary school students. Soc Psychol Educ 27(2):435–459. https://doi.org/10.1007/s11218-023-09841-8
Everitt BS, Landau S, Leese M, Stahl D (2011) Cluster analysis. John Wiley and Sons, New York
Fahrmeir L, Kneib T, Lang S, Marx B (2013) Regression—models, methods and applications. Springer, Berlin
Fencl H, Scheel K (2005) Research and teaching: engaging students—an examination of the effects of teaching strategies on self-efficacy and course climate in a nonmajors physics course. J Coll Sci Teach 35(1):20–24
Fraenkel JR Wallen NE and Hyun HH (2012) How to design and evaluate research in education. Mc Graw Hill, New York

Galano S, Liccardo A, Amodeo AL, Crispino M, Tarallo O, Testa I (2023a) Endorsement of gender stereotypes affects high school students' science identity. Phys Rev Phys Educ Res 19:010120. https://doi.org/10.1103/PhysRevPhysEducRes.19.010120

Galano S, Palazzo L, Testa I (2023b) A latent profile analysis of students' attitudes towards astronomy across grades 9–13. Int J Sci Educ 46(14):1480–1516

Gould SJ (1981) The mismeasure of man. W. W. Norton & Company, New York

Gungor A, De Cock M (2021) Validity of the physics affective characteristics scale for Flemish pharmacy and biology majors. Phys Rev Phys Educ Res 17:010132. https://doi.org/10.1103/PhysRevPhysEducRes.17.010132

Hamaker EL, Mulder JD, van IJzendoorn MH (2020) Description, prediction and causation: methodological challenges of studying child and adolescent development. Dev Cogn Neurosc 46:100867. https://doi.org/10.1016/j.dcn.2020.100867

Handley MA, Lyles CR, McCulloch C, Cattamanchi A (2018) Selecting and Improving Quasi-experimental designs in effectiveness and implementation research. Annu Rev Public Health 39:5–25. https://doi.org/10.1146/annurev-publhealth-040617-014128

Hazari Z, Sonnert G, Sadler PM, Shanahan M-C (2010) Connecting high school physics experiences, outcome expectations, physics identity, and physics career choice: a gender study. J Res Sci Teach 47:978–1003. https://doi.org/10.1002/tea.20363

Hazari Z, Chari D, Potvin G, Brewe E (2020) The context dependence of physics identity: examining the role of performance/competence, recognition, interest, and sense of belonging for lower and upper female physics undergraduates. J Res Sci Teach 57(10):1583–1607. https://doi.org/10.1002/tea.21644

Hestenes D, Wells M and Swackhamer G (1992) Force concept inventory. Phys Teach 30:141–158. https://doi.org/10.1119/1.2343497

Hestenes D, Halloun I (1995) Interpreting the force concept inventory: a response to March 1995 critique by Huffman and Heller. Phys Teach 33(8):502. https://doi.org/10.1119/1.2344278

Ho R (2013) Handbook of univariate and multivariate data analysis with IBM SPSS. CRC Press Taylor & Francis Group, Boca Raton

Huffman D, Heller P (1995) What does the force concept inventory actually measure? Phys Teach 33(3):138–143. https://doi.org/10.1119/1.2344171

Jackson DL, Gillaspy JA, Purc-Stephenson R (2009) Reporting practices in confirmatory factor analysis: an overview and some recommendations. Psych Methods 14(1):6–23. https://doi.org/10.1037/a0014694

James A, Waring M, Coe R, Hedges LV (2012) Research methods and methodologies in education. SAGE, Thousand Oaks, CA

Johnston FE (1974) Cross-sectional versus longitudinal studies nutrition and malnutrition. In: Roche AF, Falkner F (eds) Advances in experimental medicine and biology, vol 49 Springer, Boston, MA. https://doi.org/10.1007/978-1-4684-3252-7_13

Kalender ZY, Marshman E, Schunn CD, Nokes-Malach TJ, Singh C (2019) Gendered patterns in the construction of physics identity from motivational factors. Phys Rev Phys Educ Res 15:020119. https://doi.org/10.1103/PhysRevPhysEducRes.15.020119

Kalender ZY, Marshman E, Schunn CD, Nokes-Malach TJ, Singh C (2020) Damage caused by women's lower self-efficacy on physics learning. Phys Rev Phys Educ Res 16:010118. https://doi.org/10.1103/PhysRevPhysEducRes.16.010118

Kampa N, Neumann I, Heitmann P, Kremer K (2016) Epistemological beliefs in science—a person-centered approach to investigate high school students' profiles. Cont Educ Psych 46: 81–93. 0.1016/j.cedpsych.2016.04.007

Kim Y, Steiner P (2016) Quasi-experimental designs for causal inference. Ed Psych 51(3–4):395–405. https://doi.org/10.1080/00461520.2016.1207177

Kirschner S, Borowski A, Fischer HE, Gess-Newsome J, Von Aufschnaiter C (2016) Developing and evaluating a paper-and-pencil test to assess components of physics teachers' pedagogical content knowledge. Int J Sci Educ 38(8):1343–1372. https://doi.org/10.1080/09500693.2016.1190479

Kleinbaum DG, Klein M (2010) Logistic regression: a self-learning text. Springer, New York
Krijtenburg-Lewerissa K, Pol HJ, Brinkman A, van Joolingen WR (2018) Int J Sci Educ 41(3):349–366. https://doi.org/10.1080/09500693.2018.1550273
Lau F (2017) Methods for Correlational Studies. In: Lau F, Kuziemsky C (eds) Handbook of eHealth evaluation: an evidence-based approach. University of Victoria, Victoria
Maloney D, O'Kuma T, Hieggelke C, Van Heuvelen A (2001) Surveying students' conceptual knowledge of electricity and magnetism. Am J Phys 69:S12–S23. https://doi.org/10.1119/1.1371296
Martins JS, Lindsay WE (2022) Evaluation of high school student responses to the Colorado learning attitudes about science survey. Phys Rev Phys Educ Res 18:010132. https://doi.org/10.1103/PhysRevPhysEducRes.18.010132
Mason A, Singh C (2010) Surveying graduate students' attitudes and approaches to problem solving. Phys Rev ST Phys Educ Res 6:020124. https://doi.org/10.1103/PhysRevSTPER.6.020124
Matteucci M, Mignani S (2021) Investigating gender differences in mathematics by performance levels in the Italian school system. Stud Educ Eval 70(3):101002
McDermott LC, Redish EF (1999) Resource letter: PER-1: physics education research. Am J of Phys 67(9):755–767. https://doi.org/10.1119/1.19122
McMillan JH, Mohn RS, Hammack MV (2013) Quantitative research designs in educational research. Oxford Bibliographies, Oxford
Mercer N (2008) The seeds of time: why classroom dialogue needs a temporal analysis. J Learn Sc 17(1):33–59. https://doi.org/10.1080/10508400701793182
Neumann K, Viering T, Boone WJ, Fischer HE (2013) Towards a learning progression of energy. J Res Sc Teach 50(2):162–188. https://doi.org/10.1002/tea.21061
Noyens D, van Daal T, Coertjens L, Van Petegem P, Donche V (2019) Assessing students' perceptions of fit between secondary and higher education: a validation study of the SPFQ. High Ed Res Dev 39(2):273–287. https://doi.org/10.1080/07294360.2019.1662377
Pfundt H, Duit R (1994) Students' alternative frameworks and science education: bibliography. Institut fur die Padagogik der Naturwissenschaften an der Universitat, Kiel
Psillos D, Kariotoglou P (2016) Theoretical issues related to designing and developing teaching-learning sequences. In: Psillos D, Kariotoglou P (eds) Iterative design of teaching-learning sequences. Springer, New York, pp 11–34
Redish E, Saul J, Steinberg R (1998) Student expectations in introductory physics. Am J Phys 66:212–224. https://doi.org/10.1119/1.18847
Sadaghiani H, Pollock S (2014) Quantum mechanics concept assessment: development and validation study. Phys Rev ST Phys Educ Res 11:010110. https://doi.org/10.1103/PhysRevSTPER.11.010110
Schmid M, Brianza E, Petko D (2020) Efficient self-report measures for technological pedagogical content knowledge (TPACK): constructing a reliable and valid short-scale among pre-service teachers. Comp Ed 157:103967. https://doi.org/10.1016/j.compedu.2020.103967
Scott TF, Schumayer D, Gray AR (2012) Exploratory factor analysis of a force concept inventory data set. Phys Rev ST Phys Educ Res 8:020105. https://doi.org/10.1103/physrevstper.8.020105
Scotti di Uccio U, Colantonio A, Galano S, Marzoli I, Trani F, Testa I (2019) Design and validation of a two-tier questionnaire on basic aspects in quantum mechanics. Phys Rev Phys Educ Res 15:010137. https://doi.org/10.1103/PhysRevPhysEducRes.15.010137
Scotti di Uccio U, Colantonio A, Galano S, Marzoli I, Trani F, Testa I (2020) Development of a construct map to describe students'reasoning about introductory quantum mechanics. Phys Rev Phys Educ Res. 16:010144. https://doi.org/10.1103/PhysRevPhysEducRes.16.010144
Singh C, Rosengrant D (2003) Multiple-choice test of energy and momentum concepts. Am J Phys 71:607–617. https://doi.org/10.1119/1.1571832
Steegh A, Höffler T, Höft L, Parchmann I (2021) Exploring science competition participants' expectancy-value perceptions and identification: a latent profile analysis. Learn and Instr 74:101455. https://doi.org/10.1016/j.learninstruc.2021.101455

Taasoobshirazi G, Bailey M, Farley J (2015) Physics metacognition inventory Part II: confirmatory factor analysis and Rasch analysis. Int J Sci Educ 37(17):2769–2786. https://doi.org/10.1080/09500693.2015.1104425

Taasoobshirazi G, Farley J (2013) Construct validation of the physics metacognition inventory. Int J Sc Ed 35(3):447–459. https://doi.org/10.1080/09500693.2012.750433

Teo T (2013) Handbook of quantitative methods for educational research. Sense Publishers, Rotterdam

Teo T, Tsai LT, Yang CC (2013) Applying structural equation modeling (SEM) in educational research. In: Myint SK (ed) Application of structural equation modelling in educational research and practice. Sense Publishers, New York, pp 3–21

Testa I, Capasso G, Colantonio A, Galano S, Marzoli I, Scotti di Uccio U, Trani F, Zappia A (2019) Development and validation of a university students' progression in learning quantum mechanics through exploratory factor analysis and Rasch analysis. Int J Sci Educ 41:388. https://doi.org/10.1080/09500693.2018.1556414

Testa I, Colantonio A, Galano S, Marzoli I, Trani F, Scotti di Uccio U (2020) Effects of instruction on students' overconfidence in introductory quantum mechanics. Phys Rev Phys Educ Res 16:010143. https://doi.org/10.1103/PhysRevPhysEducRes.17.020130

Testa I, De Luca PR, Scotti di Uccio U (2022) Patterns of Italian high school and university students' attitudes towards physics: an analysis based on semiotic-cultural perspective. Eur J Psych Ed 37(1):785. https://doi.org/10.1007/s10212-021-00563-z

Testa I, Galano S, Scotti di Uccio U, Marzoli I, Giuliana G, Catena D, Leccia S, Puddu E (2023) Learning progressions: an overview and how-to guide for researchers in physics education. Nuovo Cim C 6(6):202. https://doi.org/10.1393/ncc/i2023-23202-y

Testa I (2024) Quantitative methods in physics education research: overview and examples of studies. In: Aydiner E, Sidharth BG, Michelini M, Corda, C (eds) Frontiers of fundamental physics FFP16. FFP 2022. Springer, Cham, pp 475–494. https://doi.org/10.1007/978-3-031-38477-6_26

Thornton R, Sokoloff D (1998) Assessing student learning of Newton's laws: the force and motion conceptual evaluation and the evaluation of active learning laboratory and lecture curricula. Am J Phys 66:338–352. https://doi.org/10.1119/1.18863

Tschisgale P, Steegh A, Kubsch M, Petersen S, Neumann K (2024) Towards a more individualised support of science competition participants—identification and examination of participant profiles based on cognitive and affective characteristics. Int J Sci Educ 46(16):1757–1781. https://doi.org/10.1080/09500693.2023.2300147

Wieman CE, Adams WK (2015) On the proper use of statistical analyses. In: Evaluation of colorado learning attitudes about science survey" by Douglas et al. https://arxiv.org/ftp/arxiv/papers/1501/1501.03257.pdf

Williamson K, Willoughby S, Prather E (2013) Development of the Newtonian gravity concept inventory. Astron Educ Rev 12:1–20. https://doi.org/10.3847/AER2012045

Yao JX, Guo YY, Neumann K (2017) Refining a learning progression of energy. Int J Sci Educ 39(17):2361–2381. https://doi.org/10.1080/09500693.2017.1381356

Yong AG, Pearce S (2013) A beginner's guide to factor analysis: focusing on exploratory factor analysis. Tutor Quant Meth Psych 9:79–94. https://doi.org/10.20982/tqmp.09.2.p079

Pre-school and Primary Education

Is It Possible to Teach Nanoscience-Nanotechnology Content to Kindergarten Students?

Giorgos Peikos, Anna Spyrtou, Giorgos Antoniadis, Katerina Karategou, Maria Tzioli, Olga Kremmyda, and Spyridoula Gkiremezi

Abstract This paper focuses on the introduction of Nanoscience-Nanotechnology (NST) content to kindergarten students. Four educational interventions were developed to explore kindergarten students' ability to use linear causal reasoning based on NST-related agents concerning the lotus effect, gecko effect, protection masks against coronavirus and NST for restoring marble sculptures. Data were collected through semi-structured interviews and kindergarten students' drawings. The results provide evidence that educational interventions can facilitate the development of linear causal reasoning when students attempt to explain NST-related processes, incorporating essential elements of the scientific model.

Keywords Nanoscience-nanotechnology education · Early childhood science education · Linear causal reasoning

1 Introduction

Exposure to Science Education during early childhood is essential, since this involves the enjoyment of exploring nature, promotes positive attitudes towards science, while early engagement with scientific phenomena can enhance children's understanding of scientific concepts that are formally studied in higher grades (Delserieys et al. 2018; Eshach and Fried 2005). In line with this, integrating Nanoscience-Nanotechnology (NST) concepts across all educational levels, starting from kindergarten, could promote the scientific literacy of modern citizens (Dorouka and Kalogiannakis 2023; Spyrtou et al. 2021).

Research focusing on teaching and learning science content shows that kindergarten students (KSs) often use causal reasoning to explain natural phenomena (Koliopoulos 2013), characterising them as "cause seekers" (Varela et al. 2022).

G. Peikos (✉) · A. Spyrtou · G. Antoniadis · K. Karategou · M. Tzioli · O. Kremmyda · S. Gkiremezi
University of Western Macedonia, Florina, Greece
e-mail: giorgospeikos@gmail.com

Z. Ješková and J. Hanč (eds.), *Advancing Physics Education*, Challenges in Physics Education, https://doi.org/10.1007/978-3-031-94207-5_5

They tend to attribute phenomena either to supernatural or magical agents or naturalistic causes, like mechanical contact (Varela et al. 2022). Educational interventions appear to help KSs construct explanations about natural phenomena that incorporate certain key elements of the scientific model using linear causal reasoning, such as a precursor form of the energy chain model (Koliopoulos 2013). Therefore, there is evidence supporting the notion that KSs can develop explanations based on an agent that causes an outcome or based on simple causal chains like A causes B causes C (Perkins and Grother 2005). Research examining students' conceptions of NST-related concepts at primary and kindergarten levels is still in its infancy, while a study conducted with primary school students (5th and 6th graders) revealed misconceptions about NST phenomena prior to any systematic instruction. More specifically, regarding the lotus effect, students tended to attribute the formation of spherical droplets on a superhydrophobic leaf's surface to its visible characteristics (Peikos et al. 2023). However, after relevant educational interventions students' explanations were more sophisticated, attributing the lotus effect either to a single NST-related agent, i.e. the leaf's nanostructure or describing it as an emergent process caused through the interaction of various agents, such as the nanostructure and the air (Peikos et al. 2022). Similarly, recent research involving early primary school students (2nd graders) showed that, after instruction, students were able to describe viral infection as a sequence of stages, where a nanoscale agent (virus) affects the micro-world (cells), which in turn affects the macro-world (infected person) (Dorouka and Kalogiannakis 2023). Additionally, a study focusing on KSs and their understanding of the gecko effect indicated that, prior to instruction, most students attributed the gecko's adhesive property to perceptually-based agents, such as suction cups, while after instruction their explanations included scientific elements, such as the nanostructure of the gecko's foot (Manoloudi and Lefkos 2023).

Given that linear causal reasoning is often used among young students when explaining natural phenomena (Koliopoulos 2013), the research question of this paper is: Can an educational intervention facilitate the development of KSs linear causal reasoning based on an NST-related agent?

2 Methods

2.1 *The Context of the Research: The Educational Interventions*

In this paper, we present four educational interventions on NST concepts, having been conducted within the framework of a postgraduate program in Science Education. These interventions were developed and implemented by postgraduate students, who were kindergarten teachers, in collaboration with Science Education researchers in Greece.

Concerning the content of the educational interventions qualitative approach was adopted (Peikos et al. 2022) and the essential concept was the size of the objects, since factors related to size and scale are crucial in describing matter and predicting its behavior, given that materials at the nanoscale exhibit unique properties enabling novel applications (Blonder and Yonai 2020; Stavrou et al. 2018). Regarding the size of objects, a qualitative approach includes the categorisation of objects of different sizes to macro, micro and nanoworld, based on qualitative criteria, i.e. the observation tools that render them accessible: the unaided eye, the optical microscope and the electron microscope (Peikos et al. 2022). In the context of explaining NST-related processes, i.e. phenomena and applications, we argue that objects from the nanoworld can be utilized as agents affecting various NST-related processes. All four educational interventions emphasised the size of objects and the observation tools, while each intervention focused on one of the following NST-related processes: the lotus effect, NST applications in restoring sculptures, the gecko effect and protection against coronavirus using face masks. The pedagogical approach was inquiry-based, emphasising active student engagement in experimental and modeling activities (Peikos et al. 2022). In parallel, a variety of methods suitable for KSs were employed, including puppets, storytelling, dramatic play (Fragkiadaki et al. 2021) and digital technologies (Dorouka and Kalogiannakis 2023). The educational interventions are presented in the following sections.

2.1.1 Educational Intervention 1: Lotus Effect

Scientific content: Lotus effect is called the superhydrophobicity and water repellent property of some plants, e.g. the lotus. Lotus leaf's surface is covered by microstructures and waxy nanostructures and air pockets are trapped within the interstitial spaces. Therefore, the water droplet has a larger surface area in contact with the air than with the nanostructures, resulting in the formation of the spherical water droplets (Kim et al. 2018).

Aim of the educational intervention: Despite the fact that the lotus effect, from a cognitive perspective, is interpreted as an emergent process caused through the collective interaction of agents, such as the micro and nanostructures and the air (Peikos et al. 2022) for preschool students, our aim was to facilitate the use of linear causal reasoning that is age-appropriate, by introducing an agent deriving from the essential concept of size, namely the leaf's nanostructure.

Indicative activities: The educational intervention (duration: 360 min) centered around a storytelling, titled "Snow White returns to nanoworld", based on the classic fairy tale "Snow White and the Seven Dwarfs" which included both narration and illustrations. The prefix "nano" is a Greek word that means "small" or "dwarf". Greek children use this prefix in their everyday communication, providing a vehicle for introducing NST-related concepts. During the narration, students were introduced to the observation tools for the macroworld (i.e., the naked eye), the microworld (i.e., the optical microscope), and the nanoworld (i.e., the electron microscope), while the teacher also guided them to engage in experimental activities related to the story.

For instance, Fig. 1 depicts an illustration, showing Snow White's tears fall onto a lotus leaf, visually demonstrating the lotus effect. Additionally, students conducted experimental activities related to the lotus effect by dropping water droplets onto cabbage leaves in order to observe the formation of spherical water droplets (Fig. 2). To explain the lotus effect, the storytelling included representations of the surface of a lotus leaf observed under an electron microscope, revealing the nano-bumps (Fig. 3), whose illustrations were also observed by students using a model of an electron microscope (Fig. 4). Furthermore, the students engaged in creating models, e.g. by using their bodies and balloons to represent nano-bumps and spherical water droplets on the lotus leaf (Fig. 5).

Fig. 1 Storytelling illustration

Fig. 2 Exploring lotus effect

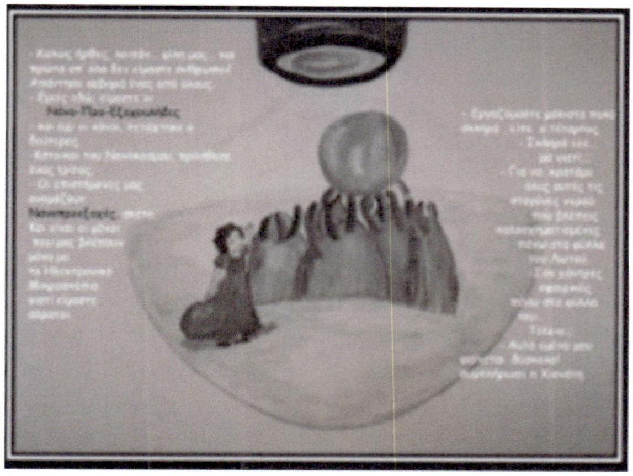

Fig. 3 Representation of the nano-bumps in the storytelling

Fig. 4 Lotus leaf under an electron microscope (artefact)

Fig. 5 Students representing nano-bumps

2.1.2 Educational Intervention 2: NST for Restoring Sculptures

Scientific content: Concerning sculptures made from natural stone, there is often a loss of cohesion due to microcracks. NST applications achieve restoration of lost cohesion by applying consolidants. A technique includes the deposition of nanolimes that are dispersions of $Ca(OH)_2$ (calcium hydroxide) nanoparticles with hexagonal structure within the stone. In turn, calcium hydroxide nanoparticles undergo carbonation by atmospheric CO_2, forming $CaCO_3$ (calcium carbonate) that has binding capacity (Sassoni et al. 2020).

Aim of the educational intervention: KSs were intended to explain how NST contributes to the restoration of sculptures, describing the role of nano-hexagons (calcium hydroxide nanoparticles) in the process. More specifically, these nano-hexagons are applied to the sculpture's wounds (cracks), with the aid of water (dispersion medium) and air (CO_2) in order to repair non-visible cracks.

Indicative activities: The educational intervention (duration: 65 min) was based on a puppet (Fig. 6) and an artefact concerning the marble sculptures (Fig. 7). The puppet narrated a story about a sculpture's problem: its wounds are too small that they are invisible with an unaided eye. Additionally, students were prompted to explore a "treatment" for these wounds, having been encouraged to observe the sculptures' cracks using a model of a microscope included in the artefact.

A discussion with a scientist (Fig. 8), supported by a model (Fig. 9) helped students describe how nano-hexagons, water and air contribute to heal the wounds and finally students created models of the nano-hexagons, using clay (Fig. 10). Last but not least, participating in a dramatic play, some students assumed the role of water, transferring the nano-hexagons to the sculptures' wounds (Fig. 11), while other students performed the role of air, taking part in the consolidation process by "blowing".

2.1.3 Educational Intervention 3: Gecko Effect

Scientific content: Geckos can attach to and detach from various surfaces at will. This attachment system consists of hierarchy of structures: macroscale structures, called lamellae and microscale structures, called setae extending from the lamellae. At the end of each setae a lot of nanoscale structures, known as spatulae, branch and form the points of contact with the surface, about three billion spatulae on two feet. When the spatulae come into contact with a solid surface, van der Waals forces help gecko adhere onto the surface (Bhushan 2010).

Aim of the educational intervention: The aim of the instruction was to encourage students to explain the gecko's adhesive property by attributing it to the nanoscale structures on its feet and the extensive contact area between a gecko's foot and the surface.

Indicative activities: The instruction (duration: 360 min) started with a video of a gecko climbing a ceiling, leading to a discussion on its climbing mechanism. Students were given three wooden models of a gecko's foot, each representing a common misconception: the presence of glue, claws or suction cups on the gecko's toes.

Fig. 6 Puppet story: sculptures' wounds

Fig. 7 Students exploring sculptures using an artefact

Fig. 8 Students discussing with a scientist

Fig. 9 Model of treatment

Through testing these models on various surfaces and recording their observations (Fig. 12), the students attempted the explanatory power of these alternative ideas.

After having explained to students that the gecko's climbing ability lies in the invisible characteristics of its foot, they observed objects like hairs under digital microscopes (Fig. 13) and then used a simulation of an electron microscope to observe the same objects with enhanced magnification. Later, they watched videos and images

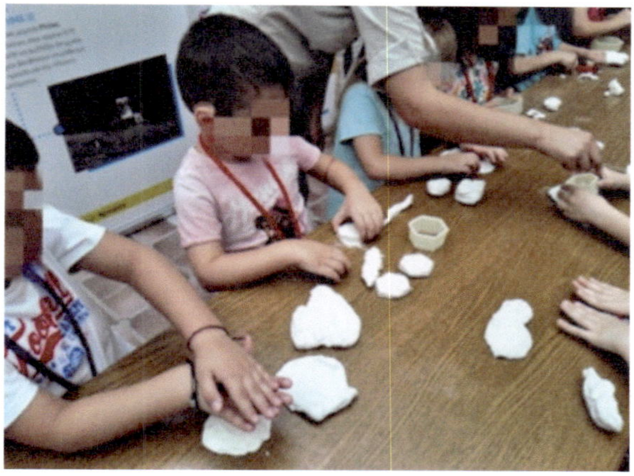

Fig. 10 Students creating models of nano-hexagons

Fig. 11 Dramatic play: healing the sculptures' wounds

Fig. 12 Students testing their alternative conceptions about the cause of the gecko effect

Fig. 13 Students using digital microscopes

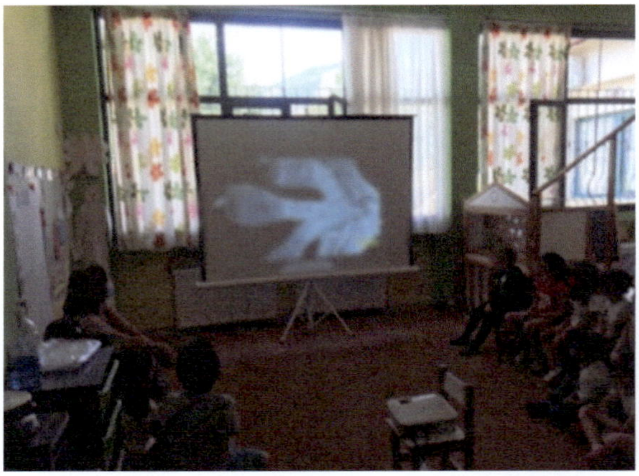

Fig. 14 Video that zoomed in on the gecko's foot

magnifying the gecko's foot, showing setae and spatulae under an optical and electron microscope respectively (Fig. 14). Students drew models representing the gecko's foot structure (Fig. 15) and participated in activities exploring applications mimicking the gecko effect, such as using gecko-inspired adhesive tape (Fig. 16).

2.1.4 Educational Intervention 4: Protection Masks Against Coronavirus

Scientific content: SARS-CoV-2 is potentially transmitted through respiratory droplets, while studies indicate that masks reduce its spread. An N95 mask, notable for its small pores' size (about 30 μm) and dense nanoscale fibres, offers a high filtration efficiency against coronavirus (Du et al. 2021).

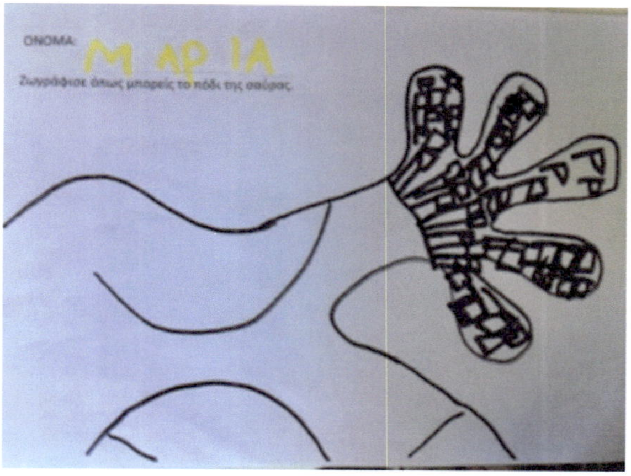

Fig. 15 Students' model of the gecko's foot

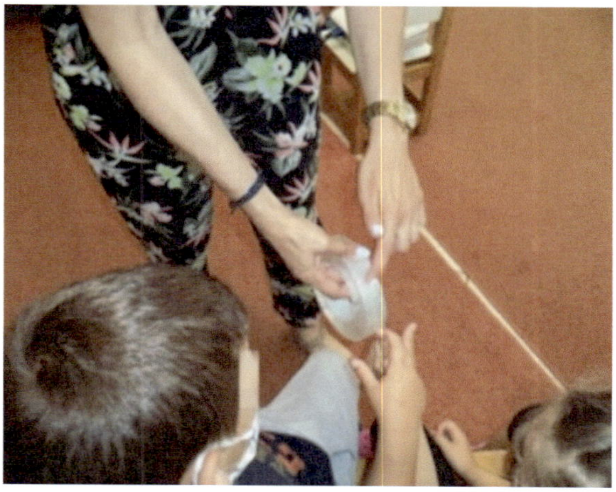

Fig. 16 Students exploring gecko-inspired adhesive tape

Aim of the educational intervention: The aim of the instruction was to help students explain the mask's filtration efficiency against coronavirus by attributing it to the structure of the mask, i.e. its non-visible pores and fibres.

Indicative activities: The intervention (duration: 270 min) featured a puppet assuming the role of a scientist who asked students questions like: "Why can't we see the coronavirus?" and "How do face masks protect us?" (Fig. 17). Students examined objects from the macro, micro and nanoworld, aided by a Scratch 3 software visualization (Fig. 18). They created a classification poster that displayed landmark objects from each world alongside their corresponding observation tools (Fig. 19). Then, students used digital microscopes and a microscope model to observe mask pores (Fig. 20) and explored how they prevent virus transmission using tangible models (Fig. 21). Another model was later used to explain mask protection against

Fig. 17 The puppet scientist

coronavirus (Fig. 22). In a dramatic play titled "Coronaviruses try to find people," they demonstrated how masks protect individuals from infection.

2.2 *Participants*

Data were collected from forty-seven (47) KSs (22 girls and 25 boys, 4–6 years old) from four different kindergartens in Greece. These KSs were selected as they attended all the lessons of their respective educational interventions. Four KSs participated in all the lessons of the educational intervention about the lotus effect, 13 KSs in the intervention on NST for restoring sculptures, 20 KSs in the intervention for the gecko effect, and 10 KSs in that for the protection masks against coronavirus.

2.3 *Data Collection*

Data were collected through semi-structured interviews. The aim was to encourage students to express their conceptions about NST-related processes, similarly to

Fig. 18 Scratch 3 visualization

previous research (Manoloudi and Lefkos 2023; Peikos et al. 2023, 2022). Indicative questions for the educational interventions (Ed.Ins.) were: (a) Ed.In.1: How does water become spherical on a lotus leaf? (b) Ed.In.2: How can nanotechnology help sculptures' wounds heal? (c) Ed.In.3: How can a gecko climb without falling? (d) Ed.In.4: How does the mask protect us? Regarding Ed.In.1, the interview process was facilitated by an artefact incorporating a tangible user interface, namely the Makey-Makey device (Fokides and Papoutsi 2020), which helped students answer basic interview questions by allowing them to interact with physical objects. Regarding Ed.Ins.2–4, students were asked to create drawings related to the above-mentioned questions, which could represent a student's expressed model of the particular concept. For Ed.In.3 and Ed.In.4, data were collected both before and

Fig. 19 Classification poster

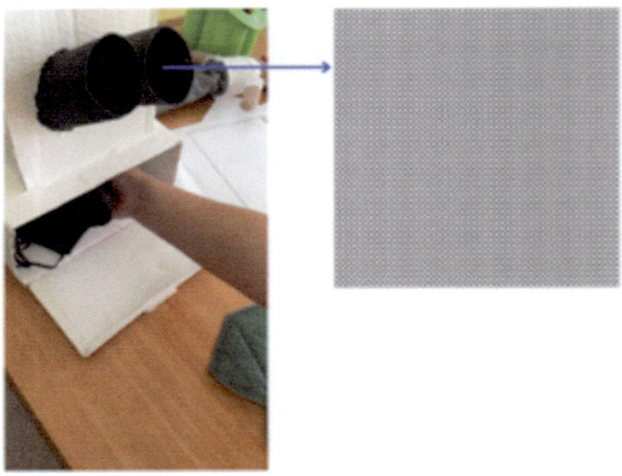

Fig. 20 Observing masks under the microscope

after the intervention, while for Ed.In.1 and Ed.In.2 data were collected only after the implementation of the educational materials.

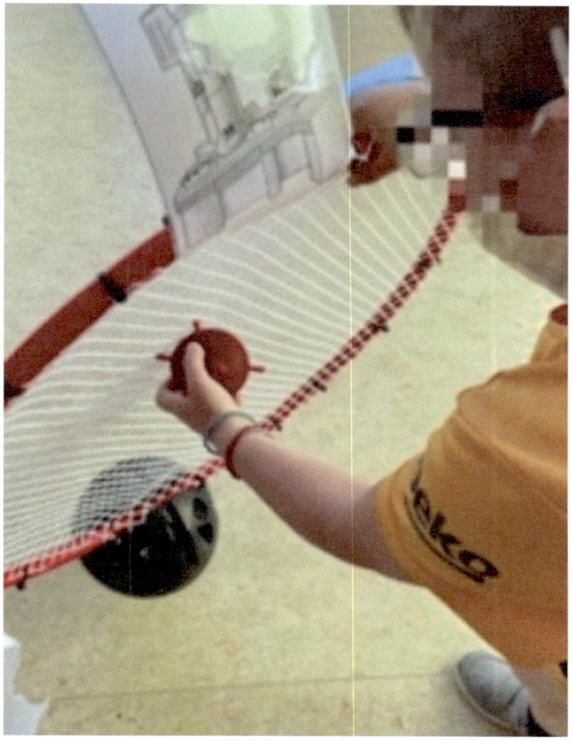

Fig. 21 Model: testing mask effectiveness

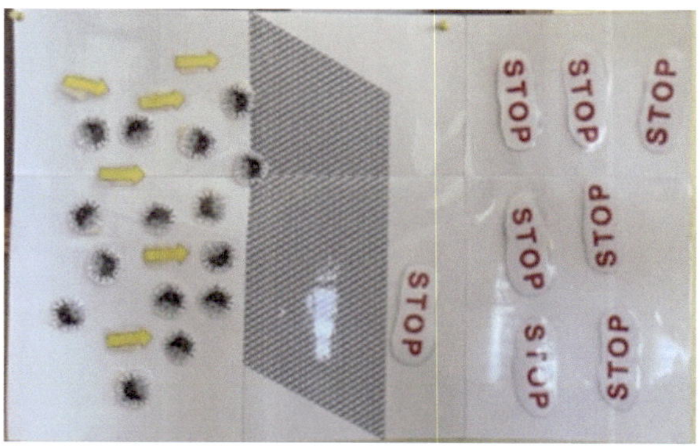

Fig. 22 Model: mask filtration

2.4 Data Analysis

Data were analyzed based on a content analysis approach (Mayring 2014) utilizing Atlas.ti software. In examining students' responses using linear causal reasoning model for explaining NST-related processes, they were classified into four categories (C): C1 included answers without any causal reasoning about the NST related process, while C2 consisted of responses using linear causal reasoning based on a "magical" agent. Moreover, C3 included answers through linear causal reasoning being grounded in a perceptual-based agent. Finally, C4 referred to answers that were formed using linear causal reasoning based on an NST-related agent.

3 Results

Figure 23 shows that, prior to the instruction a large percentage of responses was formed using linear causal reasoning based on perceptual-based agents (C3: 86.67%). Conversely, a small subset (C1: 13.33%) did not incorporate a specific agent.

Figure 24 illustrates that, after the intervention, the majority of responses shifted towards linear causal reasoning based on an NST-related agent (C4: 65.96%). Responses utilizing perceptual-based agents accounted for a smaller percentage (C3: 25.53%), while a lesser extent of responses either did not specify any agent (C1: 6.38%) or employed linear causal reasoning based on a magical agent (C2: 2.13%). In the following sections qualitative examples of students' answers and drawings are presented for each category.

Category 1: Non-causal reasoning responses. This category refers to students' answers that do not use linear causal reasoning, as they do not mention any specific agent to explain an NST-related process. In the following example from the interview, conducted prior to instruction, about mask protection against coronavirus, a student admitted that a face mask offers protection, without identifying any specific agent contributing to a mask's protective function.

3:2 Interviewer: Can you draw how the mask protects us?
Student: [The student drew the Fig. 25].
Interviewer: Can you describe your drawing?
Student: She is a lady wearing a mask.

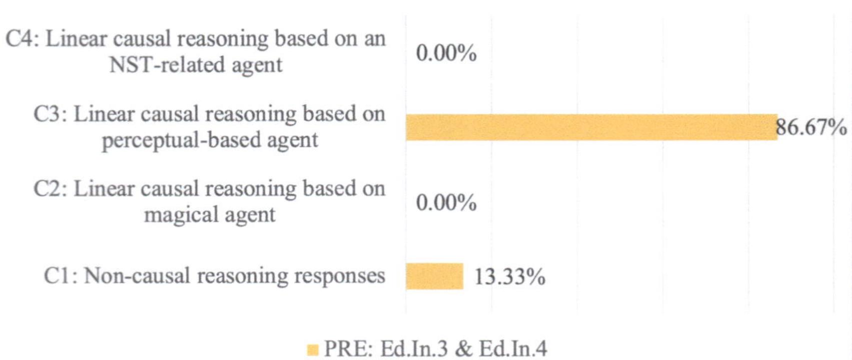

Fig. 23 Percentage of KSs responses in each category, prior to educational interventions 3 and 4

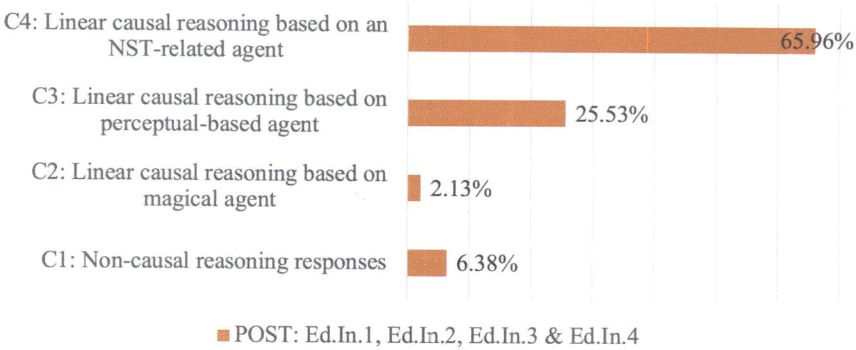

Fig. 24 Percentage of KS responses in each category, after educational interventions 1, 2, 3, and 4

Fig. 25 A student's drawing in category C1 regarding how a face mask protects us

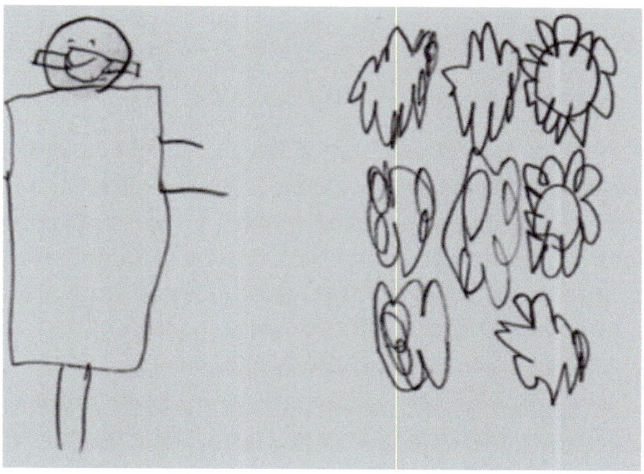

Interviewer: And what are all these things over here?
Student: Viruses.
Interviewer: How does the mask protect us from the coronavirus?
Student: The mask keeps the coronavirus away.

Category 2: Answers using linear causal reasoning based on a magical agent. This category includes answers, explaining NST-related processes based on a magical agent (Varela et al. 2022). After the instruction, when a student was asked to explain the lotus effect, the spherical droplet on the lotus leaf was attributed to "a prince", a character from the storytelling of the educational intervention.

7:2 Interviewer: This is a lotus leaf. What shape does the droplet take when it falls on the lotus leaf? Please show me, placing the droplet with the correct shape on the leaf.

Student: [The student immediately chooses the spherical droplet].
Interviewer: How did the droplet become spherical?

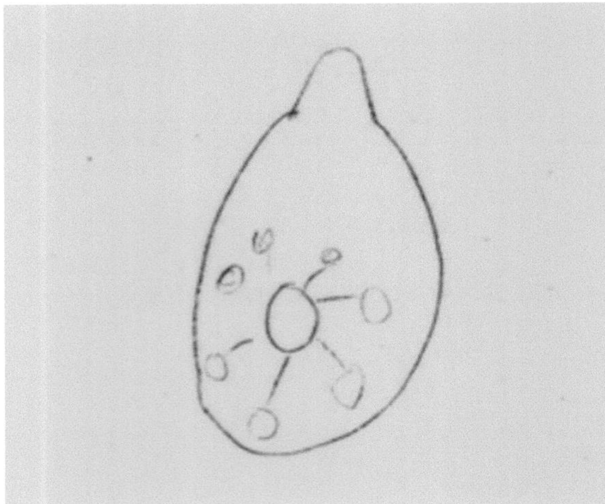

Fig. 26 A student's drawing in category C3 regarding how a face mask protects us

Student: A prince made it spherical.

Category 3: Answers using linear causal reasoning based on perceptual-based agents. In this category, the answers are associated to perceptual-based agents that are perceived with an unaided eye. Concerning mask protection, the mask's thick fabric is interpreted as an agent, as seen in the following example prior to instruction:

3:6 Interviewer: Can you draw how the mask protects us?

Student: [The student drew the Fig. 26].

Interviewer: Can you describe your drawing?

Student: You can see the coronavirus and the mask. The coronavirus is outside and cannot penetrate the mask due to the mask's thick fabric.

Regarding the gecko effect, according to students' answers, the gecko's ability to climb on various surfaces is attributed to perceptual-based agents, since they think that a gecko's foot has features such as claws, hairs, glue or suction cups. An indicative student's response from the interview prior to instruction is presented below:

6:12 Interviewer: Can you draw how the gecko can climb without falling?

Student: [The student drew the Fig. 27].

Interviewer: Can you explain how the gecko can climb without falling?

Student: [Because] the gecko has claws.

Category 4: Answers using linear causal reasoning based on NST-related agents. In this category, the NST-related agents are addressed to the structure of nanoscale objects concentrating on their size.

For the gecko effect, in students' answers of this category, the gecko's ability to climb on various surfaces is attributed to the nanoscale structure of their feet. Below there is an indicative student's response from the interview after instruction:

5:15 Interviewer: Can you draw how the gecko can climb without falling?

Student: [The student drew the Fig. 28].

Interviewer: Can you describe your drawing?

Fig. 27 A student's drawing in category C3 regarding the gecko effect

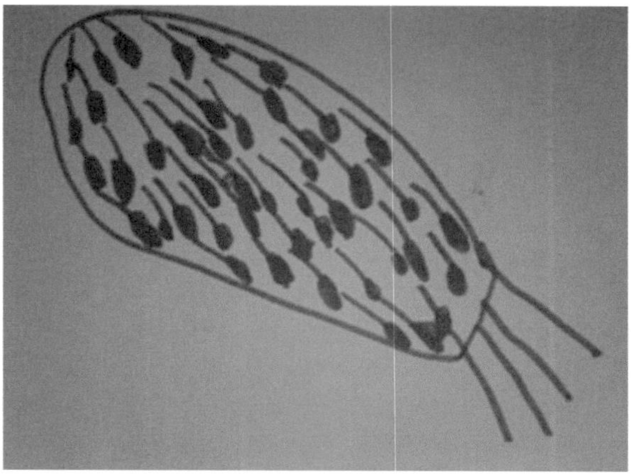

Fig. 28 A student's drawing in category C4 regarding the gecko effect

Student: It has hairs that end in spatulae. They attach to the wall and the lizard walks. We can observe them [the spatulae] under an electron microscope.

Concerning the mask protection against coronavirus, in this category the mask's protective function is explained based on the size of the mask's pores. An indicative student's response from the interview after instruction is the following:

8:9 Interviewer: Can you draw how the mask protects us?

Student: [The student drew the Fig. 29].

Interviewer: Can you describe what you drew?

Student: This is a mask that has small round pores and the coronavirus cannot penetrate the mask because it is larger than the small round pores. The arrows indicate that the coronavirus approaches the mask, without penetrating it because the pores are too small.

Regarding NST for restoring sculptures, students' explanations about the restoration of sculptures included terms related to the nanostructures of the solution used.

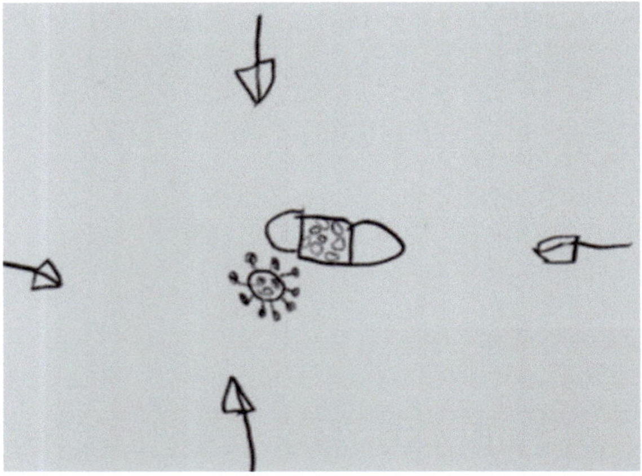

Fig. 29 A student's drawing in category C4 regarding face mask protection

An indicative student's response from the interview after instruction is presented below:

4:9 Interviewer: Can you draw how nanotechnology can help the sculptures' wounds heal?

Student: [The student drew the Fig. 30].

Interviewer: Can you describe your drawing?

Student: I drew the nanohexagons, the air and the water that travel into a sculpture's wound and treat it.

Concerning the lotus effect, students explain the spherical water droplets onto the lotus leaf based on the nanostructures of the leaf. An indicative student's response from the interview after instruction is as follows:

7:3 Interviewer: What shape will the droplet take when it falls on the lotus leaf?

Student: [The student shows the spherical shape].

Interviewer: How did the droplet acquire a spherical shape?

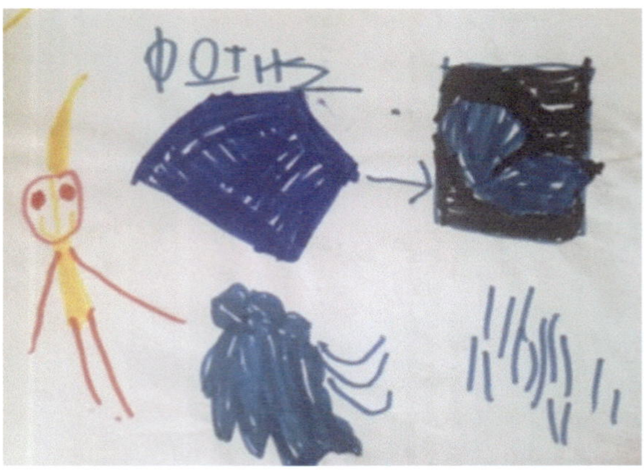

Fig. 30 A student's drawing in category C4 regarding NST for restoring sculptures

Student: It is that in which we raise our hands. [This refers to the model that the students created using their bodies to represent the nano-bumps during the instruction].

Interviewer: How does the surface of the lotus leaf appear, when being observed under an electron microscope?

Student: [The student shows the bumpy leaf].

Interviewer: Why did you choose this leaf?

Student: Because it has nano-bumps.

4 Discussion

This study aimed to examine students' explanations of NST-related phenomena and applications from the perspective of causal reasoning models. The results concerning students' preconceptions provide evidence that KSs tended to use linear causal reasoning when explaining NST-related processes, often attributing them to perceptual-based agents, revealing their misconceptions. For instance, they attributed the gecko's adhesive property to its foot claws. These results are in line with previous findings in the literature, suggesting that both KSs and primary school students tend to attribute NST phenomena to perceptual-based agents before receiving any formal instruction (Manoloudi and Lefkos 2023; Peikos et al. 2023). Furthermore, following the educational interventions, it is evident that students were able to use linear causal reasoning that incorporated key elements of the scientific model. More specifically, as presented in Fig. 31, they attributed: (a) the lotus effect to the nanoscale structure of the surface, i.e. the nano-bumps, (b) the role of NST in repairing microcracks in sculptures to nanoparticles, i.e. the nanohexagons, (c) the gecko effect to the nanoscale structures on the feet, namely the spatulae, and (d) the protective efficacy of masks against coronavirus to the pores' size of the mask. These findings confirm previous results in the literature. For example, KSs as well as primary school students, after educational interventions, can attribute phenomena of NST to nanoworld-related agents (Dorouka and Kalogiannakis 2023; Manoloudi and Lefkos 2023; Peikos et al. 2022). The above-mentioned results led us to conclude that educational interventions can facilitate the development of linear causal reasoning, incorporating essential elements of scientific explanations, when students attempt to explain NST-related processes.

In future studies, we plan to implement a professional development program for teachers, focusing on teaching NST concepts in early grades by adapting the educational materials developed during the four educational interventions. Additionally, we intend to support teachers in implementing these educational materials in their classrooms and to collect data on both their perspectives regarding the implementation and the students' learning outcomes.

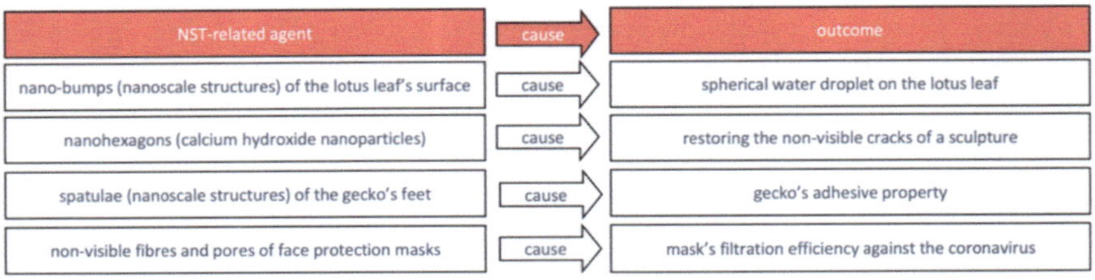

Fig. 31 Kindergarten students' linear causal reasoning including NST-related agents

References

Bhushan B (2010) Gecko feet: natural hairy attachment systems for smart adhesion. In: Bhushan B (ed) Springer handbook of nanotechnologym, 3rd edn, Springer, Berlin, Heidelberg, pp 1553–1594. https://doi.org/10.1007/978-3-642-02525-9_44

Blonder R, Yonai E (2020) Exposing school students to nanoscience: a review of published programs. In: Sattler K (ed) 21st century nanoscience—a handbook: public policy, education, and global trends, CRC Press, London, p 9

Delserieys A, Jégou C, Boilevin JM, Ravanis K (2018) Precursor model and preschool science learning about shadows formation. Res Sci Technol Educ 36(2):147–164. https://doi.org/10.1080/02635143.2017.1353960

Dorouka P, Kalogiannakis M (2023) Teaching nanotechnology concepts in early-primary education: an experimental study using digital games. Int J Sci Educ 46(13):1311–1338. https://doi.org/10.1080/09500693.2023.2286299

Du W, Iacoviello F, Fernandez T, Loureiro R, Brett DJL, Shearing PR (2021) Microstructure analysis and image-based modelling of face masks for COVID-19 virus protection. Commun Mater 2:69. https://doi.org/10.1038/s43246-021-00160-z

Eshach H, Fried MN (2005) Should science be taught in early childhood? J Sci Educ Technol 14:315–336. https://doi.org/10.1007/s10956-005-7198-9

Fokides E, Papoutsi A (2020) Using makey-makey for teaching electricity to primary school students. A Pilot Study Educ Inf Technol 25:1193–1215. https://doi.org/10.1007/s10639-019-10013-5

Fragkiadaki G, Armeni A, Zioga S, Ravanis K (2021) Dramatic play as a means to explore and support preschool children's thinking about thermal insulation. J Child Educ Soc 2(3):220–234. https://doi.org/10.37291/2717638X.20212395

Kim W, Kim D, Park S, Lee D, Hyun H, Kim J (2018) Engineering lotus leaf-inspired micro- and nanostructures for the manipulation of functional engineering platforms. J Ind Eng Chem 61:39–52. https://doi.org/10.1016/j.jiec.2017.11.045

Koliopoulos D (2013) Is it possible to teach energy in preschool education? In: Tasar F (ed) Proceedings of the WCPE conference. Gazi Üniversitesi, Ankara, pp 451–455

Manoloudi M, Lefkos I (2023) Nanotechnology in kindergarten. Is there any learning gain using an ICT-based approach? Int J Digit Soc 14(1):1868–1672. https://doi.org/10.20533/ijds.2040.2570.2023.0234

Mayring P (2014) Qualitative content analysis: theoretical foundation, basic procedures and software solution, Beltz, Klagenfurt

Peikos G, Spyrtou A, Pnevmatikos D, Papadopoulou P (2022) A teaching learning sequence on nanoscience and nanotechnology content at primary school level: evaluation of students' learning. Int J Sci Educ 44(12):1932–1957. https://doi.org/10.1080/09500693.2022.2105976

Peikos G, Spyrtou A, Pnevmatikos D, Papadopoulou P (2023) Nanoscale science and technology education: primary school students' preconceptions of the lotus effect and the concept of size. Res Sci Technol Educ 41(1):89–106. https://doi.org/10.1080/02635143.2020.1841149

Perkins DN, Grotzer TA (2005) Dimensions of causal understanding: the role of complex causal models in students' understanding of science. Stud Sci Educ 41(1):117–165. https://doi.org/10.1080/03057260508560216

Sassoni E, Ugolotti G, Pagani M (2020) Nanolime, nanosilica or ammonium phosphate? Laboratory and field study on consolidation of a byzantine marble sarcophagus. Constr Build Mater 262:120784. https://doi.org/10.1016/j.conbuildmat.2020.120784

Spyrtou A, Manou L, Peikos G (2021) Educational significance of nanoscience–nanotechnology: primary school teachers' and students' voices after a training program. Educ Sci 11(11):724. https://doi.org/10.3390/educsci11110724

Stavrou D, Michailidi E, Sgouros G (2018) Development and dissemination of a teaching learning sequence on nanoscience and nanotechnology in a context of communities of learners. Chem Educ Res Pract 19(4):1065–1080. https://doi.org/10.1039/C8RP00088C

Varela VS, Flores ML, Gayoso IGR (2022) Encouraging the construction of a precursor model about air through experimental activities in preschool. In: Boilevin JM et al (eds) Precursor models for teaching and learning science during early childhood contemporary trends and issues in science education. Springer, Cham, pp 111–129. https://doi.org/10.1007/978-3-031-08158-3_7

Theatrical Activities on Magnetic Phenomena for Scientific Learning in Primary School

Marisa Michelini and Irene Sualdin

Abstract Scientific education needs to be started early, with explorative experiences and representations of the physical world. Implicit conceptual obstacles (knots) are often present in pupils, so scientific education requires different kinds of integrated activities to produce conceptual change from the common ideas to the scientific ones. Besides, active learning generates motivation and curiosity, two fundamental elements that contribute to effective learning. This contribution presents the implementation of a Teaching Intervention Module (TIM) on magnetic phenomena, using integrated strategies of active learning and arts (drama and role-play). The activities aim to produce improvement not only in the scientific learning process but also in transversal competencies. The final product is a theatrical performance designed, prepared and implemented by children, during and for their own learning of the main basic concepts on the magnetic phenomena, beforehand explored by an Inquiry Based Learning approach. This research, with twelve 4th and 5th grade children, uses pre-post tests, tutorials and an observation grid to monitor the children's learning and understanding, as well as their transversal competencies, through both the scientific inquiry and the theatrical activities.

Keywords Drama · Active learning · Theatre · Magnetism · Arts integration · Role play

1 Introduction

Magnetic phenomena are in children's hands every day and this is an opportunity for scientific education. Previous studies investigated children's naive conceptions (Bradamante et al. 2005; Michelini and Vercellati 2014a, b; Stefanel 2008) and main

M. Michelini · I. Sualdin (✉)
Physics Education Research Unit, University of Udine, Udine, Italy
e-mail: irene.sualdin@gmail.com

M. Michelini
e-mail: marisa.michelini@uniud.it

Z. Ješková and J. Hanč (eds.), *Advancing Physics Education*, Challenges in Physics Education, https://doi.org/10.1007/978-3-031-94207-5_6

knots (obstacles) (Vercellati et al. 2010) about magnetic phenomena interpretation, and in particular the magnetic interactions, the role of poles, the idea of magnetic field and its representation (Challapalli et al. 2012; Fedele et al. 2005; Stefanel 2008). The direct experience with the phenomena is crucial for an effective understanding, and the literature underlines the need to introduce gradually the various conditions until the phenomenon occurrence, by means of the active exploration of the phenomena themselves (Fedele et al. 2005; Vartiainen and Kumpulainen 2020). The goal is to individuate the physics quantities involved and relative relationship and to lead the children to an interpretative perspective. For this scope, we have evidence that it is important to implement a micro-step path that provides explorative contexts, targeted questions, cooperative learning and reinforcement of the correct intuitions (Fedele et al. 2005). The children's personal engagement promotes the learning process, and this is the principle on which the Inquiry Based Learning (IBL) is based (Attard et al. 2021; Friesen and Scott 2013; Ord 2012; Rundgren 2018). People learn better when they personally build their own learning, being involved in learning experiences; besides knowledge is meaningful and can become competence if discovered by the individuals themselves (Panasan and Nuangchalerm 2010a, b). Therefore we designed an intervention module, based on previous research and paths already developed (Michelini and Vercallati 2014; Michelini et al. 2014), to lead the children to discover step by step the main concepts concerning magnetic phenomena. This method uses tasks that are authentic, significant for the children and linked to their experience; the children work and inquire for a long time on the same topic, they ask questions, formulate hypothesis, observe phenomena and find conclusions; this creates a deep understanding, critical thinking and scientific reasoning (Barrow 2006; Ernst et al. 2017; Panasan and Nuangchalerm 2010a, b), by means of Creative Inquiry too (Montuori 2011; Hardiman 2016). Literature underlines how modeling and representation have a strong significance in understanding scientific concepts, so in our path the children are invited to create different kinds of representations of the phenomenon (as drawings and role-play).

Creativity plays an important role in the industrial economy and technological in-novation, so the interest in introducing a creative development in schools has increased in the last decades (Sawyer 2006; Zhou 2018). Literature reveals that creativity is not just a cognitive effort but an affective one as well. A creative learning model has three levels: divergent functions; processes of complex thinking and feelings; involvement in real challenges (Fasko 2001; Hout and Krug 1995). Artistic education, diversification of experiences and improvisation have been found to be effective in reinforcing creative processes (Zhou 2018). In this work we used methods and techniques from the Arts Integration approach (Chemi 2016; Cornett 2015), drama (Chemi and Kastberg 2015; Hendrix et al. 2012; Najami et al. 2019; Michelinin and Vercellati 2014; Peleg and Baram-Tsabari 2011; Stagg and Verde 2019) and role-play (Aubusson et al. 1997; Braund and Ahmed 2019). Literature points out that the arts are excellent vehicles to reinforce understanding (Lerman 2003) and drama has an important role in influencing ideas and stimulating the critical thinking (Sloman 2012) as well as helping children to gain awareness of a complete and complex form of communication and expression, that they already possess (Beneventi and Conati

2006). The theatrical play can connect a narration, a story between characters to a scientific concept; it creates empathy, that helps cognitive learning. Therefore storytelling can be weaved to the subject to be learnt (Peleg and Baram-Tsabari 2011; Walan and Enochsson 2019). Besides, the role-play allows children to reproduce a phenomenon or any kind of topic with their own body. According to the embodied cognition theory (Amin et al. 2015; Shapiro and Stolz 2019; Wilson 2002), imagination is headed to the same cerebral mechanism used for perception and action, and abstract concepts are understood in terms of generalizations on sensorimotor experiences through metaphorical maps (Anderson and Shattuck 2012). This means that representing a concept with the body and the movement produces an authentic learning and a clear understanding of what is represented, that may be even more effective than a purely theoretical study.

We designed, implemented and monitored a Teaching Intervention Module (TIM) that aimed to overcome the main learning knots (obstacles) in the magnetic phenomena interpretation and to help children produce conceptual appropriation of the scientific concepts, so as to enable them to design and perform a theatrical representation of the main phenomena.

2 Research Questions

The research questions are:

RQ 1: How do children deal with the interpretation of magnetic phenomena and which are the main conceptual knots that they encounter?

RQ 2: How can a teacher prepare children for a theatrical activity that contributes to scientific education?

RQ 3: Through the planning of a theatrical play, how do children consolidate the principal concepts with explorative activities, and how do they strengthen learning and build a scientific and social identity?

3 Research Methodology

The educational reconstruction of the concepts (Fedele et al. 2005) was implemented adopting a Design Based Research (DBR—planning, implementation, data analysis and revision) (Anderson and Shattuck 2012; Reimann 2010) and Learning Analytics approach (learners, data, metrics/analytics, intervention) (Lockyer and Dawson 2011; Mangaroska and Giannakos 2018) with a sample of 12 children of 4th and 5th grade (one with disability certificate), participating in an exploration-based TIM of 40 h of post-school activities (distributed into lessons of about 1.5 h each) over 4 months in Trieste (Italy), from December 2021 to March 2022.

We adopt the Conceptual Labs for Operative Exploration (CLOE) strategy developed for magnetic phenomena (Challapalli et al. 2012), where the main research steps are the following:

Exploration of children ideas on their experience on magnetic phenomena (Bradamante et al. 2005; Michelini and Vercellati 2014a, b; Stefanel 2008).

Planning of an IBL path about magnetic phenomena (we based it on the CLOE labs experiences) (Challapalli et al. 2012; Michelini et al. 2023).

Building of tutorials and a pre-post test (based on the path we elaborated).

Selection/invention of theatrical activities (based on literature) (Beneventi and Conati 2006; Micheli and Papi 2007).

Building of an observation grid for transversal competencies.

Individuation of a priori expectations and creation of evaluation tables.

Implementation of in-class activities.

Data analysis of tutorials and pre-post tests by means of qualitative methods of operative classes of answers.

Comparison of data emerging by the different instruments used for data collection, data discussion and conclusions.

The CLOE strategy offers the opportunity of "a learning environment in which children have an operational role, proposing problem-situations: a context in which one is left with free exploration, then asked to make a prediction, and finally to test and compare the results of the tests with the predictions" (Gibbs, 2005). During the activities, the children are guided by worksheets (tutorials) with stimuli questions that ask children to do a prediction, then to perform the experiment, and to compare observations with their expectations, drawing their conclusions. Table 1 shows an example of the research protocol in the context.

A test of 19 open questions about magnetic phenomena experience and interpretation was submitted to each child at the beginning of the activity. The same test was submitted at the end of the entire path, to monitor the children learning evolution. Qualitative data analysis was carried out in two ways: (a) by classes of meaning, (b) by classes of typology: descriptive, conditional, interpretative.

The in-class activities were developed in four phases:

exploration of magnetic phenomena (7 interventions);

preliminary theatre activities (6 interventions);

role-play on the magnetic phenomena (4 interventions);

planning and performing a theatrical play about magnets (22 interventions).

The first exploratory phase involved actively all children in 20 experiments. The Inquiry Based Learning process went through the following concepts: identification of a magnet; existence of poles; ferromagnetic, diamagnetic and paramagnetic materials behavior; reciprocity of interaction; compass as an explorer; magnetic field existence and characteristics.

In parallel to the IBL experimental exploration, the children took part in preliminary theatrical activities and started to train competences as: body, voice, stage and time self management; work organization; communication; self expression; team work.

Table 1 Example of the stimuli questions in the tutorials

CLOE protocol	Stimuli questions	Research questions
Interactions between magnets	I have two cylindrical magnets. I want to study how they interact. I put one of them on the table and I keep the other one in my hand, approaching it to the magnet on the table. How do you expect the one on the table to behave, so how do the two magnets interact? Explain Do it and describe what you observe Now I put the magnet I have in my hand near the other end of the magnet on the table. How do you predict it will orient itself? Turn the magnet in the hand and repeat the experience. Describe what you observe What's the conclusion? How do the magnets behave with each other?	RQ (1.1) Which kinds of reasoning are used and how does reasoning play the role of conceptual referent? RQ (1.2) Are poles taken into consideration to predict the interaction and to describe the observation? RQ (1.3) Is a magnet's rotation considered as a possible outcome of the interaction and is it observed as a phenomenon? RQ (1.4) Are generalization and/or conceptualization made? Or is there just a contingent description of the phenomena?

During the theatrical activities, we observed the children behavior by means of an Observation Grid based on a Likert scale in 5 levels for the following aspects: work organization, time management, problem-solving, finding creative solutions, effectiveness of the communication, body management, voice management, stage management, courage in self-expression, motivation, team work (divided into active collaboration and respect for the others).

The same grid was filled in three different moments: at the beginning of the preliminary activities (phase 2); during the role-play (phase 3); at the end of the final performance of the play (phase 4).

The Arts Integration approach (phase 3) focused on making children represent each magnetic phenomenon with a role-play performance. In that phase the Creative Inquiry and the embodied cognition theory were operationalized. The children were divided into small groups and they had a limited time to decide how to perform the phenomenon and rehearse before performing in front of the class mates. In order to represent the phenomenon correctly, the children had to recall and explicit the past exploration and the correlated concept, deepening their understanding and reinforcing their memory (according to the embodied cognition theory (Amin et al. 2015; Shapiro and Stolz 2019; Wilson 2002)); moreover they needed to find creative solutions, training their creativity, problem solving abilities and divergent thinking (according to the creative inquiry approach (Hardiman 2016; Montuori 2011)). After a group performed their scene, the audience evaluated if the scientific concept was clear and explained why it was well or badly represented. In this way, the children went over the phenomenon again and developed critical thinking.

Phase 4 was dedicated to the planning and preparation of a theatrical play about magnetic phenomena. The children selected the phenomena to represent and then chose a way to perform them (selecting an option from the types of performances elaborated during phase 3). A full script was built by a series of brainstorming, symbolic representations in the blackboard, trials and discussions with the teacher. Step by step, this entire process required us to find many creative solutions, to evaluate alternatives, to solve problems, to deepen the scientific concepts, to practice the theatrical abilities, to handle a team work, to organize the work in the given time, to make costumes and props. The children were motivated because there was a concrete final product to produce (the final performance in front of an external audience of youngsters and adults) and this stimulated them to commit to their tasks.

After the public performance, the children filled in the post-test (the same as the pre-test) and a survey of 15 questions focused on their involvement in the theatrical activities, their perception of their own learning and of the entire experience.

4 Data Analysis

All 12 children submitted the pre-post test at the beginning and at the end of the activities. Data analysis was carried out according to predefined categories by means of literature and conceptual goals, completed a posteriori including examples of observations, representations and interpretations by children. For each question we used a qualitative methodology of analysis based on reasoning categories.

We chose to show here a few sample tables used to categorize the answers of the pre-post test (see Tables 2 and 3).

Table 2 shows that in the post-test two children gave both description and explanation. Moreover, it disappears the possibility for the magnets non to get attracted, and the explanations are more articulated and precise.

Table 3 shows that in the pre-test only 3 children generically declared that the nail sticks; while in the post-test 9 children gave the correct answer, in 4 cases together with the explanation and interpretation of the phenomenon. The temporary magnetization was one of the most loved experiments and many children demonstrated to have a clear understanding of it during the preparation of the theatrical play.

The tutorials and pre-post test analysis show that most of the concepts (Identification of a magnet by the interaction with other materials; The magnetic interaction occurs at a distance; Every magnet has two poles; The magnet's poles are indivisible; Not all the metals are ferromagnetic; A compass needle is a magnet; Temporary magnetization is possible) were learnt by the majority of students, while some concepts (By approaching homologous poles of unconstrained magnets, a rotation occurs with consequent attraction; Interaction is reciprocal; Difference between the interaction between magnets, and between a magnet and a ferromagnetic material; Earth has a magnetic North Pole and South Pole) would need further work.

Table 2 Categorized answers of the pre-post test question: "You have a magnet with the shape of a small cylinder. You cut it in two and take one piece. You put that piece near another whole magnet. You see that they stick together, as before. How do you explain this?"

Categories/ subcategories	Role	Pre test—original sentences	Post test—original sentences	Pre n° answers	Post n° answers
Phenomenon	Descriptions	– No, it doesn't stick as before – It still sticks – They stick together		4	0
	Interpretations		– They attract each other as before – They attract each other anyway	0	6
Explanation	Interpretations	– It stays magnetic	– Even if you divide a magnet in many pieces, they still have a South Pole and a North Pole – Another North or South Pole is generated – It still has two poles	4	6
No answer				4	2

The pre-post test analysis shows also an increase of the total answers the children gave and a significant increase of the interpretations.

Figure 1 shows how the cognitive categories of answers emerged in the pre-post tests: experience recall, recall of an experiment, description of an observation (contingent descriptive rules regarding the specific situation observed in that moment), generalization of a phenomenon (generalization of the rules and conditions observed, always in a descriptive manner), interpretation of a process, strategies (indications about what to do to obtain a certain result).

In the pre-tests the descriptive answers (47.43%) prevail, while conditional (24.35%) and interpretative answers (10.25%) are often confused or deriving from common sense ideas. In the post-test the answers related to the personal experiences diminish (5.9%) (4% are related to the theatrical show); a few experiment recalls appear (3%); there is a small incrementation of the number of descriptive answers, in fact reduced with respect to pre-test (43%); the enunciation of conditions diminishes (15.56), probably because this type of answer in many cases is replaced by interpretations, that have an incrementation (24.73%); finally we see a little incrementation of the strategies (8.06%).

Table 3 Categorized answers of the pre-post test question: "I let the paper clip stick to the magnet. Now I put the nail near the paper clip. What do you expect to see? Explain your forecast"

Categories/ subcategories	Pre-test—operative definitions	Post-test—operative definitions	Pre n° answers	Post n° answers
Phenomenon			7	6
Attraction	– It sticks	– The nail sticks to the paper clip – It sticks	1	5
No attraction	– It doesn't stick	– The nail doesn't stick to the paper clip	6	1
Explanation	– They don't move, because they are simple pieces of iron	– It sticks because the paper clip has temporarily become magnetic – They stick together because in that moment the magnet has a sufficient force to keep them both – It sticks because there is the magnet – They both sticks, because the magnetic field reaches the paper clip	1	4
Other	– They both attract each other – The nail falls down		3	0
No answer			1	2

During the work we noticed that the request of written answers in tutorials and tests do not restitute the richness of reasoning orally obtained by children. In fact, several children had writing issues, so they wrote very few words, instead of deeply explaining their reasoning. Some other students openly refused to answer, just because they felt lazy.

Figure 2 shows just three of the 12 transversal competencies observed (listed in the section Research Methodology of this document). In every competency there is evidence of a general improvement. The majority of children increased their motivation and other competencies in the last phase, the preparation of the theatrical show. The highest general increase is in the areas of effective communication, stage management and active collaboration, while the absolute highest score is in motivation.

Analyzing the final questionnaire about theatre, we see that all the participants rated the activities 10/10 or more. The only thing a student specified was not appreciated was the filling of the questionnaire, because many children didn't like to write. At the question: "What have you learnt from the theatrical activity?" Some children focused on the science content ("many things about magnets", "the magnetic field"), others on the theatrical competences ("to act", "to smile even when I am angry", "that you are not supposed to show your back to the audience"), others referred to

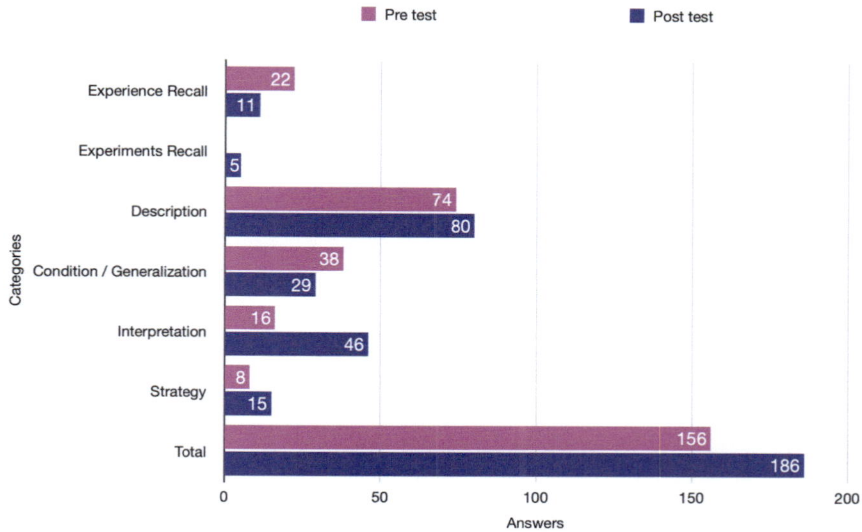

Fig. 1 Distribution of the kind of answers per cognitive categories in pre-test (156, number of answers collected considering all the pre-tests together) and post-test (186, number of answers collected considering all the post-tests together)

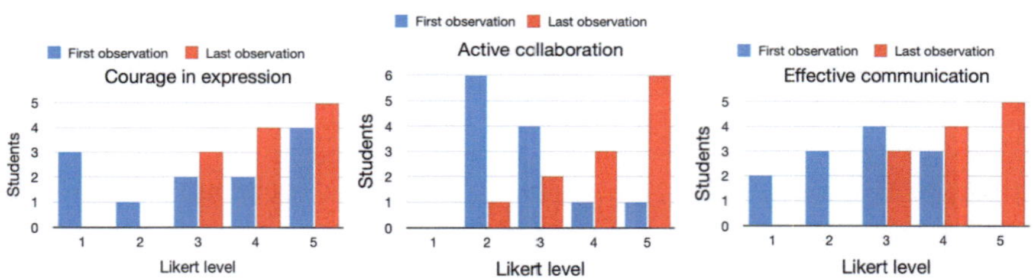

Fig. 2 Number of students for each Likert level. Data from the observation done by means of the Rubric to monitor the transversal competencies gained during the theatrical activities

less specific elements ("it is a wonderful activity", "everything", "what was supposed to be learnt"). To the specific question "Did the theatrical activity help you learn/understand/remember the magnetic phenomena better?" all 12 students answered positively. To the question "In what way has theatre been useful to learn concepts?" some children focused on the pleasant emotions the activities gave them ("You have fun") and literature makes clear that pleasure increases motivation; while some other children focused on the specific activities they found useful in order to learn ("I like to see things and experience them", "To put it into practice", "Learning the lines by heart"). In the questionnaire, a student also specified that "I learn things better with activities than with yelling teachers", suggesting that practice is helpful for effective learning. 10 students out of 12 declare that they would prefer to prepare and perform a play in order to learn a subject instead of approaching it with more traditional school methods.

5 Conclusions

We found evidence that each child builds their own learning path by reworking the experience and personal analysis of it by means of the CLOE strategy (Challapalli et al. 2012; Michelini et al. 2023); the learning process is also influenced by communication and physical elements, by emotions, sensations and by the peer interactions. The integration of expressive, communicative and experiential activities with the scientific ones generates a mutual enhancement. The implementation of this educational proposal based on active learning methodologies, as the Inquiry Based Learning (Attard et al. 2021; Rundgren 2018; Friesen and Scott 2013; Ord 2012), focusing on conceptual understanding, and artistic theatrical activities, using the Creative Inquiry (Hardiman 2016; Montuori 2011) and the Arts Integration (Ødegaard 2003; Stagg and Verde 2019) approaches, has very good outcomes.

The children started to familiarize with the scientific inquiry method and to adopt it correctly; they familiarized also with the basic techniques of theatre and trained several transversal competencies. Practical activities are motivating and make the subject clearer and more concrete to the children.

Observation, tests, tutorials and the questionnaire show that the children face every experimental exploration in a positive, active and responsible way. The biggest conceptual knots they encounter are related to the reciprocity of the interactions and the distinction between interactions between two magnets or a magnet and a ferromagnetic material (**RQ1**). The planned goal requires that the teachers: provide preparatory exercises on theatrical activities, prepare the theatrical script, organize rehearsals. The learning open environment provides opportunities for the children to: reflect and reconstruct the core concepts; individuate representation methods; select the proposals; create the theatrical context and characters; prepare the costumes and props; perform in front of an audience (**RQ2**). Children show a good appropriation of the main concepts (7/12 identify a magnet from interactions with other materials, 9/12 individuate poles phenomenon and 8/12 that poles cannot be divided, 12/12 the role of distance in the magnetic interactions, 7/12 identify the compass needle as a magnet, 9/12 explain how not all metals are ferromagnetic, 10/12 are aware of the temporary magnetization). The theatrical performance also confirms the gain of soft skills and an identity planning ability (the children chose a way to perform and correctly acted out the phenomena, even when they had to improvise out of the script); children were enriched by expression; communication; body, voice and space management; collaboration; cooperation; respect and work organization (the observation grid shows an overall improvement across all these skills) (**RQ3**).

We have evidence of a way to carry out a TIM by means of IBL and drama able to produce effective conceptual learning: the majority of children gained a clear understanding of the main concepts and chosen phenomena, and enhanced their motivation, communication and social skills.

The main limits of this research—that would need further study—are the small sample, the long time needed to implement the theatre play preparation, the demanding work due to many written tutorials to be filled by the children: oral interviews appear to work better. Then, the extra-curricular activity gave a lot of freedom for the research: now a feasibility study for the curricular integration of the proposal is needed. The research aimed to the individuation of modalities and potentialities for the scientific and expressive learning by means of a theatrical production: this needed a lot of time. Phases 1 and 3 are easily implementable in a school context, while phase 4 may be implemented one or twice a year, probably not more, as an integrative activity. The preparation of our theatrical play took 22 interventions of about 1 h in this research.

References

Amin TG, Jeppsson F, Haglund J (2015) Conceptual metaphor and embodied cognition in science learning: introduction to special issue. Int J Sci Educ 37(5–6):745–758. https://doi.org/10.1080/09500693.2015.1025245

Anderson T, Shattuck J (2012) Design-based research: a decade of progress in education research? Educ Res 41(1):16–25. https://doi.org/10.3102/0013189X11428813

Attard C, Berger N, Mackenzie E (2021) The positive influence of inquiry-based learning teacher professional learning and industry partnerships on children engagement with STEM. Front Educ 6:693221. https://doi.org/10.3389/feduc.2021.693221

Aubusson P, Fogwill S, Barr R, Perkovic L (1997) What happens when students do simulation role-play in science? Res Sci Educ 27:565–579. https://doi.org/10.1007/BF02461481

Barrow LH (2006) A brief history of inquiry: from Dewey to standards. J Sci Teach Educ 17(3):265–278. https://doi.org/10.1007/s10972-006-9008-5

Beneventi P, Conati D (2006) Nuova guida di animazione teatrale. Sonda Casale, Monferrato—AL, Italy

Bradamante F, Fedele B, Michelini M (2005) Children's spontaneous ideas of magnetic and gravitational fields. In: Pinto R, Couso D (eds) The ESERA selected contributions book, Barcellona

Braund M, Ahmed Z (2019) Drama as physical role-play: actions and outcomes for life science lessons in South Africa. J Biol Educ 53(4):412–421. https://doi.org/10.1080/00219266.2018.1490799

Challapalli SRCP, Michelini M, Vercellati S (2012) Informal learning in CLOE labs to build the basic conceptual knowledge of magnetic phenomena. In: Taşar MF (ed) Proceedings of the WCPE. Pegem Akademi, Ankara, pp 741–745

Chemi T (2016) Science and the arts: Curriculum integration, learning, and emotions in schools. In: Bellocchi A, Quigley C, Otrel-Cass K (eds) Exploring emotions, aesthetics and wellbeing in science education research, vol 13. Springer, Cham, pp 203–218

Chemi T, Kastberg P (2015) Education through theatre: typologies of science theatre. Appl Theatr Res 3(1):53–65. https://doi.org/10.1386/atr.3.1.53_1

Cornett CE (2015) creating meaning through literature and the arts: arts integration for classroom teachers. Pearson, Allyn & Bacon, Boston, MA. Ch. 1-2-3

Ernst DC, Hodge A, Yoshinobu S (2017) What is inquiry-based learning. Not AMS 64:570–574. https://doi.org/10.1090/NOTI1536

Fasko D (2001) Education and creativity. Creat Res J 13(3–4):317–327. https://doi.org/10.1207/S15326934CRJ1334_09

Fedele B, Michelini M, Stefanel A (2005) 5–10 years old pupils explore magnetic phenomena in cognitive laboratory (CLOE). In: Pinto R and Cuso D (eds) Proceedings of the fifth international ESERA conference on contributions of research to enhancing students' interest in learning science, Barcelona

Friesen S, Scott D (2013) Inquiry-based learning: a review of the research literature. Alberta Ministry of Education, vol 32

Gibbs G (2005) Improving the quality of student learning. Doctoral Thesis. University of South Wales, United Kingdom

Hardiman MM (2016) Education and the arts: educating every child in the spirit of inquiry and joy. Creat Educ 7(14):1913–1928. https://doi.org/10.4236/ce.2016.714194

Hendrix R, Eick C, Shannon D (2012) The integration of creative drama in an inquiry-based elementary program: the effect on student attitude and conceptual learning. JSTE 23(7):823–846. https://doi.org/10.1007/s10972-012-9292-1

Houtz JC, Krug D (1995) Assessment of creativity: resolving a mid-life crisis. Educ Psychol Rev 7:269–300. https://doi.org/10.1007/BF02213374

Lerman ZM (2003) Using the arts to make chemistry accessible to everybody. 2002 James Flack Norris Award, sponsored by ACS Northeast section. J Chem Educ 80:1234–1242. https://doi.org/10.1021/ed080p1234

Lockyer L, Dawson S (2011) Learning designs and learning analytics. In: Proceedings of the 1st international conference on learning analytics and knowledge. Association for Computing Machinery, New York, pp 153–156. https://doi.org/10.1145/2090116.20901

Mangaroska K, Giannakos M (2018) Learning analytics for learning design: a systematic literature review of analytics-driven design to enhance learning. IEEE Trans Learn Technol 12(4):516–534. https://doi.org/10.1109/TLT.2018.2868673

Micheli S, Papi S (2007) Libri in scena. Giochi e attività teatrali con i libri. Erickson, Gardolo—TN, Italy

Michelini M, Vercellati S (2014a) The IWB as a bridge between phenomena exploration and interpretation of electromagnetic phenomena in construction of formal thinking. In: Dormido S, de la Torre L (eds) Proceedings of the MPTL18, vol 37 European Physical Society, Madrid, pp 101–106

Michelini M, Vercellati S (2014b) Pupils' discussion to understand magnetic interactions. In: Constantinou CP, Papadouris N, Hadjigeorgiou A (eds) E-Book Proceedings of the ESERA 2013 confernce: science education research for evidence-based teaching and coherence in learning (Nicosia, Cyprus). European Science Education Research Association, Nicosia, Cyprus, pp 457–463

Michelini M, Santi L, Stefanel A and Vercellati S (2014) Building vertical paths in exploring magnetic phenomena developing formal thinking. In: Constantinou CP, Papadouris N, Hadjigeorgiou A (eds) E-Book Proceedings of the ESERA 2013 conference: science education research for evidence-based teaching and coherence in learning (Nicosia, Cyprus). European Science Education Research Association, Nicosia, Cyprus, pp 974–982

Michelini M, Santi L, Stefanel A, Chiofalo M (2023) Entangled research methods for the building of coherent conceptual thematic learning paths and connecting research with praxis. Il Nuovo Cimento 46 C: 196. https://doi.org/10.1393/ncc/i2023-23196-4

Montuori A (2011) Reproductive learning. In: Seel NM (ed) Creative inquiry, encyclopedia of the sciences of learning. Springer, Boston MA

Najami N, Hugerat M, Khalil K, Hofstein A (2019) Effectiveness of teaching science by drama. Creat Educ 10(1):97–110. https://doi.org/10.4236/ce.2019.101007

Ødegaard M (2003) Dramatic science: a critical review of drama in science education. Stud Sci Educ 39:75–101. https://doi.org/10.1080/03057260308560196

Ord J (2012) John Dewey and experiential learning: developing the theory of youth work. Youth Policy 108:55–72

Panasan M, Nuangchalerm P (2010a) Learning outcomes of project-based and inquiry-based learning activities. J Soc Sci 6(2):252–255. https://doi.org/10.3844/jssp.2010.252.255

Panasan M, Nuangchalerm P (2010b) Learning outcomes of project-based and inquiry-based learning activities. J Soc Sci 6(2):252–255. https://doi.org/10.3844/JSSP.2010.252.255

Peleg R, Baram-Tsabari A (2011) Atom surprise: using theatre in primary science education. J Sci Educ Tech 20:508–524. https://doi.org/10.1007/s10956-011-9299-y

Reimann P (2010) Design-based research. In: Markauskaite L, Freebody P, Irwin J (eds) Methodological choice and design: scholarship, policy and practice in social and educational research. Springer, Dordrecht, Netherlands, pp 37–50. https://doi.org/10.1007/978-90-481-8933-5_3

Rundgren CJ (2018) Implementation of inquiry-based science education in different countries: some reflections. Cult Stud of Sci Educ 13:607–615. https://doi.org/10.1007/s11422-016-9787-8

Sawyer RK (2006) Educating for innovation. Think Ski Creat 1(1):41–48. https://doi.org/10.1016/j.tsc.2005.08.001

Shapiro L, Stolz SA (2019) Embodied cognition and its significance for education. Theory Res Educ 17(1):19–39. https://doi.org/10.1177/1477878518822149

Sloman A (2012) Using participatory theatre in international community development. Community Dev J 47(1):42–57. https://doi.org/10.1093/cdj/bsq059

Stagg BC, Verde MF (2019) Story of a seed: educational theatre improves students' comprehension of plant reproduction and attitudes to plants in primary science education. Res Sci Technol Educ 37(1):15–35

Stefanel A (2008) Disciplinary knots and learning problems in electromagnetism. AIP Conf Proc 1018:231–235. https://doi.org/10.1063/1.2947691

Vartiainen J, Kumpulainen K (2020) Playing with science: manifestation of scientific play in early science inquiry. Eur Early Child Educ Res J 28(4):490–503. https://doi.org/10.1080/1350293X.2020.1783924

Vercellati S, Michelini M, Lambourne R, Mathelisch L (2010) A discussion of disciplinary knots on electromagnetism and superconductivity using a web environment in the context of an EU project (MOSEM) for research-based in service teacher training. Il Nuovo Cimento, Multimedia in Physics Teaching and Learning. 33(3):189–193. https://doi.org/10.1393/ncc/i2010-10637-6

Walan S, Enochsson AB (2019) The potential of using a combination of storytelling and drama, when teaching young children science. Eur Early Child Educ Res J 27:821–836

Wilson M (2002) Six views of embodied cognition. Psychon Bull Rev 9:625–636. https://doi.org/10.3758/BF03196322

Zhou K (2018) What cognitive neuroscience tells us about creativity education: a literature review. Glob Educ Rev 5(1):20–34

Secondary Education

Introduction to Measurement Uncertainties for Middle School Classes

Hannah Loidl and Clemens Nagel

Abstract Introducing measurement uncertainties aims to foster middle school students' abilities in critical thinking, conducting experiments, and properly analysing results. In the first author's master thesis, a teaching concept for the beginning of natural science classes was developed and evaluated in a middle school in Vienna. The concept was evaluated and improved in three cycles using the design-based research method. Students should perform a simple experiment that leads them to measurement uncertainties. Sources for uncertainties and the trustworthiness of the experiment are discussed. Using simple methods of statistics such as the arithmetic mean, the experiment will be analysed. The tested class revealed positive outcomes. The students' discussions during and after conducting the designed experiment provided valuable insights. Furthermore, linking the word trustworthiness to the technical term measurement uncertainties seems to be a promising way to introduce students to the concept of measurement uncertainties.

Keywords Measurement uncertainties · Experimental skills · Nature of science · Secondary curriculum · Design-based research

1 Introduction

Acquiring experimental skills is part of every physics curriculum. While the hands-on part of experiments raises the students' interest in physics, the part of calculating and discussing measurement uncertainties (MU) sometimes has an unsatisfactory character for students, also at the university level. However, interpreting measurement uncertainties opens engaging discussions about conducting experiments. It may also

H. Loidl (✉)
Department of Research and Development, KPH Vienna/Lower Austria, Vienna, Austria
e-mail: hannah.loidl@kphvie.ac.at

C. Nagel
Basic Experimental Physics Training and University Didactics, University of Vienna, Vienna, Austria
e-mail: clemens.nagel@univie.ac.at

Z. Ješková and J. Hanč (eds.), *Advancing Physics Education*, Challenges in Physics Education, https://doi.org/10.1007/978-3-031-94207-5_7

deepen the understanding of the examined phenomena and allows a more nuanced view on how scientists derive results. Therefore, a concept for introducing measurement uncertainties was developed and evaluated in a Viennese middle school as a part of the first author's master thesis. The method of design-based research was applied. Materials were evaluated with methods such as probing acceptance interviews, film- and audio recordings during class and written assessments. Finally, an interview with an experienced physics teacher revealed potential to improve the existing teaching concept (Loidl 2021).

1.1 Theory of Measurement Uncertainty in GUM

The formulated teaching goals are in line with the theory of measurement uncertainty proposed by the Guide to Expression of Uncertainty (GUM). Every result of a measurement has an uncertainty consisting of different components. The GUM groups them into two categories, namely "those which are evaluated by statistical methods" (type A) and "those which are evaluated by other means" (type B) (BIPM 2008). As recommended by the GUM, we differentiate between errors and uncertainties. Errors should be avoided whenever possible, however if avoidance is not possible, errors lead to systematic effects which must be corrected. Depending on the mathematical problem and depth of the analysis, different mathematical methods for the evaluation of uncertainties must be chosen. Since the teaching concept was developed for students at the age of eleven to thirteen, we chose low level statistic values e.g., the range (or standard deviation for advanced students) for calculating type A uncertainties. Also, the teaching materials ask students to reflect on the uncertainty of the applied measurement device. Type B uncertainties are in general determined deductively by other than statistic means. Even an individual measurement has an uncertainty due to the nature of the measurement itself, for instance caused by the production or function of the measuring device used. Measurement instruments with a measuring scale typically consist of at least three main components, which arise due to the calibration of the device and the process of graduation of the scale (linearity uncertainty). The third main component is the reading uncertainty met in the process of measuring when the value is in the middle of two scale divisions (Nagel 2021, 2025a). Therefore, the reading uncertainty equals half of the smallest scale division. In the teaching concept for assessing type B uncertainty, we used reading uncertainty as a simple tool to introduce the concept of type B MU.

1.2 *Why Should We Teach Measurement Uncertainty in School?*

At the first glance teaching MU does not seem like an exciting topic for middle school classes. This might be influenced by the fact that teachers themselves struggle in their academic education with analysing data in experimental courses, including measurement uncertainties (Nagel et al. 2021). However, introducing MU in physics classes provides a fruitful learning opportunity for students as discussing results and their meanings has the potential to support critical thinking. Also, engaging more deeply with the potential poor quality of data may improve data literacy.

Besides the above-mentioned considerations, acquiring experimental and evaluation skills for data analysis is part of every physics curriculum. Conducting experiments and analysing self-collected data by including MU is fundamental in the learning process of the Nature of Science. As a result, students gain insights in the scientific process of generating knowledge and competencies in data literacy (Kok 2020).

Moreover, a proficient use of data is crucial in today's society, as seen during the Covid-19 pandemic, where absolute numbers were widely cited without sufficient discussions of uncertainties (Priemer 2022). Furthermore, students are constantly exposed to (fake-)experts and (fake-)data in social media. Therefore, it seems necessary to confront children with this topic from a young age to prepare them for evaluating the reliability of (scientific) data.

1.3 *Research Gap and State of the Field*

Considering uncertainties is necessary to obtain reliable results. Conversely, data without the expression of their uncertainty are not trustworthy. By introducing students to the theory of measurement uncertainty, they may be able to distinguish between scientific and non-scientific data. Nevertheless, most physics teachers resort to straightforward experiments when trying to prove a concept. The lack of well-established (teaching) strategies may hinder teachers from including MU in their physics classes, as suggested by Holz and Heinicke (2019). However, in recent years some ideas and concepts were published e.g., in the Austrian physics didactics journal "Plus Lucis" which dedicated a whole edition to measurement uncertainties (Nagel and Neumann 2021). Also, Bärenthaler-Pachner (2022) developed and evaluated a teaching concept for introducing middle school students (age 15) to MU at the beginning of science classes in his master thesis. In his concept, he focuses on the method of type B evaluation of uncertainty and is using the bridging-concept of "trustworthiness" as the main domain specific design principle. Priemer (2022) published an entire book, explaining not only the importance of uncertainties but also providing ideas for discussing uncertainties in everyday life examples. In his dissertation, Kok

(2022) developed a digital learning environment to teach MU in high school and discussed at what grade level MU can successfully be introduced in school.

Another way to teach uncertainty in elementary school was presented by Munier et al. (2013). The authors invented two teaching/learning-sequences and examined "to what degree elementary school can foster understanding and skills related to measurement, measurement uncertainty, and the analysis of actual measurement data" (Kok 2022). However, there is a lack of teaching units introducing MU in middle school and most of the research concerns higher education and laboratory courses at the university level.

1.4 Design-Based Research and Test Instruments

The teaching concept was invented by using the method of design-based research, following the methodology introduced by Bakker (2019). In three cycles the teaching materials were probed and improved. In the first cycle, ideas were collected, and the invented experiment was tested in initial trials with critical friends using the action research method (Altrichter et al. 2018). The resulting experiment, along with the instructions, was incorporated into the teaching materials. Afterwards, the developed materials were tested by the method of probing acceptance interviews (Wiesner and Wodzinski 1996) with two students in a middle school in Vienna. Additionally, an online survey was conducted to determine which statistical key figures are taught in math classes up to the 8th grade (students aged 14) in Austria. Encountered problems during the probing acceptance interviews as well as the results from the online survey were used to refine the teaching materials.

In the third cycle, the teaching concept comprising two school lessons (each 50') was tested in a physics class, in the same school as the above-mentioned, with students aged 13–14. Due to the Covid-19 pandemics the materials were tested in one class only and conducted by an experienced physics teacher in the field of MU. For the analysis, the teaching units were filmed and four working groups were audio-recorded during the whole session. All recordings were transcribed and analysed using qualitative content analysis (Lamnek and Krell 2016). Also, the class's physics teacher attended both lessons and recorded noteworthy observations using an observation guide. After the intervention, an expert interview (Lamnek and Krell 2016) with the performing teacher was conducted to discuss any difficulties concerning the teaching concept. Three days after the intervention, the students completed a written test. In contrast to the teaching concept, the written test (Krüger et al. 2014) focused on an experiment related to time measurement to assess whether the students could apply their newly acquired knowledge on a different topic. In the last cycle, the teaching materials were once more refined, and the results presented in the author's master thesis.

2 Aim of the Teaching Concept

The aim of the study was to create a teaching concept for students at the beginning of science/physics classes. It was designed to consist of only two 50-min lessons, serving as a starting point for further experiments and data analysis processes. After introducing MU to middle school classes, MU should be integrated into every experiment to establish a practical routine in the scientific evaluation of data.

As a result of the developed teaching units, students go through an incremental process: starting with the simple task of identifying sources of MU, then advancing to incorporate MU in their calculations, and ultimately being able to evaluate the reliability of measurements by integrating MU into their analytical thinking strategies, see also Fig. 1.

In an initial approach to the concept of MU, students utilise their existing mathematical skills to calculate statistical parameters e.g., arithmetic mean, range and, eventually, standard deviation within their physics class. Furthermore, students shall recognise the necessity of MU for comparing different results and understand its importance in evaluating the trustworthiness of data. The presented didactic ideas of MU and subject-specific theory by GUM provided the base for formulating the following teaching goals and key ideas (see Table 1). Also, preliminary studies as part of a bachelor thesis at the University of Vienna and insights from Heinicke's dissertation (Holz and Heinicke 2019) were incorporated into the teaching concept.

2.1 Main Ideas of the Concept

Preliminary studies in the author's bachelor thesis revealed that high school students (aged 15–16) were able to discuss the relevance of considering MU in scientific contexts. However, when dealing with uncertainties in experimental settings, students often resorted to simply copying already learned mechanisms without critically examining them. Therefore, the concept should focus on conducting a simple experiment with intentionally integrated flaws. The sub-optimal experimental setup—using a measuring device that provides imprecise readings and following an unsatisfactory method (explained in the following chapter)—exposes students to sources of MU, errors, and potential improvements in experimental design. Based on the teaching method "activating and confronting misconceptions" inspired by Wiesner

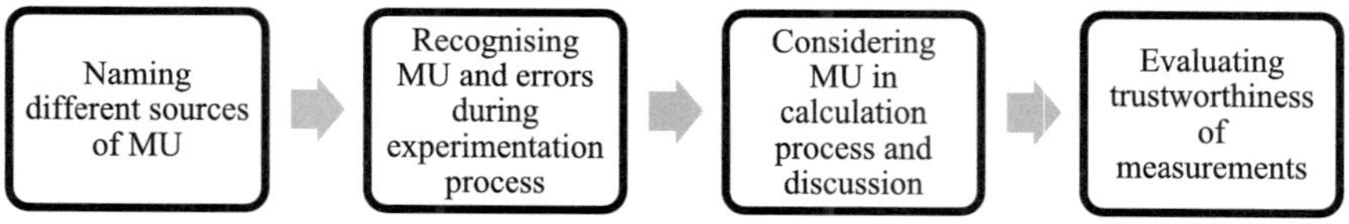

Fig. 1 Stages in the learning process of measurement uncertainties (MU)

Table 1 Overview of teaching units and associated goals, as well as key ideas

<table>
<tr><th></th><th>Superior goal</th><th>Immersing teaching goals</th><th>Key ideas</th></tr>
<tr><td>Unit 1</td><td>Thematising the trustworthiness of a measurement (Introduction to the concept of MU)</td><td>• Being able to explain why every measurement has an uncertainty
• Being able to evaluate and compare results by considering measurement uncertainties</td><td rowspan="2">✓ The individual measurements in a series of measurement are usually not identical.
✓ The mean of a series of measurement is the best estimate of the true value. The more measurements are taken, the closer the mean value gets to the true value. The mean of a series of measurement is the best estimate of the true value. The more measurements you take, the closer the mean value gets to the true value.
✓ The smaller the range (standard deviation of the mean) of a series of measurement the more trustworthy the result.
✓ There are different causes for MU: Randomness of measured values and uncertainty of the measuring device.</td></tr>
<tr><td>Unit 2</td><td>Consolidate gathered knowledge.</td><td>• Being able to analyse self-collected data and state results including their measurement uncertainty
• Being able to name different sources of measurement uncertainties and consider MU for evaluating the reliability of data</td></tr>
</table>

et al. (2018), students should reflect on the reliability of their values and results by performing a seemingly simple task and grappling with its unexpected outcomes.

Additionally, the experiment should get along with usual measurement devices, making it cost-effective and easily accessible for teachers. Some of the key elements of the teaching concept e.g., a scale for evaluating the trustworthiness, can be used for every single future experiment. However, this does not replace a reflection or discussion of the results by considering MU.

Furthermore, there must be a strict differentiation between type A and type B evaluation of uncertainty and errors, as stated in the GUM (BIPM 2008). Whenever students speak of errors in measurement, we have to be certain they do not confuse them with MU. Besides that, teachers should be aware of the terms and vocabulary they use when teaching MU.

Before implementing the technical term MU, students should evaluate the trustworthiness of their measurements. The explicit distinction of childrens' everyday language and technical terms should facilitate the introduction of a new concept in science class. The following teaching concepts were primarily developed for the beginning of science/physics classes (children at the age of 11–13 years).

2.2 *Main Structure of the Curriculum*

2.2.1 First Teaching Unit

The concept is supposed to take place in two teaching lessons and was created to first introduce MU in physics classes. Students shall work in small groups of up to three people without the teacher's help. Each group finds a prepared box with a precisely weighed amount of water, a small measuring cup and another empty box for scooped water, as well as a worksheet with a table for the measured values. The teacher introduces the work-phase by stating: "How much water is in the box in front of you? Please scoop the water by using the small measuring cup, read the amount of water and note the value in the table on your worksheets. However, there is one condition: You are not allowed to lift the box, completely, at least one corner of the box must touch the table at any time!".

Now students shall start to execute the experiment in groups. This might lead to students taking turns in the process of scooping and measuring and presents a starting point for the following discussion: Did every student perform the measurement in the exact same way? Was every student reading the values from the same angle?

As soon as a group scooped as much water as possible, they calculate the scooped amount of water. Afterwards, the groups assess the trustworthiness of their measurement by using a scale on the worksheet, see Fig. 2.

Then, the students mark their group-result (measured amount of water) with an X on a number line. The number line is prepared in advance by the teacher on a flipchart poster (or board). Preferably, the poster is placed outside of the classroom, so that slower groups do not get influenced by the quicker groups' results.

By comparing their own result with the ones of other groups on the number line (the first group gets an input by the teacher, like: "Imagine there would be some other results scattered randomly all over the number line") the students think about possible reasons and sources for the different results, as instructed on the back of the worksheet. This leads students to reflect on factors that influence the measurement and sets the beginning of considering MU and errors in (scientific) experiments. The sentence "The bigger the trustworthiness, the smaller the measurement uncertainty of a measurement" should be completed by the students on the worksheet. This connects the trustworthiness (reliability) with the concept of MU.

In the next step, students assess the MU of the measuring cup by thinking of the smallest amount of water that can be measured with the respective instrument. The MU of the instrument is therefore half the step width of the scale. For this task, students can take a prepared hint on a card from the teacher's table, see Fig. 3.

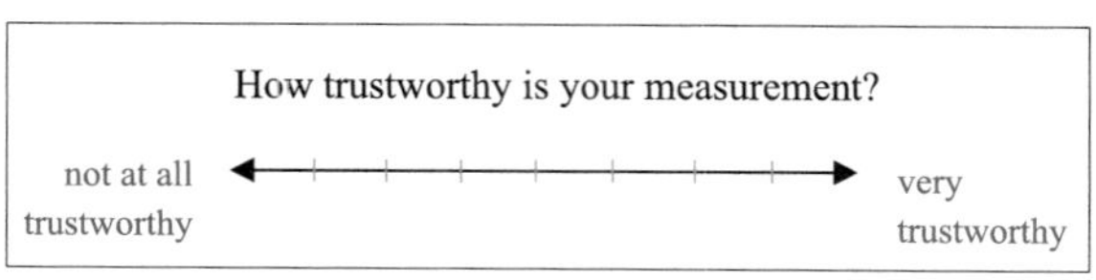

Fig. 2 By using a scale, students should think of the trustworthiness of their own measurement

Hint 1

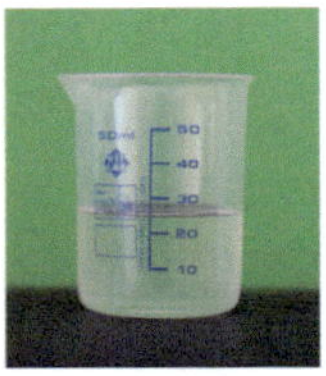

What if the water is between two marks?

Hint 2

What should be done if there is still water remaining in the box at the end?

Hint 1

You can only measure as accurately as your measuring instrument allows. Therefore, you must decide whether the water is closer to the upper or lower line. The largest uncertainty in your measurement is then plus or minus 5 ml.

Hint 2

Can the remaining amount of water in the box be estimated? If so, how would you proceed? What do you notice when examining the results on the number line and including the estimated amount of water in the container in your considerations?

Fig. 3 Students can find hint cards on the teacher's table when confronted with one of the two following situations. Hint 1: The water is between two marks of the measuring cup. Hint 2: Students are wondering how to measure the remaining water in the box

As an extra task, groups that finished assess the remaining amount of water in the box, which is linked to the systematic effect of the conducted experiment. Also in this case, students can find a prepared hint-card on the teacher's table. Furthermore, students shall think about the meaning of the found systematic effect and conclusions for the marked results on the number line.

To conclude the first unit, experiences and findings are exchanged in a subsequent plenary session guided by the teacher. The number line with the results of all groups provides the fundament of the following discussions. At this point, the teacher reveals that every group had the exact same amount of water in their boxes. The students now think of sources and reasons for the scattered results, possible outliers, and, if applicable, systematic effects. The following questions might be of use for the teacher's guiding position:

- Why are there x different results?

- Whose result is correct? Which group is right? Does anyone have "the" right answer?
- Where do you think might be the "actual" result of the experiment? Do you have an idea, how we could determine the "actual" answer?
- Do you have suggestions for improving this experiment?

If in some groups the systematic error was discovered, it could be thematised by using the number line. In this case, the scale of the number line should be corrected accordingly. The following questions might be posed:

- Did someone think about the remaining water in the box?
- Did some group encounter a systematic effect during or after the measuring process?
- How could we correct this systematic effect error?

A picture of the class's results on the number line provides the starting point for the next unit.

2.2.2 Second Teaching Unit

For repetition, the second unit starts with comparing the varying group's results with the aid of a number line (the poster from the last unit should be fixed on the board visible for all students). Also, each student should be provided with a copy of the photo and glue it on the distributed worksheets from the today's lesson. Thereafter, the students calculate the arithmetic mean (depending on the group's mathematical skills in form of self-study or together with the teacher's assistance) and then depict the result on their number lines. This way, one can see the varying values in a series of measurement, which leads students to the key idea: "The individual measurements in a series of measurement are usually not identical". Now, the group's results are part of one series of measurement, again one could argue that different groups might have measured in slightly different ways. Depending on the class, one can find a lot of different clues for fruitful discussions of how to experiment properly and improve the design of the conducted experiment. Next, the class calculates the range, which functions as a first and very simple statistical key figure for the discussion of the trustworthiness of the measurement. Hereby, students achieve the key idea "The smaller the range, the more trustworthy the result". In classes with advanced mathematical skills, the range can be replaced by the standard deviation.

On the worksheet, students find further exercises for practicing and dealing with MU for comparison of different measurement results. For example, results and MU of different groups, which conducted the same experiment under the same conditions shall be compared by reading the respective data out of a table (see Table 2) and ordered according to their trustworthiness. Also, scattered values, their arithmetic means and ranges on number lines can be used for practising the evaluation of the trustworthiness of different results. It is important to train in dealing with measurement uncertainties using various creative methods to avoid merely copying learning

Table 2 Exercise on dealing with measurement uncertainties

By using measurement uncertainties, scientific results can be compared. Imagine other classes conducted the same experiment as yours. Compare your class's result with those of the other classes, taking measurement uncertainties into account.

Class	Result (mean) in ml	Uncertainty of the series of measurement (range) in ml	Uncertainty of the measuring device in ml
Class X	498	100	5
Class Y	505	70	5
Class Z	505	80	5
Your class			

a. Arrange the measurements of the classes according to their measurement uncertainty.

b. Which class had the most trustworthy result and why?

mechanisms. For instance, concept cartoons have been developed and provided for discussing results by considering MU amongst students. However, it is important that students find (rightful) arguments for their presumptions.

The materials for the presented teaching concept can be found on the Website of the Austrian Educational Competence Centre Physics (AECCP) (https://aeccp.univie.ac.at/lehrer-innen/fuer-den-unterricht/). Unfortunately, so far, the materials are only available in German language, as further research on a proper translation for an English-speaking context is needed.

3 Evaluation and Findings

With the method of qualitative content analysis, the transcripts of the probing studies as well as the transcripts of the audio and video recordings during the intervention in class were analysed. Three categories were identified (see Fig. 4 for a better overview). The first category, *strategies*, included students who were able to apply methods such as the calculation of the arithmetic mean, range and/or standard deviation. Also, students had interesting ideas in discussing the trustworthiness of their measurements by comparing their results with other groups' results, some even included the MU in discussion. The subcategory, *trial and error*, included students which tried creative and different strategies to improve their measurements and results. One example is a group using a sponge to transfer spilled water into the measuring cup to reduce the error of their measurement. However, such ideas must lead to further discussion in the plenary session.

The second category, *thinking strategies*, included ideas generated during the group discussions, e.g., identifying sources of MU and suggesting improvements

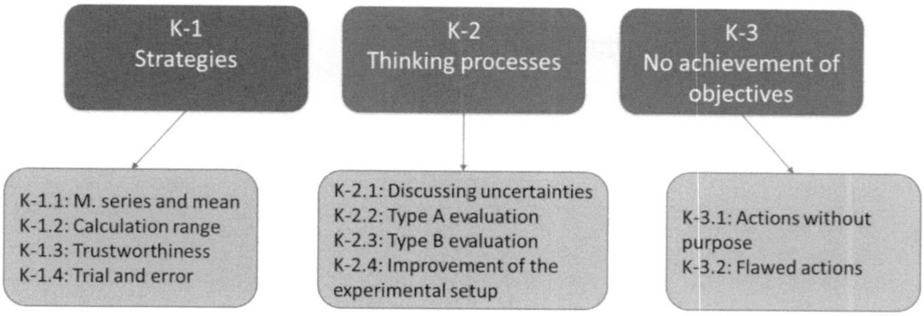

Fig. 4 Identified categories and subcategories

for the experiment. Some groups had intense discussions concerning the amount of water in their measuring cups before writing down their assessed values. In the plenary session, students offered good ideas for improving the experimental setup (e.g., using more appropriate measuring tools, comparing the amount of water to other groups, cooperative work, …) and showed intuitively already well-established competencies in experimental skills.

The last category, *no achievement of objectives*, was assigned in situations and actions that did not lead to a deeper understanding of the concept of MU or led to misconceptions and flawed thinking strategies. Altogether, the teaching concept revealed first promising results to introduce students to the concept of MU. The intended key ideas, shown in Table 3, were found in the discussions of students during group work as well as in the plenary sessions during observation in class, but also in the written test three days after the intervention. However, the third key idea was only half checked, because there was only one student, mentioning, that one should repeat the measurement and then calculate the mean for a reliable result. Nevertheless, in the written test, this key idea was successfully achieved by most of the students.

Interestingly, it seemed that students often relate MU to themselves, rather than seeing MU as an important part of every single measurement in scientific settings in order to compare and assess the trustworthiness of results. The fact that students relate MU to themselves as experimenters might be a sign of low self-esteem when it comes to their experimental skills. So, teaching MU in school might not only improve the self-conception concerning the own competencies in natural sciences but is also part of acquiring scientific literacy and empirical knowledge every student should reach when attending science classes. Heinicke (2012) found similar statements in interviews she conducted for her dissertation. Further, students had their difficulties in differentiation between errors and measurement uncertainties, this should be addressed when introducing to MU. The introduction of MU through the concept of trustworthiness of a measurement seemed to be a successful way of linking the everyday language of the children to the technical term of MU.

Table 3 Reached key ideas in the different cycles of evaluation

	Key ideas	Probing acceptance study	Observation in class	Written test
1	The measured values of a series of measurements are in general not identical.	✓	✓	✓
2	The mean of a series of measurement is getting close to the true value of a measurand.	✓	✓	✓
3	The more measurements you take the closer you get to the true value.		(✓)	✓
4	The smaller the range the better the trustworthiness of the measurement.		✓	✓
5	There are different sources of uncertainties: randomly varying measurements (type A) and others, caused by the measuring instrument (type B).	✓	✓	✓

4 Conclusion and Prospects

The presented teaching concept was developed in three cycles and tested in one class of a Viennese middle school (age 13–14 years) with the method of Design-Based research. An explanation of the method can be found within Bakker's Framework (2019). For evaluation, different methods of empirical research, such as probing acceptance interviews (Wiesner and Wodzinski 1996), participating observation (Lamnek and Krell 2016), an expert interview, video and audio recordings and a written test after the intervention were used.

The simple experiment resulted in a high motivation for students, independent of their performance levels, to participate in the proposed activities. The collaborative work led to discussions about the exact measurement values and to the desired start for students to reflect on the existence of MU. By assessing the trustworthiness of one's own measurements, MU were successfully implemented as a necessary information regarding scientific results. By using a number line, the mean, range and scattering of values the trustworthiness of different results could be easily and vividly compared.

As there is practically no need of any prior knowledge in experimenting, the teaching units can easily be used in lower classes e.g., at the beginning of science/ physics classes. The use of mathematical methods can be supported by the teacher according to the competencies of the students. Further, the units might also be interesting for older students if MU have not been discussed yet. In this case, the standard deviation should replace the range as a better measure of statistical dispersion. For a deeper understanding and introduction of type B MU (age 14–15 years) one should consider the teaching concept from Bärenthaler-Pachner (2022), which can also be found on the website of the AECCP.

A delayed performance test, approximately two years after the intervention revealed positive results concerning the long-term stability of the teaching concepts from Loidl (2021) and Bärenthaler-Pachner (2022). Students reached 9 out of 15 points (60%) (Nagel 2025b). However, the testing scale included both grade levels. This might result in a lower total score of the younger students.

Meanwhile the concept was tested and evaluated in 23 more classes and more than 450 students, held by different teachers in Vienna. As expected, the learning effects are lower than in the development stage, but the outcomes are still satisfactory. One reason for the smaller learning effects might be due to the precise conduction of the teaching concept through the responsible teachers or due to the individual age-related conceptual understanding of trustworthiness.

For a successful implementation of the teaching concept a good intuition of the teacher seems to be essential for a fruitful discussion of the students' ideas. Also, we recommend the use of simple mathematical tools such as the arithmetic mean and range of a series of measurement to concentrate on the topic of MU and reduce distracting difficulties concerning mathematical competencies of the students. Once students got introduced to MU a repeated application of acquired strategies, critical examination and interpretation of results should be part of future experiments in science classes.

References

Altrichter H, Posch P, Spann H (2018) Lehrerinnen und Lehrer erforschen ihren Unterricht, 5th edn. Verlag Julius Klinkhardt, Bad Heilbrunn

Bakker A (2019) Design research in education: a practical guide for early career researchers. https://doi.org/10.4324/9780203701010

Bärenthaler-Pachner R (2022) Entwicklung und Evaluation einer Lernumgebung zum Thema Messunsicherheit in der Sekundarstufe II. Master Thesis. University of Vienna, Vienna

BIPM (2008) JCGM 100:2008 Evaluation of measurement data - Guide to the expression of uncertainty in measurement (GUM). International Organization for Standardization 1995, Geneva. ISBN 92–67–10188–9 p ix

Heinicke S (2012) Aus Fehlern wird man klug. Eine Genetisch-Didaktische Rekonstruktion des Messfehlers. Logos, Berlin

Holz C, Heinicke S (2019) Messunsicherheit – ein ungeliebter Gast im Physikunterricht? In: Maurer C (ed) Naturwissenschaftliche Bildung als Grundlage für berufliche und gesellschaftliche Teilhabe, vol 39. Universität Regensburg, Kiel, pp 89–92

Kok K, Boczianowski F, Priemer B (2020) Messdaten im Physikunterricht auswerten—wann sind Messunsicherheiten wichtig? MNU-J 4:292–5. https://doi.org/10.18452/27175

Kok K (2022) Certain about uncertainty—what students need to know about measurement uncertainties to compare data sets. Dissertation. Humboldt-Universität zu Berlin, Berlin

Krüger D, Parchmann I and Schecker H (2014) Methoden in der naturwissenschaftsdidaktischen Forschung. Springer, Berlin Heidelberg). https://doi.org/10.1007/978-3-642-37827-0

Lamnek S, Krell C (2016) Qualitative Sozialforschung: mit Online-Material 6th ed. Beltz, Weinheim Basel
Loidl H (2021) Entwicklung und Evaluation von Unterrichtseinheiten zum Thema Messunsicherheiten. Master Thesis. University of Vienna, Vienna
Munier V, Merle H, Brehelin D (2013) Teaching scientific measurement and uncertainty in elementary school. Int J of Sci Educ 35:2752–2783. https://doi.org/10.1080/09500693.2011.640360
Nagel C (2021) Sicher ist sicher! Fachliche Klärung für didaktische Rekonstruktion von Messunsicherheiten im Unterricht. Plus Lucis 4/2021:7–11
Nagel C (2025a) Sicher ist sicher! Fachliche Klärung für die didaktische Rekonstruktion von Messunsicherheiten im Unterricht. In: Priemer B, Kok K (eds) Messunsicherheiten im Physikunterricht. Berlin Universities Publishing, Berlin, pp 53–65. https://doi.org/10.14279/depositonce-21608
Nagel C (2025b) Trustworthiness as central design principle for introducing uncertainties of measurements to students. J Phys: Conf Ser 2950:012016. https://doi.org/10.1088/1742-6596/2950/1/012016
Nagel C, Lux B, Steindl S (2021) Die Thematisierung von Messunsicherheiten im Physikunterricht—Eine Umfrage. Plus Lucis 4/2021:4–6
Nagel C, Neumann S (2021) Messunsicherheiten—Sicher ist sicher! plus Lucis 4/2021
Priemer B (2022) Unsicherheiten, aber sicher! Vom kompetenten Umgang mit ungenauen Daten. Springer, Berlin, Heidelberg
Wiesner H, Wodzinski R (1996) Akzeptanzbefragungen als Methode zur Untersuchung von Lernschwierigkeiten und Lernverläufen Lernen in den Naturwissenschaften ed R Duit (Kiel: IPN), pp 250–74
Wiesner H, Schecker H, Hopf M (2018) Physikdidkatik kompakt 4th ed. Aulis, Freising

Conceptual Outcomes Implementing Different Approaches to Optical Spectroscopy in Secondary School

Daniele Buongiorno, Danilo Catena, Marisa Michelini, and Lorenzo Santi

Abstract An educational proposal on light spectroscopy was designed according to the theoretical framework of the Model of Educational Reconstruction (MER) by means of Design-Based Research (DBR) methods. The proposal was implemented several times according to three different approaches, focused on the Light Sources exploration (LS), the Light Emission process (LE) and the Spectra Exploration (SE), respectively. This study illustrates the results coming from a pretest–posttest aimed at measuring the students' conceptual understanding after several implementations of the proposal, with a total number of 257 upper secondary school students (17–19 years old) in northern Italy. Some items were specific for each approach, therefore here we focus on the results related to the common aspects. The data were analyzed with qualitative methods, categorizing the student's answer for each item, according to an iterative process. The results show that, although each approach reports some specific benefits, the LS produce a greater improvement of students' conceptual understanding. For this approach, the results about students' ideas of the colors' nature and the spectra formation and interpretation are reported. Implications for teaching and further research steps are also discussed.

Keywords Conceptual learning · Optical spectroscopy · Design-based research

D. Buongiorno · D. Catena (✉) · M. Michelini · L. Santi
Department of Mathematics, Computer Science and Physics, University of Udine, Udine, Italy
e-mail: catena.danilo@spes.uniud.it

M. Michelini
e-mail: marisa.michelini@uniud.it

L. Santi
e-mail: lorenzo.santi@uniud.it

Z. Ješková and J. Hanč (eds.), *Advancing Physics Education*, Challenges in Physics Education, https://doi.org/10.1007/978-3-031-94207-5_8

1 Introduction

Spectroscopic phenomena play a crucial role in several fields, especially in astrophysics and in providing an experimental basis for the structure of matter (Bailey et al. 2012). Therefore, in several European curricula optical spectroscopy is recognized among the topics of modern physics to be introduced in secondary school (Michelini 2001). Nevertheless, the lack of its teaching practice and the many learning obstacles which characterize this set of phenomena require the elaboration and implementation of innovative educational proposals, developed through based-research methods.

The Physics Education Research Group at the University of Udine (Italy) has developed a teaching intervention that we are going to illustrate in the next section: despite a common conceptual referent, it shows several alternative approaches. Indeed, starting from the phenomenology, it is possible to base the educational path on the sources of light, or on the emission process. Furthermore, moving the focus directly on the spectra exploration, we obtain the third approach. In this way, it is possible to investigate if there is significant difference in the students' conceptual understanding after implementing the proposal following different approaches.

2 Theoretical Framework and Literature Review

The educational proposal was developed according to the Model of Educational Reconstruction (Duit et al. 2012), by means of Design-Based Research methods (Duit et al. 2005). This theoretical framework requires a reconstruction of the contents from an educational point of view, an analysis of the main conceptual knots and the main interpretative problems that have emerged from the history of physics and the design of conceptual micro-steps, in which active learning strategies make it possible to overcome the main learning difficulties.

The conceptual knots that we identified from the literature review can be classified in three macro-categories:

1. *Spectral lines and atomic energy levels*: there is a very common belief regarding the association of each line in a discrete spectrum with a single energy level of the source (Korhasan and Wang 2016). Students hold the idea that the amount of radiation emitted is related only to the final or initial level, not considering the ground level as an energy level (Ivanjek et al. 2015a).
2. *Functional role of the spectroscope components* and characteristics of the observed spectrum: students often fail to attribute a functional role to each component: according to their reasonings a prism always produces a continuous spectrum, regardless of the nature of the emitting source, or the idea that a discrete spectrum is always visible—even without a dispersive element, such as a prism or a diffraction grating—are present among the students (Ivanjek et al. 2015b).
3. *Distinction between a diffraction pattern and a discrete spectrum*: students often think that a discrete spectrum can only be observed in the case of monochromatic

sources or using a diffraction grating (regardless of the nature of the source). The same problem arises when students are unable to distinguish a diffraction pattern from a discrete spectrum, confusing diffraction maxima with spectral lines (Ivanjek 2012). These conceptual knots seem to be related to the model of light that students adopt to explain the light-matter interaction.

Thus, each category is related to specific learning difficulties, that can be considered important obstacles both for upper secondary school and undergraduate students. We also built a match between these categories and the different approach we proposed, as specified in the next section.

3 The Educational Proposal

The educational path that we proposed analyses the connection between the macroscopic world, namely, it aims to correlate the spectral emission with the energetic structure of matter. As we briefly pointed out in the introduction, three different approaches are proposed, based on the Light Sources exploration (LS), the Light Emission process (LE) and the Spectra Exploration (SE). All these approaches exploits energy as a common conceptual referent to describe the color of light. To get insight of the proposal, we report the conceptual steps that we identified for the LS approach:

1. *The light sources*: a review of different natural and artificial light sources is introduced, helping students to recognize a light source as a system able to transform energy. Thus, the problem of interpreting the emission processes is also posed. This represents a valid context to reinforce the idea that the generation of light is a process that has to be interpreted in energetic terms. The different light sources as incandescent, fluorescent lamps, LEDs and gas-discharge lamps are examined from a structural point of view, according to the emission processes and the emitted light, also discussing their technological characteristics.
2. *The light emitted by different sources*: the quantities characterizing the light emitted by the sources are discussed: intensity and color. Concerning the color, the exploration of how different colored lights can be obtained leads to revisiting the additive and subtractive mechanisms of production of different colors, exploring if there are intrinsically colored light sources. The presence of an energy exchange within a light source emerges from the analysis of the emission of an incandescent bulb: as the electrical power supplied to the system increases, light is obtained with different intensities and colors. The phenomenological laws of Stefan-Boltzmann, the identification of the emissive and absorbing power and the fact that their relationship is only a function of temperature, directs the examination of the emission process towards the increase of the power supplied to a source. The discovery of infrared radiation is introduced in this context, in which the idea that the emission of radiation occurs at any temperature emerges. The historical discovery of ultraviolet radiation confirms this idea experimentally

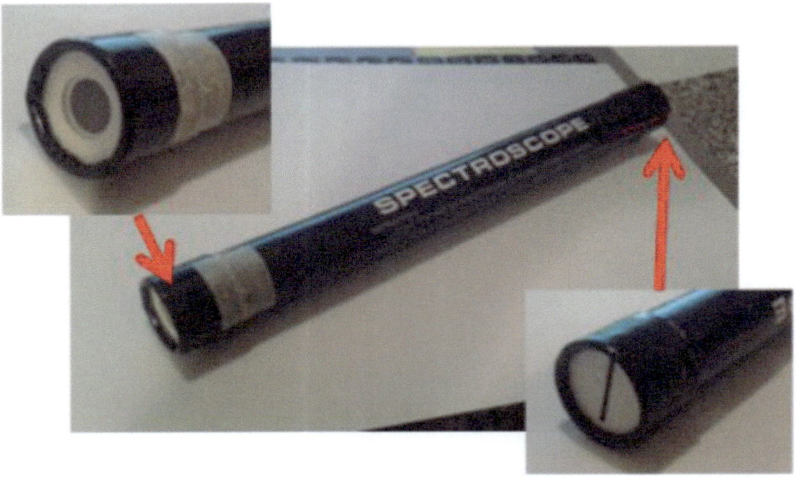

Fig. 1 A spectrometer with its grating and slit. We asked some students to build a similar homemade version providing them some instructions, in order to let them aware about the roles of different parts of the spectroscope

and allows to see the colors of the visible spectrum as different radiations. The problem of describing a white light is addressed recognizing Newton's double prism experiment, showing that white light is composed of different colors.

3. *The structure of a spectroscope*: the use of a grating spectroscope–as the one illustrated in Fig. 1—allows to characterize a light source depending on the emitted spectrum. After classifying spectra in three categories–continuous, band and discrete–depending on the characteristics of the sources, the structure of the spectroscope is analyzed. This analysis is carried out considering the functional role of its components (slit, reticulum, tube).
4. *The energetic interpretation of colors*: the analysis of the nature of light is based on the photoelectric effect, where the photons' energy corresponds to the color of light, and their number corresponds to the light intensity.
5. *The analysis of Balmer and Rydberg formulas*: The measurement of color was made using a photon model of light, so a spectrum can be considered a graph with arbitrary units of measurement, with consistent placement of lines on the energy axis. On this basis, students receive the spectrum of the hydrogen atom and are invited to find a law that describes it, retracing the Balmer's experience. We were inspired by his original article to build the activity, in which the prerequisites are only the 4 arithmetic operations, but whose experience is of great value in building a physical identity. Thus, students are able to identify the law of the phenomenon following the proposal by Rydberg. They can recognize that the description of a spectrum is linked to the difference of two squared quantities. These are interpreted as the difference between energy states, aligning with the Einstein's photon hypothesis.

6. *The interpretation of emission spectra*: The emission process is illustrated through the link between energy levels and spectral lines. The transition from energy levels to spectral lines is a different methodological and conceptual challenge than the reconstruction of the energy structure of an emitting system starting from its spectrum: both aspects were considered in the process, using practical examples.

The other approaches differs from the LS one about some aspects. In particular, in the LE approach, the emission of light in macroscopic terms is studied starting from the Stephan-Boltzmann law according to the interpretation of color as a physical quantity, alongside the analysis of light-matter interactions. In the SE approach the students use a spectroscope, exploring its functional characteristics to identify spectra coming from different sources.

Following this educational proposal, the main research questions that guided our study are:

1. How does each approach help overcome the main learning difficulties emerging from literature?
2. How does the approach integrating analysis of technological light sources and technological aspects of the spectroscope (LS approach) improve the learning outcomes?

4 Methods

4.1 Participants

For this study, we involved a convenience sample consisting of 257 students. They came from the last two years (17–19 years old) of upper secondary schools in the north of Italy, whose curriculum is focused on mathematics and natural sciences. This study was carried out within an orientation program, in collaboration with the University of Udine. The schoolteachers did not provide any specific information about the addressed topics before the educational path was implemented, as the school curriculum does not focus on optical spectroscopy. At the moment of the implementation, the students were not introduced to modern physics at all.

The students were subdivided in 4 groups, depending on the implemented approach: they experienced the teaching intervention in different periods, but the amount of the teaching interventions was the same for all groups (6 h). The number of participants for each group was: 52 students for the LE approach, 63 students for the SE approach, and 142 for the LS approach. The last cluster consisted of two different groups (82 + 60 participants) as they experienced the educational paths in different moments. In the next section we specify the reason for this subdivision.

4.2 Instruments

A conceptual survey was designed to investigate all the addressed conceptual knots in the educational path. The survey consisted of different versions depending on the group it was submitted to, but the central core was the same. It consisted of 18 questions: some of them were full open-ended, some others were asking for a justification after a multiple choice. Furthermore, some items were posed in the form of a dialogue between two or three students, who claim different opinions. These are related to common-sense ideas reported in literature, as pointed out above, so in some cases they are all wrong. Students are asked to agree/disagree with these opinions or to argue their own thesis. In this way we include the common-sense ideas not only in the path's constructions, but also in the final instrument measuring the efficacy of the teaching interventions. Several questions for the same topic are proposed: this allow us to investigate the level of coherence in students' reasonings, which cannot be analyzed by a single-item instrument. The items they had were already validated in a previous research study (Buongiorno and Michelini 2019).

4.3 Data Analysis

To answer the research questions, the collected data were analyzed using qualitative methods. We used an iterative process of categorization (for full open-ended questions), looking for interactions for categories of different items. The categorization was based on the structure of the answer itself, but also on the understanding of the conceptual knots we identified a-priori. Thus, the qualitative methods are justified by the fact that we were searching for insights about the mental models exploited by the students in order to account for the different aspects characterizing the spectroscopic phenomena. Data were collected online some weeks after the path implementation, and the students filled in it under the supervision of their schoolteachers.

5 Results

5.1 Differences Between the Approaches

Here we report the main results in order to answer the first research question. In Fig. 2 three questions with answers' categorization are reported. As each of them refers to a different approach, it provides an example of how they improve different conceptual aspects. The categories' names directly reflect their conceptual meaning.

As far as the light emission process is concerned, it emerges from the post-tests that the students who have followed an approach starting from light sources (LS)

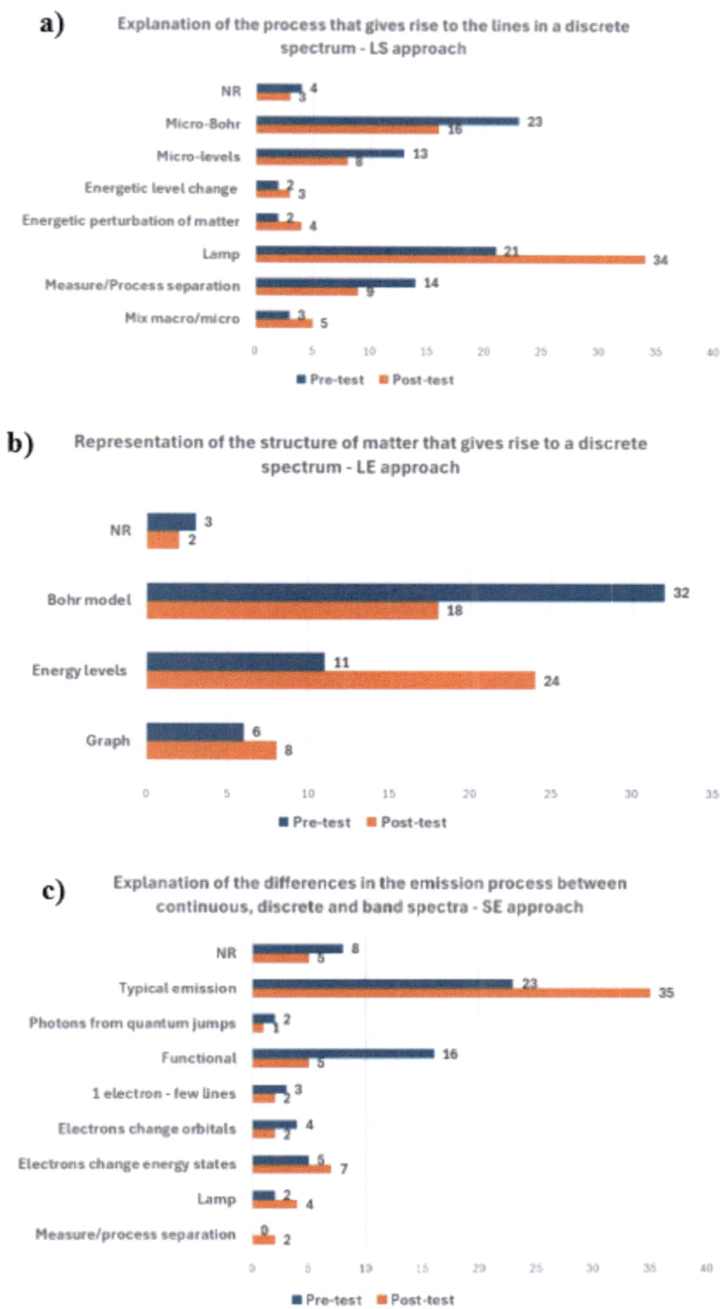

Fig. 2 (**a**) Categories about the explanation of the process that gives rise to the spectral lines in the LS approach (N = 82). (**b**) Categories about the representation of the structure of matter giving rise to a discrete spectrum in the LE approach (N = 52). (**c**) Categories about the differences in the emission process between different kind of spectra in the SE approach (N = 63). The differences between the pre-test and the post-test are reported. In all cases NR identifies blank answers

mainly describe it from a macroscopic point of view: we named this as “Lamp” category, with respect to the technological functional process implemented (~25% in the pre-test, ~40% in the post-test). About the LE-approach, we observed a shift in terms of description of the structure of matter from the Bohr model (~20% in the pre-test) to a model based on energy levels (namely, in which electrons or matter change levels) or a model in which the light is emitted as a result of an energy transformation, avoiding the descriptive spatial representation of the emitting system (~45% in the post-test). The idea according to which the energies emitted are those which characterize the source is not observed anymore. The interpretation in terms of photon emission becomes predominant after education (~55% in the post-test compared to ~35% in the pre-test) in the SE group. Exploration through the spectroscope in this phenomenological approach fostered students in describing a source in functional terms when its emission process is taken into account, promoting the description of the observed spectra in terms of energy. The number of students describing spectra when asked to account for the emission process decreases after instruction (~10%), showing a slight shift towards a microscopic interpretation.

As regards the conceptual model linking discrete emissions to energy levels, the analysis of other items reveals that all approaches and perspectives are indifferent to obtain conceptual understanding of the addressed situations: ~60% of participants were able to properly identify this link in the post-test for all approaches. The negative value of the energy levels appears to be justified by the students in any case using the semi-classical Bohr model. Nevertheless, considering other aspects the LS approach seems to be more effective in improving students’ conceptual understanding, although each approach can report different specific benefits.

For instance, a changing perspective about a description of the atom in terms of energy levels and of the spectra as a transition between them—rather than in terms of spatial representation by orbits—emerges for the LS approach (~75% compared to 45–50% for the other groups). It is interesting to note that most of the participants in the LS group (~60%) changed the conceptual referents (energy instead of wavelength), while for the other groups the percentage is significantly lower (35–40%). Interestingly, the participants who experienced the LS approach express themselves mainly in terms of “energy perturbation” or “transformation of matter”, thus preferring the perspective of light-matter interaction. Thus, the photon as an indefinite entity that carries energy is the main conceptual referent.

The previous results induced us to implement the educational path according to the LS approach in a second experimentation with N = 60 (new) participants, in order to assess the stability of the conceptual understanding produced by the teaching intervention. In this case we only used a post-test, as the results of the pretest we already submitted were homogeneous for the various groups, thus providing a stable point of reference for students’ previous knowledge. The results for this new implementation, providing a measure for the effectiveness of the educational path, are reported in the next paragraph.

5.2 *The Effectiveness of the LS Approach*

Answering the first research question, the LS approach resulted to be the most effective for our educational path in fostering students' conceptual understanding about spectroscopic phenomena. For its new implementation, in order to answer the second research question, we involved a new group of 60 upper secondary school students. In this case we added some other questions to the post-test, and the corresponding students' understanding was evaluated taking into account the coherence between the answers provided within the same class of items. Indeed, depending on the topic addressed, we classified the post-test's items in three main groups. They are related to, respectively: the nature of colors, the spectra formation, and the spectra interpretation. Here we report the main results coming from the answers' categorization.

In Fig. 3 the results related to the nature of colors are reported. The large majority of students (53 out of 60) recognize that a specific color can be obtained both from addition synthesis and diffraction phenomena. This is coherent with the fact that they conceive color (our questions are related to violet) as a specific and measurable physical quantity (51 out of 60 students), but just 15 of 60 participants were able to apply this information in the specific context proposed by the question. Indeed, half of the students claimed to find not only violet, but also red and blue lines looking with a spectroscope at the violet color coming from a prism hit by white light.

The two main questions related to spectra formation are reported in Fig. 4. They both refer to a continuous spectrum. Similar percentages of students stated that

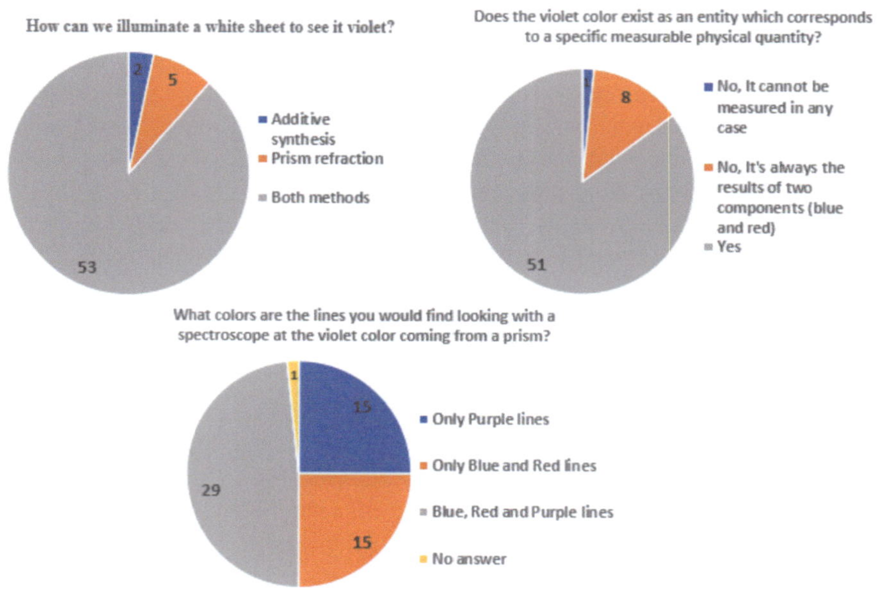

Fig. 3 Items about the nature of colors (N = 60). The categories were predetermined (multiple choice), but students' explanations were consistent with their meaning

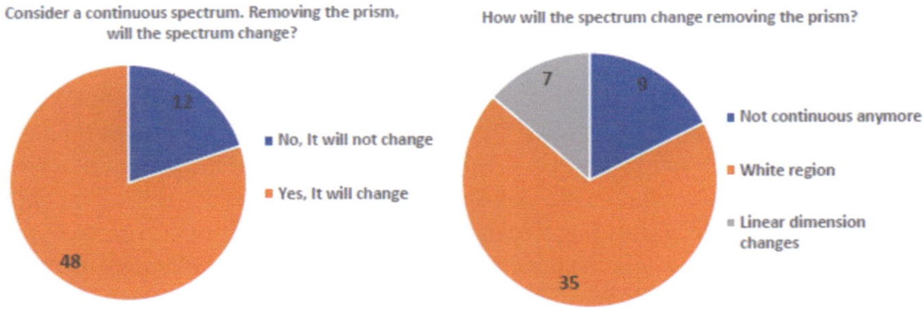

Fig. 4 Item about the spectra formation (N = 60). The question was full open-ended and the categories are not mutually exclusive. In the second graph only the answers of the participants stating that the spectrum is changing are reported

removing the prism (48 out of 60) or replacing it with a diffraction grating (46 out of 60), the continuous spectrum would change. These questions are reformulated from another instrument (Ivanjek et al. 2020), in order to capture possible association between the categories of the two items. Interestingly, we found that the majority of students (35 out 60 students) recognized that removing the prism we would obtain a continuous white light region, consistently with the fact that just 6 students gave the same answer when the prism is replaced a diffraction grating. Their explanations suggest that they attribute the same role to these objects in determining the shape of the spectrum, mostly without specifying it. Furthermore, the two items share other two categories, given by previsions that the spectrum would not be continuous anymore, and that the linear dimensions would change. As we found other two categories in the second question–(1) replacing the prism with the diffraction grating we would see the spectrum as we could see it at the mirror; (2) replacing the prism with the diffraction grating we would see the diffraction band that are rainbow-colored, while the one in the center would be larger and white–prior results about confusing spectra observation with diffraction pattern are reinforced.

There are six main questions related to spectra interpretation. In this group of items, distractors are selected from the main common-sense ideas found in literature (Buongiorno and Michelini 2019; Ivanjek et al. 2015a, 2020). In Fig. 5, we reported the graphs related to some items' categorization. Less than a quarter of participants (12–13 students) recognize that there is not an intermediate state between two energy levels, and that the ionization process takes place if the energy of the system assumes a positive value. The other results show that students find easier starting from energy levels (right answer for 51 out of 60 participants) than starting from differences between levels (right answer for 39 out of 60 participants) in linking the two aspects. Most of the students were able to individuate the correct number of energy levels necessary to get a specific emission spectrum. Among who report a complete and correct justification, they adopt two different strategies: the former consists in trying attempts with different number of levels, the latter consists in applying combinatorial calculations.

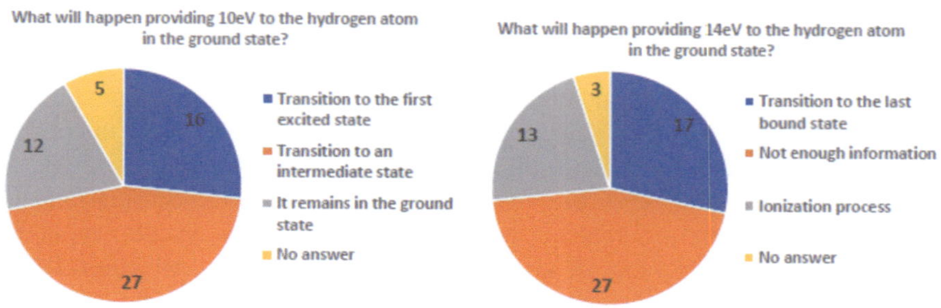

Fig. 5 Items about the spectra interpretation. Both the questions were posed in the form of a dialogue between two students, where they both were wrong

6 Conclusion

This study reports the results of different implementation of an educational path about optical spectroscopy for upper secondary school students. Their resulting conceptual understanding was mainly measured through open-ended questions as pretest and posttest, and the data were analyzed in a qualitative way through answers' categorization. Three different approaches were applied to the same core of the educational proposal, with different groups of students experiencing them. The results show that, despite each approach induce its specific benefits, overall the one focused on light sources (LS approach) produced a greater conceptual gain with respect to the ones focused on the light emission process and the spectra exploration.

Thus, a final implementation of the LS approach was carried out in order to measure students' conceptual understanding. Our results show that, even the large majority of the students recognize that it is possible to obtain a color both from and color synthesis, only 25% of the students understand that its nature can only be represented by one of these two alternatives. These results consider the nature of colors from a different point of view with respect to the one prevalent in literature (Ward et al. 2007). Furthermore, the majority recognize the role of the prism in providing the spectra, but they (expect for 10% of the participants) did not consider properly the role of the diffraction grating in the experimental set-up. This confirms prior results about the confusion between discrete spectra and diffraction pattern (Ivanjek et al. 2020). Finally, about 80% of the students were able to link energy levels and spectra emissions, despite the difficulties in dealing some specific aspects (as the ionization process).

These results demonstrate that stimulating reasoning about the intrinsic meaning of data and its macroscopic limits allows students to gain interpretive mastery: they acquire physical identity in terms of the role of a scientific model in the development of scientific ideas for the explanation of natural phenomena. The educational approach that builds interpretation starting from phenomenology can also be extended to other scientific contents, as support for planning teaching activities for science teachers: it aims at encouraging the transition from traditional to innovative teaching, able to face the challenges of conceptual and operational appropriation that learning poses.

Despite this, some limitations have to be discussed. First of all, some non-scientific conceptions persist in student's reasoning about the nature of colors and the role of dispersive elements in the spectra's formation. Furthermore, we have no measures about the previous content knowledge of students, which can affect the results through the interaction with the different approaches. Finally, we have no insights about the mental models of students about light, which could be considered as the origin of their alternative conceptions. On the other hand, we consider this research study as a first step to give a contribution to the research about students' conceptual understanding of optical spectroscopy and modern physics.

This educational proposal can be considered an example about the activities related to modern physics that could be implemented at school. The key aspects for the curriculum lie in the link between the macroscopic and the microscopic world, and in attributing a new meaning to a fundamental entity of physics: the light. As its study is already implemented in the contexts of optics and electromagnetic waves, an interesting aspect to be considered as fruitful for research is how the different models of light (i.e., ray, wave, photon) coexist or exclude each other in the context of students' mental models, depending on the kind of phenomenon they are studying: this will be the next step of our research.

References

Bailey JM, Johnson B, Prather EE, Slater TF (2012) Development and validation of the star properties concept inventory. Int J Sci Educ 34(14):2257–2286. https://doi.org/10.1080/09500693.2011.589869

Buongiorno D, Michelini M (2019) Research-based proposals on optical spectroscopy and secondary students' learning outcomes. J Phys Conf Ser 1287:012004. https://doi.org/10.1088/1742-6596/1287/1/012004

Duit R, Gropengießer H, Kattmann U (2005) Towards science education research that is relevant for improving practice: the model of educational reconstruction. In: Fischer HE (ed) Developing standards in research on science education, Taylor & Francis, London, pp 1–9

Duit R, Gropengießer H, Kattmann U, Komorek M, Parchmann I (2012) The model of educational reconstruction—a framework for improving teaching and learning science. In: Jorde D, Dillon J (eds) Science education research and practice in Europe, vol 5. Sense Publishers, Rotterdam:), p 13–37. https://doi.org/10.1007/978-94-6091-900-8_2

Ivanjek L, Shaffer PS, Planinic M, McDermott L (2020) Probing students understanding of spectra through the use of a typical experiment used in teaching introductory modern physics. Phys Rev Phys Educ Res 16:010102. https://doi.org/10.1103/PhysRevPhysEducRes.16.010102

Ivanjek L, Shaffer PS, McDermott LC, Planinic M, Veza D (2015a) Research as a guide for curriculum development: an example from introductory spectroscopy. I. Identifying student difficulties with atomic emission spectra. Am J Phys 83(1):85–90. https://doi.org/10.1119/1.4901977

Ivanjek L, Shaffer PS, McDermott LC, Planinic M, Veza D (2015b) Research as a guide for curriculum development: an example from introductory spectroscopy. II. Addressing student difficulties with atomic emission spectra. Am J Phys 83(2):171–178. https://doi.org/10.1119/1.4902222

Ivanjek L (2012) An investigation of conceptual understanding of atomic spectra among univeristy students. Master's thesis, University of Zagreb

Korhasan ND, Wang L (2016) Students' mental models of atomic spectra. Chem Educ Res Pract 17:743–755. https://doi.org/10.1039/C6RP00051G

Michelini M (2001) Supporting scientific knowledge by structures and curricula which integrates research into teaching. In: Pinto R, Surinachi S (eds) GIREP book—selected contributions of the physics teacher education beyond 2000 international conference. Elsevier, Paris, p 77

Ward RB, Sadler PM, Shapiro II (2007) Learning physical science through astronomy activities: a comparison between constructivist and traditional approaches in grades 3–6. Astron Educ Rev 6(2):1–19. https://doi.org/10.3847/AER2007015

Effectiveness of MiniPIX Detector Activities Dispelling High School Students' Misconceptions About Radioactivity and Ionising Radiation

Daniel Parcerisas, Rafael Ballabriga, Francisco Barradas-Solas, Cristina Cabo, Michael Campbell, Massimiliano Fiorini, David Gonzalez Saba, Erik H. M. Heijne, Ana Rita Pinho, and Maria Cristina Trevissoi

Abstract Research indicates that students commonly retain misconceptions about radioactivity and ionising radiation, impacting their learning and civic responsibility. MiniPIX cameras, utilising CERN's Timepix hybrid pixel detectors, are useful tools

To John B. Brennan. A teacher who inspired. A friend who remained. A memory that endures.

D. Parcerisas (✉) · D. G. Saba
Sagrada Familia School, Gava, Spain
e-mail: d.parcerisas@safagava.edu

D. G. Saba
e-mail: dgonza44@xtec.cat

R. Ballabriga · M. Campbell · E. H. M. Heijne · A. R. Pinho
CERN, Genève, Switzerland
e-mail: rafael.ballabriga@cern.ch

M. Campbell
e-mail: michael.campbell@cern.ch

E. H. M. Heijne
e-mail: Erik.Heijne@cern.ch

A. R. Pinho
e-mail: ana.rita.pinho@cern.ch

F. Barradas-Solas
Educational Innovation and Teacher Training Service, Madrid Education Department, Madrid, Spain
e-mail: fbarradas@educa.madrid.org

C. Cabo
IGFAE, University of Santiago de Compostela, Santiago de Compostela, Spain
e-mail: cristina.cabo@usc.es

M. Fiorini · M. C. Trevissoi
INFN and University of Ferrara, Ferrara, Italy

Z. Ješková and J. Hanč (eds.), *Advancing Physics Education*, Challenges in Physics Education, https://doi.org/10.1007/978-3-031-94207-5_9

to introduce particle physics and radioactivity in secondary education. An analysis evaluating MiniPIX's effectiveness in dispelling these misconceptionss was conducted. Surveying 65 Spanish 14-year-old students, their ideas about radioactivity and radiation were collected before and after regular lessons. An additional 1-h experimental session using the radiation camera followed the regular lesson. Results showed a 1.8 times greater improvement in correct answers attributable to the MiniPIX activity compared to the regular lesson, especially on questions covered experimentally during the detector session.

Keywords MediPIX detector · Radioactivity education · Radioactivity and radiation misconceptions

1 Introduction

Teaching radioactivity and radiation has long posed challenges due to documented misconceptions and spontaneous beliefs among students. Various causes for these difficulties have been suggested, including lack of experimental tools and the conceptual abstraction involved.

Recently, state-of-the-art Timepix detectors, read out by the MiniPIX EDU system, have been introduced into European high school classrooms. These tools aim to enhance visual and experimental learning about radioactivity and particle physics. The authors have initiated a systematic evaluation of the MiniPIX detector's effectiveness in refining these ideas. These misconceptions, intuitive beliefs, confusions and other alternative ideas deviate from scientifically accepted knowledge, as per Griffiths and Preston's definition (Griffits and Preston 1992).

1.1 Medipix Detectors

The Timepix chip, a hybrid pixel detector with a 256×256 pixel matrix and 55 μm pixel pitch, was designed at CERN in the framework of the Medipix2 collaboration. Its technical details have been extensively documented by Llopart (2007). Functionally, the detector comprises of a matrix of radiation-sensitive units connected to individual pulse-processing electronics (Campbell 2001). When ionising radiation interacts with the sensor material, it absorbs all or part of its energy, generating charge

e-mail: massimiliano.fiorini@unife.it

M. C. Trevissoi
e-mail: mariacristina.trevissoi@unife.it

E. H. M. Heijne
Institute of Experimental and Applied Physics, Czech Technical University, Prague, Czechia

carriers (Fig. 1) that induce an electric current processed in the readout electronics. The system, which is noise-hit-free, ensures suitability for measuring low radiation levels, such as those from educational sources (Fig. 2). Timepix measures deposited energy by counting the time-over threshold of the signal, allowing new educational activities focused on particle energy measurement.

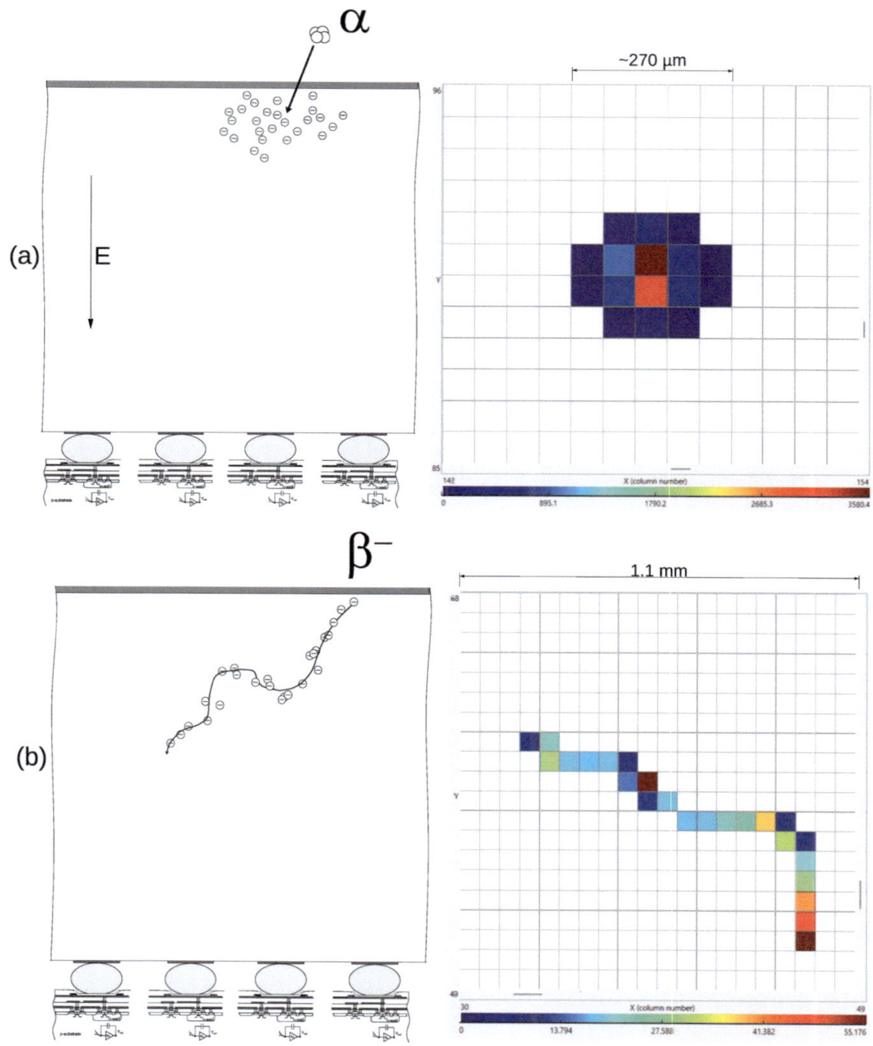

Fig. 1 Representation of the hybrid pixel detector ionisation (left) for alpha (**a**) and beta (**b**) particles and their tracks visualised with the PixetPro software (right). The sensor material (not to scale) is on the top and it is connected to the readout electronics at the bottom

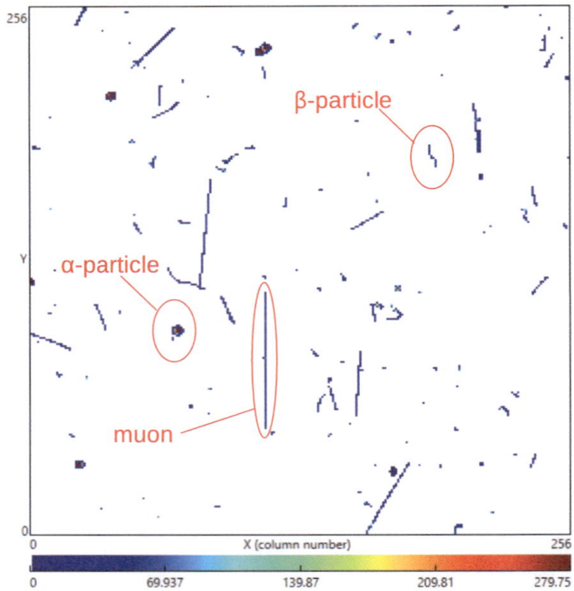

Fig. 2 Radiation at Puig Falco (altitude 2700 m) as seen with Timepix2 (20 min frame). The recording shows a few different particles, with their typical clusters of deposited energy, in keV, per pixel, on a colour scale

Timepix devices' high pixel segmentation aid in identifying various particle species within radiation environments. Different particles exhibit distinct interactions with detector materials, which is crucial for educational purposes. For instance, low-energy X- or γ photons concentrate energy in a single pixel hit or small pixel clusters (Ballabriga et al. 2016). Alpha particles create large "round" shapes by depositing intense but shallow energy (see Fig. 1a). Beta particles display trajectory deviations (Fig. 1b) in the material they interact with. In contrast, high-energy muons follow straight paths (Fig. 2). This enables students to differentiate ionising radiation types. This capability also helps comprehend radioactivity, background radiation, and distinctions between ionising and non-ionizing radiation, like those emitted by smartphones, which are invisible. The visibility of radioactivity, ease of use, and ability to measure low-activity sources suggest these devices could correct students' ideas about radioactivity.

1.2 MiniPIX Detectors in Education

The attributes detailed earlier have transformed this scientific device into an innovative pedagogical tool widely used in European educational projects. The use of the Medipix detector in secondary education began with a school visit to CERN in 2007. This led to integrating MX-10 sensors into schools through programs like LUCID and CERN@school, established by the UK's Institute for Research in Schools (Parker et al. 2019).

The Institute of Experimental and Applied Physics (IEAP) at Czech Technical University developed educational kits (Holik and Vícha 2020), supporting teacher training, workshops, and student projects. These kits are shared among IEAP, high schools, science centers, museums, and planetariums. Most Czech users trained with IEAP CTU, utilising educational videos and a tailored manual titled "Experiments Using Pixel Detector in Teaching Nuclear and Particle Physics" (Vícha 2017), covering experiments from basic to university-level complexities.

In 2017, the ADMIRA Project began in Barcelona through a collaboration between CERN's Microelectronics Section, the University of Barcelona, and the Sagrada Familia School (Parcerisas et al. 2022). This initiative, involving nearly thirty Catalonian schools, fosters teacher collaboration and integrates experimental particle physics into classrooms, developing activities and teacher training programs (ADMIRA 2021).

ADVACAM, based in Prague and holding the CERN license to commercialize Medipix and Timepix-based detectors, recently created cost-effective MiniPIX EDU cameras for schools and universities, alongside educational kits and resources (ADVACAM).

CERN's Microelectronics and Knowledge Transfer Sections have promoted technology adoption in education by participating in the GIREP 2022 conference and lending devices to new initiatives via the Medipix Collaboration. This support has fostered groups across Europe, such as the University of Ferrara in Italy, Proxecto MEDRA (Proxecto MEDRA-IGFAE) and MadPIX (MadPIX) in Spain, focusing on school detector usage.

These initiatives demonstrate that the MiniPIX detector is an effective tool for teaching radioactivity and particle physics. Its versatility facilitates work with low-activity radiation sources and encourages interdisciplinary collaboration. This widespread use raises the question of its actual impact, which the current research aims to explore.

2 Students' Misconceptions About Radioactivity and Radiation and Possible MiniPIX EDU Capabilities to Dispel Them

In the latter half of the previous century, educational researchers delved into learning progress factors, emphasising the pivotal role of students' prior subject knowledge (Griffiths and Preston 1992). Interest in misconceptions about radioactivity surged after the 1986 Chornobyl nuclear plant accident, persisting into the 21st century and extending to fields like particle physics. This section examines this research, selected beliefs for investigation, their causes, and how specific MiniPIX capabilities can address them.

2.1 *Definition and Importance of Misconceptions in the Learning Process*

Cardoso et al. (2020) define misconceptions as flawed notions formed from personal experiences, often conflicting with established scientific concepts. These alternative beliefs represent students' understanding of natural phenomena, deviating from accepted scientific consensus (Griffiths and Preston 1992). Recognised as fundamental elements, students' alternative beliefs have been emphasised in education since the 1970s (Plotz 2016; Neumann and Hopf 2012, Gutierrez et al. 2000).

Neglecting students' alternative conceptions in education can result in a teacher-centric approach, hindering meaningful learning and resulting in passive students. Understanding these alternative conceptions the teachers can take advantage in comprehending student challenges from a constructivist perspective (Neumann and Hopf 2012). Studies highlight the resilience of these ideas to traditional teaching methods, particularly evident in topics like atomic structure and radioactivity (Eijkelhof 1996). Addressing and clarifying students' beliefs on these topics is crucial for substantial learning outcomes (Ilyas and Saeed 2018).

2.2 *Types of Misconceptions*

International studies have revealed widespread misconceptions among students regarding radiation and radioactivity. Extensive research in science education consistently shows students' persistent struggles in understanding these concepts, even after instruction (Prather 2005). Research indicates that children and adults often interchangeably use terms like 'radioactivity' and 'radiation' despite physicists assigning specific meanings to each (Lijnse et al. 1990). Specifically focusing on the Chornobyl incident, Eijkelhof and Millar found in 1988 a lack of differentiation in terminology among students, reflecting an undifferentiated concept of 'radiation/ radioactive material', as cited by Boyes and Stanissteet (Boyes and Stanisstreet 1994).

Nine misconceptions (shown in Table 1) have been selected and grouped into three sets based on their relation to the origin of radioactivity, the associated hazards and benefits, and the nature of the phenomenon.

2.2.1 Sources

Some students mistakenly believe radiation is solely artificial, originating only from nuclear power plants (Eijkelhof 1990). Evidence reported to IOP Spark shows some students incorrectly think only large nuclei can be radioactive (IOPSpark).

Table 1 Selected misconceptions about radiation and radioactivity

Type	Code	Description
Sources	M1	Confusion by the nature of radiation, and some think it is artificial. (Eijkelhof 1990)
	M2	Radiation only stems from nuclear power plants (Eijkelhof 1990)
	M3	Only large nuclei can be radioactive (IOPSpark)
Risk and benefits	M4	Views about the risk of radiation and radioactive materials are based more strongly on everyday information than scientific ideas. (Lijnse et al. 1990, Eijkelhof 1990, Millar 1994, Plotz and Hopf 2016, Rego and Peralta 2006)
	M5	Aware that radiation can cause cancer, but few are aware of its beneficial uses (Rego and Peralta 2006)
	M6	See ionising radiation (and possibly 'radiation' in general) as harmful to health (Eijkelhof 1990)
Nature and properties	M7	Many are unable to distinguish between non-ionising and ionising radiation and do not think of light as 'radiation' (Rego and Peralta 2006)
	M8	Do not clearly distinguish between irradiation and contamination (Boyes and Stanisstreet 1994, Eijkelhof 1990, Millar 1994)
	M9	Confusion between radioactivity and radiation concepts (Eijkelhof 1990)

2.2.2 Risk and Benefits

Research by Millar (1994), Eijkelhof (1996), Plotz and Hopf (2016), and Rego and Peralta (2006), among others, reveals that students' perceptions of radiation risks rely more on everyday information than on scientific knowledge (Lijnse et al. 1990). They often view ionising radiation, and perhaps radiation in general, as harmful to health. While many acknowledge its association with cancer, only a few are aware of its beneficial applications.

2.2.3 Nature and Properties

Rego and Peralta (2006) highlight students' difficulty in distinguishing between ionising and non-ionizing radiation, echoed by Neuman and Hopf (2012). Millar (1994) and Eijkelhof's (1990) studies extensively examine students' challenges in differentiating between irradiation and contamination. These persistent alternative ideas and uncertainties about radioactivity and radiation's traversal and environmental impact reveal substantial confusion between these concepts.

2.3 Causes of Misconceptions and Capabilities of the MiniPIX Suitable for Dispelling Them

The study of misconceptions about radioactivity reveals complex influences, often due to their imperceptibility without specialised tools. MiniPIX capabilities, such as visual output, user-friendly design, sensitivity, and experimental functionality, discussed in Sect. 1.1, could address these issues. This section links the causes of misconceptions to the specific detector attributes that can help overcome them.

Eijkelhof notes that societal, political, and media influences shape students' understanding, while Boyes and Stanistreet highlight the topic's complexity as a source of misunderstanding (Boyes and Stanisstreet 1994). Millar and others recommend Geiger-Muller counters to address the imperceptibility of radioactivity (Millar et al. 1990). Introducing cameras to visualise radiation could help achieve this goal.

Ronneau underscores the complexity in assessing risks, considering natural and human-made uses or misuses (Ronneau 1990). Stochastic processes in radioactivity involving probability assessments add to the conceptual challenges (Slovic et al. 1981). The detector's sensitivity allows students to measure common natural sources, which could aid comprehension of associated statistical and cumulative risks.

Johnson et al. stress the difficulty of understanding atoms due to their microscopic nature, suggesting the need for bridging abstract concepts with concrete experiences (Johnson and Hafele 2010). The MiniPIX's ease of use allows focus on ongoing processes, potentially bridging macroscopic and microscopic knowledge. Hands-on experiments align with Piaget's concept of concrete reasoning and could help address the topic's complexity.

In summary, the MiniPIX's ability to visualise particle tracks can help address radioactivity's imperceptibility and the lack of direct experience. Its sensitivity and ease of use could allow students to observe radioactivity's stochastic nature, measure common sources, and better understand additional radioactive risks. The device's experimental nature brings learning to a more concrete level, possibly aiding in overcoming formal reasoning challenges in understanding radioactivity.

3 Methodology

A concise one-hour demonstrative session using MiniPIX detectors was designed to assess their efficacy in refining students' ideas. The activity targeted misconceptions, confusions, unawareness, and other alternative ideas aligned with radioactivity and ionising radiation concepts taught in Spain's third year of mandatory education (typically aged 14). The session complemented the Nature of Matter topic within the Physics and Chemistry curriculum (Generalitat 2022).

Students ($N1 = 65$) were surveyed using questions addressing misconceptions from prior research papers by Boye and Stanistreet (1994), Millar (1994), Plotz and Hopf (2016), as well as Rego and Peralta (2006), chosen to gather information. This

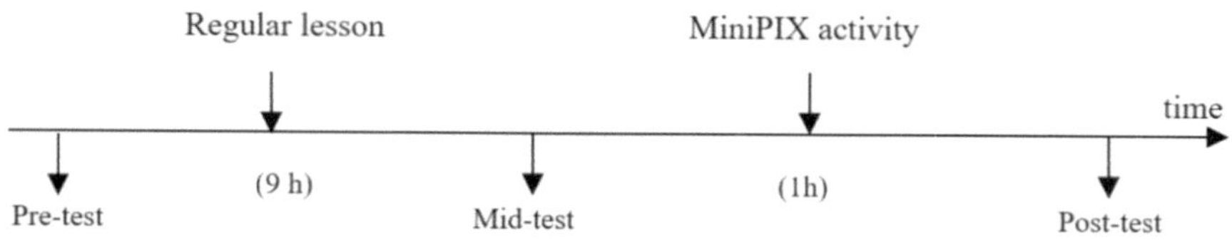

Fig. 3 Temporal sequence of the activities and the surveys for data gathering

served as the pre-test before their standard 9-h course on radioactivity, involving textbook material, presentations, and exercises. After the regular lesson, a subset (N2 = 20) served as a control group, answering the questionnaire as a mid-test to gauge improvement from standard teaching methods. Subsequently, the rest of the students (N3 = 45) participated in the MiniPIX activity before completing the survey as a post-test (Fig. 3).

By comparing results between the mid-test and post-test groups, both of which received the standard lesson, any notable improvement in responses could be attributed to the MiniPIX activity's effectiveness in enhancing understanding of radioactivity.

3.1 Design of the Activity

Due to constraints in detector availability and time, the activity aimed to measure radon radiation using one detector, replacing the original Geiger device used in the activity proposed by Austen and Brouwer (1997). Each of four or five students was provided with the necessary materials for charging a party balloon electrically and operating the detector safely. They also received a structured worksheet to annotate key ideas. A projection system allowed sharing the measurements with the class.

The session started with theoretical concepts on indirect detection and the relationship between radioactivity and radiation, including an experimental background radiation measurement. Each group prepared a charged balloon and hung it in isolation. Then, the MiniPIX detector's functionality and radioactivity applications were explained. Results from the background radiation measurement were projected, showing natural radioactivity. Students discussed differences in particle tracks (alpha, beta, gamma, muons). Following a short presentation on primary radiation sources, each group measured the radiation emitted by the particles captured by the charged balloon. Although background radiation acquisition took 15 min and the students' balloons were only 60 s, the contaminated balloon emitted notably higher radiation, particularly alpha particles, highlighting statistical risk accumulation and distinctions between contamination and irradiation. Shielding alpha radiation with paper emphasised variations in the behaviour of different types of ionising radiation.

3.2 Design of the Questionnaire

Many questions from four previous studies have been combined into a unique survey, as shown in Table 2. Some redundant questions or those unrelated to the analysed misconceptions have been discarded. However, issues arose due to intentional ambiguity in questions from Boye and Stanistreet's research, using terms like 'radioactivity' and 'radiation' together, leading to ambiguous answers. Future investigations should address this concern.

Additionally, questions from Millar's research present a specific situation regarding radioactivity or radiation uses, requiring students to answer three related questions. Despite indicating that all three questions should be answered, some students responded to only one or two questions, invalidating results related to this aspect.

The questionnaire has 28 questions and 88 different answers. Considering all the referred issues, after removing all of Millar's answers, the study has 76 different answers to be analysed.

4 Results

The analysis included only data from students who completed the pre-test ($N1 = 65$) and either the mid-test (before MiniPIX activity, $N2 = 20$) or the post-test (after MiniPIX activity, $N3 = 45$).

Each response was categorised based on how the topic was addressed during the sensor activity: experimentally (E), theoretically (T), indirectly (I), or not covered at all (N). Responses were classified into correct or wrong answers, simplifying varied question structures for statistical analysis. Doubtful or unknown responses were always treated as wrong answers.

The percentage of correct responses was computed for each answer at the pre-test, after Structure of Matter (mid-test), and after the MiniPIX activity (post-test). Comparisons between pre-test to mid-test and mid-test to post-test determined differences in correct answers attributed to the regular lesson and the sensor activity, respectively.

As expected, some answers increased the proportion of correct answers, some decreased, and few remained equal in some or both comparisons. However, the number and percentage of increase vs decrease vs equal proportion due to the standard lesson and the MiniPIX activity were utterly different, as shown in Figs. 4 and 5, suggesting a different impact from each activity.

To determine which of these variations were not attributable to the differences in the groups, both comparisons underwent a Chi-squared analysis to study the significance of the changes. Two null hypotheses were formulated. The first null hypothesis posits no difference in the proportions of correct answers before and after the didactical unit. Similarly, the second null hypothesis posits no difference in the

Table 2 Questions from the survey, including the type of answer. Some texts have been shortened to facilitate the readability of the table

Num	Questions	Answers
1	Have you ever heard of radiation?	Y/N
2	Where did you hear about radiation?	At school[a]
3	Do you know the difference between ionising and non-ionising radiation?	Y/N
4	Do all types of radiation produce the same effect on the body?	Y/N
5	Have you ever heard of natural radioactivity?	Y/N
5	Radiation used in radiography and in nuclear magnetic resonance is the same	Y/N
6	Are the same radioactivity and radiation?	Y/N
7a-h	Select the types of radiation you know about	EM, X-Rays, UV, Gamma, Cosmic, IR, Visible, Nuclear
8a-g	Select whether you think these radiations are harmful or not	Smartphones, X-rays, UV, Nuclear, Microwaves, IR, Visible[b]
9a-g	Choose whether you think these radiations are ionising or not	Smartphones, X-rays, UV, Nuclear, Microwaves, IR, Visible[c]
10a-e	Where does radiation come from?	Cosmic rays, the Sun, food, rocks under the earth, radon
11	All radiations have the same characteristics	Five levels of agreement/disagreement
12	Radiation can induce cancer but can also cure it	Five levels of agreement/disagreement
13	Radiation is used in industry, agriculture and energy production	Five levels of agreement/disagreement

(continued)

Table 2 (continued)

Num	Questions	Answers
14	The effects of radiation exposure may not be immediate	Five levels of agreement/disagreement
15	When exposed to a definite dose, the risks are independent of the radiation type	Five levels of agreement/disagreement
16	Different living beings have different sensitivities to radiation	Five levels of agreement/disagreement
17	The effect of radiation is independent of the organ absorbing it	Five levels of agreement/disagreement
18	Only heavy atoms as uranium can be radioactive	Five levels of agreement/disagreement
19a-h	Radioactivity or radiation comes from…	Outer space, rocks under the earth, rubbish tips, a hospital, nuclear power stations, factories, acid in the rain, gas from the ground
30a-h	Radioactivity or radiation can travel from one place to another:	As waves, as a cloud of gas, at very high speeds, through solid objects, as beams, as electricity, like particles, as heat
21a-1	Radioactivity or radiation can be used for…	Killing germs, measuring the speed of cars, treating cancer patients, sending radio messages, doing scientific experiments, to find out if people are lying, making food stay fresh, making lasers work, estimate age in archaeology
22a-i	Radioactivity or radiation can cause	Death, the 'greenhouse effect', mutations, heart attacks, damage to the ozone layer, newborns being abnormal, people glowing in the dark, people getting cancer, damage to buildings

[a] Only analysed option. Other possible answers were at home, at general media, at social media, scientific media

[b] Several degrees: Is harmful, Too much of it is harmful, Research Differs, Is harmless, I do not know

[c] Three answers: Ionizing, Non-ionizing, I do not know

proportions of correct answers before and after the MiniPIX activity. The p-value for both analyses was set at 0.05. Those answers which present a significant difference in proportions for either the first or second hypothesis have been gathered in Table 3, and those p-values less than 0.05 have been highlighted in bold.

Due to the small sample size, particularly in the middle test, concerns arose regarding the suitability of the chi-squared test for determining the significance of the observed improvements. To minimise the risk of Type I errors, an 'N-1' chi-squared test was conducted, following Campbell's recommendation (2007). Additionally, all

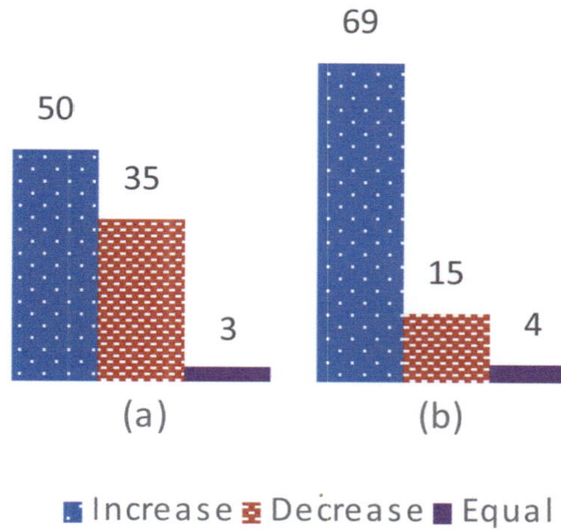

Fig. 4 Number of responses that present an increase, a decrease or remain equal in the proportion of correct answers due to the regular lesson (pre versus mid test) (**a**) and the MiniPIX activity (mid versus post test) (**b**)

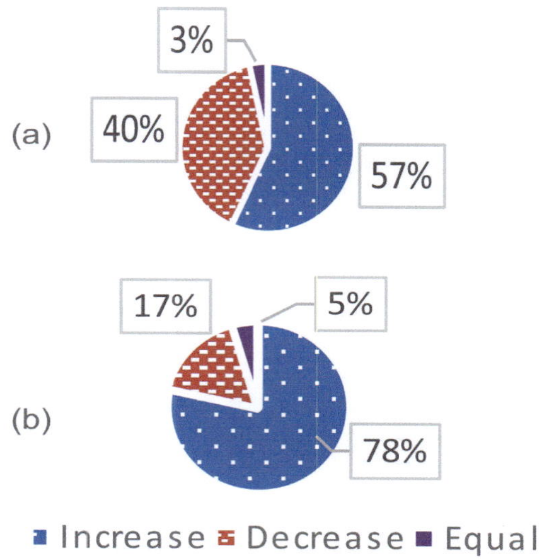

Fig. 5 Percentage of responses with an increase, decrease or no change in the proportion of correct answers due to the regular lesson (pre versus mid test) (**a**) and the MiniPIX activity (mid versus post test) (**b**)

2 × 2 tables were reviewed, and the three cells with expected values of less than five were carefully checked to ensure that the null hypothesis was not incorrectly rejected.

Table 3 Q&A with a significant improvement in right answers between pre and mid-test and/or mid and post-test

Quest.[1]	Ans.[2]	Miscon.[3]	Type[4]	Pre-test (%)	Mid-test (%)	Post-test (%)	p-val 1	p-val 2
2	–	M4	E	43	85	98	**0.0011**	**0.0433**
3	–	M7	E	5	25	91	**0.0086**	**0.0001**
5	–	M1	E	49	40	89	0.4832	**0.0001**
6	–	M9	E	85	65	98	0.0506	**0.0002**
7	d	M7	E	60	85	98	**0.0404**	**0.0433**
7	e	M7	E	14	20	49	0.5183	**0.0289**
8	a	M6	E	25	50	58	**0.0354**	0.5522
8	e	M6	T	26	20	49	0.5880	**0.0289**
8	g	M6	T	18	5	27	0.1563	**0.0432**
9	a	M7	E	31	40	73	0.4572	**0.0117**
9	b	M7	T	26	60	60	**0.0053**	1.0000
9	d	M7	E	29	55	80	**0.0342**	**0.0391**
9	g	M7	T	14	35	49	**0.0372**	0.2985
10	a	M1	E	51	35	82	0.2128	**0.0002**
10	c	M1	T	6	0	62	0.2649	**0.0001**
10	d	M1	T	25	65	89	**0.0011**	**0.0222**
10	e	M1	E	18	10	58	0.3976	**0.0003**
12	–	M5	T	43	70	80	**0.0358**	0.3808
13	–	M5	T	58	50	78	0.5308	**0.0249**
19	b	M2	E	46	85	87	**0.0023**	0.8295
20	c	M1	E	40	40	71	1.0000	**0.0187**
20	d	M1	E	26	50	58	**0.0448**	0.5522
20	g	M1	E	69	55	84	0.2516	**0.0133**
21	b	M5	T	22	10	47	0.2360	**0.0043**
21	d	M5	T	42	35	78	0.5791	**0.0009**
21	f	M4	N	89	95	73	0.4282	**0.0432**
21	i	M5	T	15	40	78	**0.0168**	**0.0030**
25	a	M8	T	75	63	89	0.2979	**0.0144**
25	b	M8	T	67	46	84	0.0925	**0.0017**

[1] Question number as labelled in Table 2. Only those answers with a significant increase are shown
[2] Answer label as shown in Table 2
[3] Misconception number as labelled in Table 1
[4] Type of work done during the activity: experimental (E), theoretical (T), indirect (I) or not covered (N)

5 Discussion

Table 3 reveals intriguing insights. Only 28 of the responses showed significant improvement after both the regular lesson and the MiniPIX activity, constituting just 32% of the total. This suggests the persistence of students' alternative beliefs, supporting previous research findings.

One answer (21f) notably decreased in correct responses after the detector activity. From Boye and Stanistreet's prior research ("Radioactivity or radiation can be used to find out if people are telling lies"), this question was not covered during either the sensor activity or the regular lesson. The proportion of correct answers started high at 89%, increased to 95% after the regular lesson, and then dropped to 73% after the MiniPIX activity. According to Boye and Stanistreet's research, the proportion of 14–15-year-old students answers is slightly below 80%, similar to the present study's final value. So, this fluctuation might not necessarily be related to the learning process but potentially due to statistical limitations. The p-value for this change is close to the significance threshold, indicating the need for further research.

Moreover, there were twelve significant improvements from the pre-test to the mid-test due to the standard lesson and twenty-two improvements from the mid-test to the post-test due to the MiniPIX activity (Fig. 6). This discrepancy suggests a greater impact of the detector activity than the standard lesson. Specifically, six improvements were solely due to the standard lesson, sixteen were exclusively due to the sensor activity, and six were attributable to both, suggesting that the MiniPIX activity's effectiveness might surpass that of the regular lesson addressing students' ideas.

Five of the six answers showing significant improvement due to the regular lesson, rather than the MiniPIX activity, were introduced theoretically in the regular lesson and the extra activity, so the improvement likely occurred during the initial introduction. Despite being demonstrated experimentally with the detector, the final question about radiation's ability to penetrate solid objects shows similar student responses to

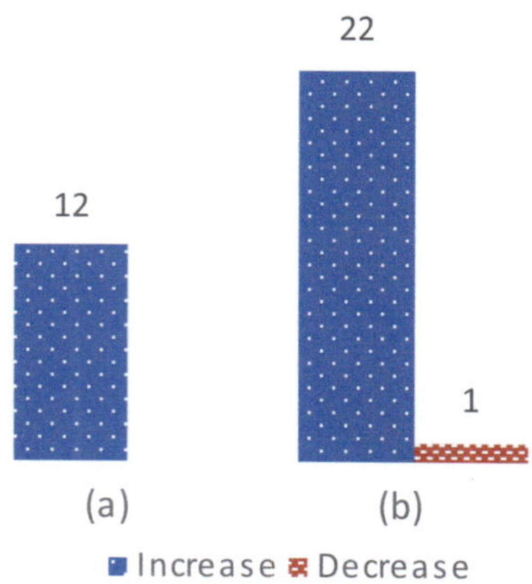

Fig. 6 The number of questions that present a significant change due to the MiniPIX activity (**b**) is 1.8 times greater to the regular lesson (**a**)

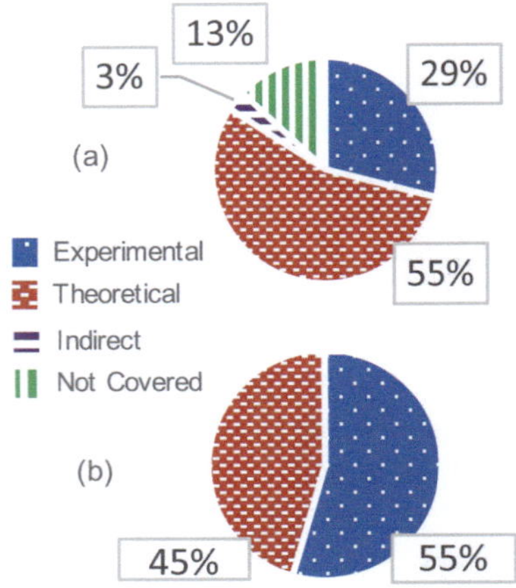

Fig. 7 Percentage of answers by type of work done during the MiniPIX activity in total (**a**) and with a significant improvement (**b**)

those covered only theoretically. The varying behaviour of different radiation types could explain these difficulties, but the test design cannot confirm this.

The observation that the MiniPIX activity potentially produced better results in dispelling misconceptions doesn't inherently prove the detector's direct impact. Focusing on answers exhibiting significant improvement due to the sensor activity could clarify this aspect.

The plotting of the proportions of each type of work, as explained in Sect. 4, during the detector activity for all survey answers showed a certain distribution (Fig. 7a). However, the plot revealed a different pattern when considering only the answers with significant improvement during the sensor activity (Fig. 7b). There's a distinct disparity between the two plots, suggesting a potential correlation between the type of work during the detector activity and the presence or absence of significant improvement.

Despite most answers being theoretically covered during the activity with the sensor, a majority of answers that exhibited a significant improvement were those where experimental work was undertaken with the MiniPIX camera. This suggests a robust association between the experimental engagement with the MiniPIX during the activity and the effective refining of students' ideas.

6 Conclusions

The study's findings corroborate earlier research, highlighting students' challenges in overcoming their previous beliefs. The effectiveness of experimental work with MiniPIX detectors stands out significantly, especially considering the study's constrained time and resources, urging further exploration in this field.

To address existing issues, the authors plan to conduct interviews to understand better how students' comprehension changes. Improvements such as resolving question ambiguities and expanding sample size and data collection methods are also on the agenda.

Moreover, intriguing questions have emerged for future investigation. Testing similar activities with a Geiger counter could discern if the MiniPIX detector or the experimental work is the critical factor. Replicating the study with different activities using the sensor or varying supervision levels aims to explore factors influencing results. Long-term retention of students' improved knowledge also deserves investigation. In any case, the authors intend to continue analysing these and other data.

Overall, this research suggests that employing MiniPIX detectors in educational activities can significantly aid in refining students' ideas about radioactivity and radiation. With the pressing challenges in energy production and solutions to the climate crisis, having a well-informed population on this subject could be crucial for Western democracies in the forthcoming decades.

References

ADMIRA (2021) Projecte ADMIRA|ServiPartícules. ICCUB. https://serviparticules.ub.edu/projectes/projecte-admira. Accessed 13 Mar 2025

ADVACAM MiniPIX EDU. https://advacam.com/camera/minipix-edu/. Accessed 13 Mar 2025

Austen D, Brouwer W (1997) Radioactive balloons: experiments on radon concentration in schools or home. Phys Educ 32:97. https://doi.org/10.1088/0031-9120/32/2/016

Ballabriga R et al. (2016) Review of hybrid pixel detector readout ASICs for spectroscopic X-ray imaging. JINST 11:P01007. https://doi.org/10.1088/1748-0221/11/01/P01007

Boyes E, Stanisstreet M (1994) Children's ideas about radioactivity and radiation: sources, mode of travel, uses and dangers. Res Sci Technol Educ 12(2):145–160

Campbell I (2007) Chi-squared and Fisher-Irwin tests of two-by-two tables with small sample recommendations. Statist Med 26(19):3661–3675. https://doi.org/10.1002/sim.2832

Campbell M (2001) Electronics for pixel detectors. In: Proceedings of the 7th workshop on electronics for LHC experiments. CERN, pp 11–16. https://doi.org/10.5170/CERN-2001-005.11

Cardoso P, Nunes M, Silva G, Braghittoni L, Trindade M (2020) Conceptions of high school students on atomic models, radiation and radioactivity. Phys Educ 55(3):035030. https://doi.org/10.1088/1361-6552/ab7fc6

Eijkelhof HMC, Utrecht R (Netherlands) (1990) Radiation and risk in physics education. INIS 22(2):22010294

Eijkelhof H (1996) Radiation risk and science education. Radiat Prot Dos 68(3–4):273–278

Generalitat 175/2022. (n.d.). XTEC—Xarxa Telemàtica Educativa De Catalunya. https://xtec.gencat.cat/ca/curriculum/eso/curriculum-175-2022/. Accessed 13 Mar 2025

Griffiths A, Preston K (1992) Grade-12 students' misconceptions relating to fundamental characteristics of atoms and molecules. J Res Sci Teach 29(6):611–628

Gutíerrez E et al (2000) ¿Qué piensan los jóvenes sobre radiactividad, estructura atómica y energía nuclear? Ensen Las Ciencias 18:247–254. https://doi.org/10.5565/rev/ensciencias.4043

Holik M, Vícha V (2020) IEAP CTU Prague—SESTRA kit. Cvut.cz. http://www.utef.cvut.cz/sestra_kit_en. Accessed 13 Mar 2025

Ilyas A, Saeed M (2018) Exploring teachers' understanding about misconceptions of secondary grade chemistry students. IJCDSE 9(1):3323–3328. https://doi.org/10.20533/ijcdse.2042.6364.2018.0444

IOPSpark (2024) Misconceptions. https://spark.iop.org/some-students-think-only-large-nuclei-can-be-radioactive. Accessed 13 Mar 2025

Johnson A, Hafele A (2010) Exploring student understanding of atoms and radiation with the atom builder simulator. AIP Conf Proc 1289:177–180. https://doi.org/10.1063/1.3515191

Lijnse P, Eijkelhof H, Klaassen C, Scholte R (1990) Pupil's and mass media ideas about radioactivity. Int J Sci Educ 12(1):67–78

Llopart X, Ballabriga R et al (2007) Programmable pixel readout chip for arrival time, energy and/or photon counting measurements. Nucl Instrum Meth A 581(1–2):485–494

MadPIX|Consejería de Educación, Ciencia y Universidades de Madrid. https://innovacionyformacion.educa.madrid.org/proyectos/otros/madpix. Accessed 12 Mar 2025

Millar R (1994) School students' understanding of key ideas about radioactivity and ionizing radiation. Public Underst Sci 3(1):53–70

Millar R, Klaassen K, Eijkelhof H (1990) Teaching about radioactivity and ionising radiation: an alternative approach. Phys Educ 25(6):338–342

Neumann S, Hopf M (2012) Students' conceptions about 'Radiation': results from an explorative interview study of 9th grade students. J Sci Educ Technol 21(6):826–834. https://doi.org/10.1007/s10956-012-9369-9

Parcerisas D et al (2022) ADMIRA project: teaching particle physics at high school with Timepix detectors. Phys Educ 57(2):025018. https://doi.org/10.1088/1361-6552/ac4143

Parker B, Thomas L, Rushton E, Hatfield P (2019) Transforming education with the Timepix detector—Ten years of CERN@school. Radiat Meas 127:106090. https://doi.org/10.1016/j.radmeas.2019.03.008

Plotz T (2016) Students' conceptions of radiation and what to do about them. Phys Educ 52(1):014004. https://doi.org/10.1088/1361-6552/52/1/014004

Plotz T, Hopf M (2016) Two concepts of radiation—a case study investigating existing preconceptions. Eur J Sci Math Ed 4(4):447–459. https://doi.org/10.30935/scimath/9484

Prather E (2005) Students' beliefs about the role of atoms in radioactive decay and half-life. J Geosci Educ 53(4):345–354. https://doi.org/10.5408/1089-9995-53.4.345

Proxecto MEDRA—IGFAE. https://igfae.usc.es/labs/proxecto-medra/. Accessed 13 Mar 2025

Rego F, Peralta L (2006) Portuguese students' knowledge of radiation physics. Phys Educ 41:259. https://doi.org/10.1088/0031-9120/41/3/009

Ronneau C (1990) Radioactivity: a natural phenomenon. J Chem Educ 67(9):736–737

Slovic P, Fischhoff B, Lichtenstein S (1981) Informing the public about the risks from ionizing radiation. Health Phys 41(4):589–598. https://doi.org/10.1097/00004032-198110000-00001

Vícha V (2017) Experiments using pixel detector in teaching nuclear and particle physics. Czech Technical University, Prague

First Outcomes of a Teaching–Learning Sequence on Quantum Information and Computation

Massimiliano Malgieri, Claudio Sutrini, Giacomo Zuccarini, and Chiara Macchiavello

Abstract In this work we report on the results of the first refinement cycle of a teaching learning sequence based on the educational reconstruction of quantum information and computation at the level of secondary school. The design of the teaching–learning sequence is based on the Model of Educational Reconstruction, relying on an analysis of the scientific subject matter as well as the perspectives and needs of teachers and students. The results served as the base for refinements and improvements of the sequence, for a new testing cycle performed starting in 2023.

Keywords Quantum mechanics education · Educational reconstruction · Quantum computation · Quantum information

1 Introduction

In recent works, a number of authors have proposed courses, tools and strategies in an effort to advance the scope of education to quantum mechanics (QM) in secondary school to include topics related to the "second quantum revolution" (Riedel et al. 2019). Following the teacher professional development course we designed and implemented in 2020 (Sutrini et al. 2022) we built a teaching–learning sequence (TLS) based on the Model of Educational Reconstruction (Duit et al. 1997) to introduce concepts of quantum information and computation. We present the results for

M. Malgieri (✉) · G. Zuccarini · C. Macchiavello
Department of Physics, University of Pavia, Pavia, Italy
e-mail: massimiliano.malgieri@unipv.it

G. Zuccarini
e-mail: giacomo.zuccarini@unipv.it

C. Macchiavello
e-mail: chiara.macchiavello@unipv.it

C. Sutrini
Istituto Superiore "Taramelli – Foscolo", Pavia, Italy
e-mail: claudio.sutrini01@universitadipavia.it

Z. Ješková and J. Hanč (eds.), *Advancing Physics Education*, Challenges in Physics Education, https://doi.org/10.1007/978-3-031-94207-5_10

three tests: one in the context of a vocational summer school, guided by researchers, and two in a context of curricular teaching, guided by the classroom teacher, formed in our professional development course. We note that, despite the smallness of the sample, these are to our knowledge the first data on student understanding of quantum computation in curricular secondary school teaching.

1.1 Previous Work on the Teaching–learning of Quantum Computer Science

While research on the teaching and the learning of quantum physics is a well-developed field within physics education, the teaching–learning of quantum technologies, and quantum information science in particular, represents still a largely uncharted territory, which only recently has received significant attention. For example, Pospiech (2021) proposes a course on QM for the German high school, in which quantum computing and quantum cryptography are introduced as rich technological contexts in which the fundamental concepts of quantum theory (e.g. superposition, entanglement, incompatibility, measurement) find their full development and application. Walsh and coauthors (2021) designed and tested a one year high school course on quantum computing based on classical wave optics, with a focus on hands-on experiments and simulation activities adopting an inquiry-based approach. Satanassi et al. (2022) developed a quantum computing course for high school students based on the general idea of leading students to follow the evolution of computational thinking in human history, from the most primitive computing machines, and ending with quantum computers and algorithms. Angara and coauthors (2020) proposed a coding approach, revolving around the introduction to basic quantum computing concepts using freely available tools such as Qiskit (IBM 2023). A comprehensive review of educational efforts aimed at expanding and improving education on quantum technology has been published in 2022 (Kaur and Venegas-Gomes 2022).

1.2 Research Question

This study concerns an intermediate step of a design-based research (DBR) process whose overarching research question (Bakker 2018) can be summarized as follows: can a TLS based on the MER be effective in leading secondary school students to achieve conceptual understanding of the fundamental ideas of quantum information and computer science? In particular, the present study directly addresses the following, more restricted sub-questions:

(a) *What were the educational outcomes of the first cycle of testing of our TLS, as measured by the evaluation instruments which are available at the present time, and with the limitations of the available sample?*
(b) *How can these data be used for a revision of the TLS and progressing to the next cycle of testing*

2 Educational Reconstruction of Quantum Information and Computation

The design of the TLS has been developed in the framework of the Model of Educational Reconstruction (Duit et al. 1997) by taking into account an analysis of the subject matter from the theoretical perspective, an analysis of the teachers' perspective, and an analysis of the learners' perspective.

2.1 Analysis from a Theoretical Perspective

The birth of quantum information science was driven by an awareness of the need to bring computation back to a physical dimension (Deutsch 1985). Among the foundational works of quantum information as a discipline, we considered as extremely relevant from an educational point of view the research on the thermodynamics of computation, carried out by Bennett (1988), Landauer (1996) and others. This literature line provided us with a first conceptually relevant element: the adoption of an integrated approach which combined physics, logic, and computation. The connection between disciplines can be highlighted by making use of a diagrammatic approach, using different, but interconnected, circuit representations, to show the logic, informational, and physical structure of a computing process; and designing activities in which students had to translate a process from one representation to another, or connect elements with corresponding functional roles in different representations. We hypothesized that, as pointed out by several authors in the past (Munfaridah et al. 2021) consistently adopting a strategy based on multiple representations of processes could also bring educational advantages in terms of learning gains.

Finally, the analysis of the structure of quantum information science highlighted for us the need for a clarification of the computational/physical structure of quantum information-processing. Indeed, our impression was that although the functioning of a quantum algorithm can be understood by inspecting the mathematical passages leading from input to output, there was a lack of deeper level analysis highlighting some general functional elements, connected to the structure of quantum theory, which led to the well-known advantages of quantum computation. Starting from the work in (Satanassi et al. 2022) where a basic principle of educational design is identified in the reconceptualization of the three main phases of foundational experiments (preparation—transformation—measurement) in terms of computation (input information—processing—output information), we worked on further clarifying of the information processing stage. In order to obtain a more fruitful *elementarization* of quantum algorithms and protocols we propose, based on common elements of the Deutsch-Josza and Grover algorithms, to tentatively decompose the information processing phase into three sequential processes: (1) the enabling of *parallelism* to generate an equal superposition of basis states; (2) the transfer of information encoded in the oracle function to the target register(s) exploiting the *product structure* of quantum compound states; (3) the enabling of *interference* by means of a network of logic gates to produce the desired state on which measurement can be performed.

2.2 Analysis from the Perspectives of Teachers

As the main source of information on teachers' perspectives on the teaching of quantum information science, we used the results of a teacher professional development course offered by the physics department of the University of Pavia in 2020–2021 (Sutrini et al. 2022). In the context of such course, teachers found a circuit approach very useful to illustrate the dialectic between the disciplines involved. When asked about their choice between a qubit implementation based on spin or on polarization, most of them showed a preference for the latter since polarization (for classical light) is comprised in the traditional secondary school curriculum and better understood by teachers. Although positive and negative elements are emphasized for both choices in the literature (synthetically: choosing spin may emphasize a discontinuity element between classical and quantum physics, and thus reduce the possibility that students form hybrid classical-quantum mental models; while choosing polarization may emphasize continuity, sometimes with unwanted educational results, but on the other hand offers the possibility of performing significant experimental activities) in this case the preference of teachers and the opportunity of performing experimental work, a need strongly perceived by students (see the next Subsection) were decisive for our choice. Thus, the initial introduction to QM, necessary to proceed forward to quantum computation, was performed by revising an existing educational path for secondary school set in the contexts of polarization and the hydrogen atom (Zuccarini and Michelini 2022). Operations that can be performed on polarization states, by means of calcite crystals, polarizers, half- and quarter- wave plates, were also the base for the initial discussion of quantum encoding and processing in terms of logic gates.

Another indication we took from the teachers' perspective was to renounce to the matrix representation of operators, since while vectors are widely covered by the secondary school program, matrices are only a peripheral topic which is often skipped entirely. Thus, in the course linear transformations are discussed exclusively by means of their vector representation (transformations of basis vectors).

2.3 *Analysis from the Perspectives of Learners*

In literature on the understanding of the quantum state in the context of polarization, students show a reluctance to think about the polarization states of a photon as a two-state system (Marshamn and Singh 2015) and a tendency to interpret the state vector as the mathematical representation of a physical quantity (Pospiech 2021). In order to help students overcome interpretive pitfalls connected with the context of polarization, e.g., the one-to-one relation between the space of properties (Euclidean polarization plane) and that of states (abstract vector space) when only real coefficients are used, and build a global knowledge structure on the relation between physical properties and state space in the encoding procedure it may be useful to guide them to encode information also in a different physical property. In addition, adding such a second property which can be used to encode one qubit of information allows us to construct two qubit circuits (e.g. the Deutsch algorithm) using only one photon, and which can be properly simulated using only one beam of classical coherent light. Thus, based also on a significant amount of scientific literature covering this possibility (Englert et al. 2001) and also on previous encouraging tests with students on the sum over paths approach to QM (Malgieri et al. 2017) we included in the course the encoding of a qubit as which-path information of a photon after a non-polarizing beam-splitter.

Another element we took in high consideration were the results of a distance-learning course for self-selected secondary school students held by us in 2020–2021 in which the small sample, despite having demonstrated a high level of understanding of the topics proposed in the final test, complained about the excessive abstractness and little attention to experimental aspects. Many of the choices we performed can be connected to the results of this small trial: enhancing visualization, by emphasizing the geometric interpretation of the gates H, X, Z as axial symmetries of the state plane; design of ideal experiments, by constantly transposing circuit diagrams into possible optical implementations (optical circuits); actual lab activities, by realizing in the lab some of the optical circuits designed during the lessons, exploiting the analogy, valid for linear optics designs, between single photon devices and classical coherent polarized light.

2.3.1 Embedding the Analysis in a Conceptual Change Framework

All the choices previously discussed were embedded in the perspective of promoting students' conceptual change from a classical to a quantum perspective, both in physics and computation jointly, based on the framework discussed in (Zuccarini and Malgieri 2022). In the initial introductory course in the context of the linear polarization of the photon, students' evolution of model-building from a classical to a quantum perspective was favored by using activities that mirror discipline-specific epistemic practices (theoretical modelling); and by requiring them to explicitly analyse change and continuity in physical concepts/constructs in theory change (observable, measurement, state, vector, superposition, interference, correlation). In the transition from classical to quantum computation, students were guided to recognize that a theory change in physics involves also a transition in logic and probability and so a transition in computation. Classical and quantum encoding strategies, gates, circuits, algorithms, and protocols were discussed in parallel, highlighting the computational role of new physical properties (e.g., entanglement). In the process of building multiple types of optical encodings (using polarization states and which-way information), learning was again facilitated through the use of activities that mirror epistemic practices related to the implementation of computing (ideal experimental design of gates and circuits) and the analysis of change and continuity in optical notions from a classical to a quantum description (from an elementary representation of electric and magnetic field vectors to quantum state vectors in both encodings). In the introduction to the Deutsch algorithm, the «problem of the coin» (determining whether a coin is genuine or defective/counterfeit) was used as a bridge from classical to quantum processing of a classically encoded database. Finally, lab activities (e.g. construction of a logic gate for qubits in polarization encoding, realization of the Deutsch algorithm with classical light (Sutrini et al. 2023) were designed to implement and discuss the transition from ideal experimental design to a real lab, with the goal to integrate theoretical and empirical perspectives.

3 Structure of the Course

The TLS lasts about 25 h, starting from a basic introduction to quantum mechanics using a two state approach, and ending with quantum algorithms and protocols (Deutsch and Grover algorithms; teleportation protocol). Instruction proceeds through a variety of activities, including inquiry-based and modelling tasks described in two–three-page worksheets which students fill during the course, always alternating the attending of a lesson with frequent moments of individual work. At the end of instruction, students are given a final test. The structure of the TLS is synthetically reported in Table 1.

Table 1 Structure of the teaching–learning sequence

Topic		Hours
A	Introduction to quantum mechanics	6
B	From classical logic to its quantum counterpart	1
C	One-qubit gates: polarization and which path encoding	3
D	Two qubit gates: polarization and which path encoding	2
E	Deutsch's and Grover's algorithms	3
F	Lab activities	3
G	Teleportation protocol	2

4 Context of the Experimentations

We discuss data from three experimentations. One experiment, henceforth called the UPV sample, was performed in the context of the summer school on quantum technologies at the University of Pavia and included 14 self-selected students bound to start the final year of secondary school. The experiment was guided by researchers. Two experiments were instead performed in curricular teaching, guided by teachers formed in our professional development courses. The first one was performed at Liceo Volta, in Castel San Giovanni (henceforth the CSG sample), including 23 students; the second one in Liceo Gramsci, Florence (henceforth the FL sample), including 18 students. Lab activities (item F in Table 1) were only performed in the UPV sample due to insufficient instrumentation in secondary school labs, and the Deutsch and Grover algorithms (item E in Table 1) were not performed in the FL sample by decision of the teacher.

5 Methods

The data considered in this study come from the worksheets individually filled by students during the course. Data on the final test was not yet available at the time of the GIREP conference.

The questions are supposed to measure students' conceptual understanding of basic elements of quantum information science. The only indication we have so far of their validity, is that they were designed through a collaboration between experts in physics education and experts in quantum information science. Note that, since the teaching of quantum information at secondary school level is a rather new research field, there is so far no validated instrument in the literature to test conceptual understanding in this sub-discipline.

Student answers in this preliminary study were analysed using a dichotomic approach (correct/incorrect). This choice was taken since the questions did not generally require sufficiently complex answers to use more sophisticated (e.g. rubrics-based) scoring approaches, nor the sample was sufficiently large to justify the adoption of phenomenographic categories, although this may be done in the future. The correctness of the answers was evaluated by three researchers separately (MM, CS, GZ) and dubious cases were resolved by consensus.

6 Results

We summarize the main strengths and weaknesses of our approach and describe how we are using them to improve our TLS.

Generally, we find that both the main educational choices of (a) subdividing the information processing phase in the composition of three sequential subprocesses, and (b) adopting a diagrammatic approach consisting of different types of diagrams with a different kind of relationship to real systems are productive for students.

However, we find a number of semiclassical/hybrid interpretations of quantum phenomena, and difficulties in precisely identifying the differences between classical and quantum. Thus, the hypothesis that the context of quantum computation can help students pinpoint the peculiar characteristics of quantum physics needs to be further scrutinized.

6.1 Logic Gates in Polarization Encoding

Results for the basic elements of the polarization encoding of qubits were very encouraging. In the item shown in Fig. 1, (effect of a π phase shifter on a qubit prepared in different states) the correct answer was proposed by all students (14/14) in the UPV sample, by 16/18 students in the FL curricular sample, and by 14/22 students in the CSG curricular sample. In the lower part of Fig. 1 we show one of the students' responses judged as correct.

Another item (Fig. 2) required students to design a Z gate in polarization coding using one or more of some optical devices proposed. The correct setup (as in Fig. 2) was proposed by all students (14/14) in the UPV sample, by 17/18 students in the FL curricular sample, and by 11/22 students in the CSG sample. Again, from most answers the productivity of representing gates as transformations of vectors in the polarization space emerges. However, an analysis of the language used by students already highlights problems with the classical/quantum interface. Not many students write consistently of a spatial separation of single photon states; many refer to a "ray", "flow", or "beam".

6.2 Logic Gates in Which Path Encoding

While students do well for simpler gates in 'which path' coding, such as a single NOT gate, they show several sources of confusion for the more complex cases as the one displayed in Fig. 3.

The question is answered correctly for both sub-items by 8/17 students in the CSG sample, and by 8/14 students in the UPV sample. The FL sample answered a slightly different question with similar content (15/18 correct). In this item we find indications that some students may find the double coding of qubits (which is essential for arriving to more complex applications such as the Deutsch algorithm and teleport protocol) still confusing at the time they compiled the worksheet, as they mix elements of the two codings (e.g. replacing birefringent crystals for beam splitters). We have indications from the final test (not discussed at the conference) that

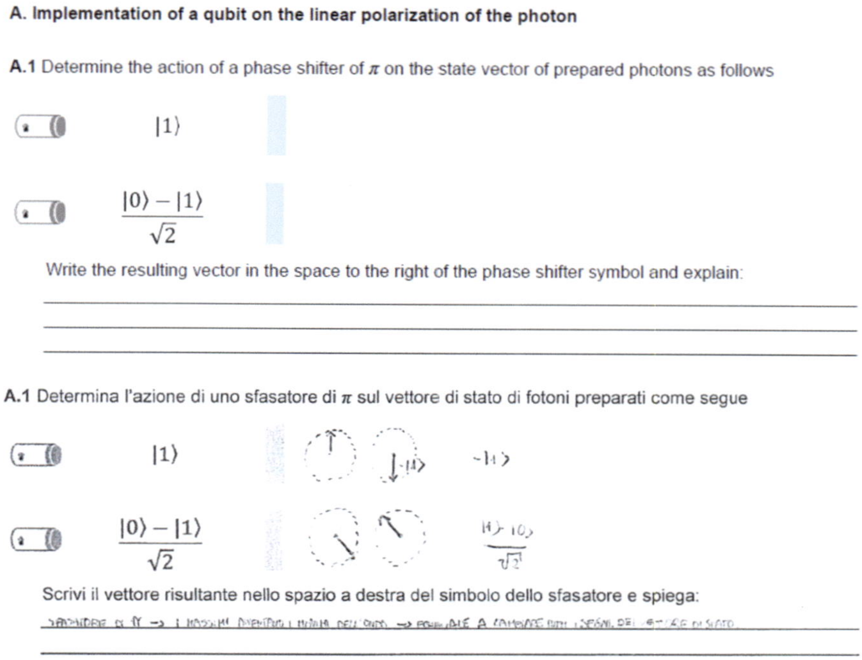
A. Implementation of a qubit on the linear polarization of the photon

A.1 Determine the action of a phase shifter of π on the state vector of prepared photons as follows

$|1\rangle$

$\frac{|0\rangle - |1\rangle}{\sqrt{2}}$

Write the resulting vector in the space to the right of the phase shifter symbol and explain:

A.1 Determina l'azione di uno sfasatore di π sul vettore di stato di fotoni preparati come segue

$|1\rangle$ $-|1\rangle$

$\frac{|0\rangle - |1\rangle}{\sqrt{2}}$ $\frac{|1\rangle - |0\rangle}{\sqrt{2}}$

Scrivi il vettore risultante nello spazio a destra del simbolo dello sfasatore e spiega:

Fig. 1 Worksheet item on polarization encoding of qubits. Example of a correct answer provided by a student. The written text can be translated as "*a π phase shift $\rightarrow$ brings wave maxima to minima $\rightarrow$ equivalent to changing the sign of the state vector*"

this confusion is mostly resolved proceeding with the sequence, as students become capable to analyse, or translate from a logic to an optical/circuital representation, moderately complex two-qubit circuits.

6.3 *Deutsch Algorithm*

Questions related to the Deutsch algorithm were only given in the CSG and UPV samples, as in the FI sample the teacher decided not to propose this topic to students. One of the questions (reported in Fig. 4) asked to write down explicitly the state of the system before and after the final Hadamard gate, and the outcome of the measurement, for all the possible oracle settings.

Most students correctly completed the table both in the CGS sample (13/16) and in the UPV sample (9/14). Some of the incorrect answers contained one or two sign mistakes. In this particular item no special criticalities were highlighted. A more complete data analysis of the educational outcomes for the CGS and the UPV samples, limited to understanding of the Deutsch algorithm, is reported in (Sutrini et al. 2023). In general, we find that the subdivision of the operation of the quantum

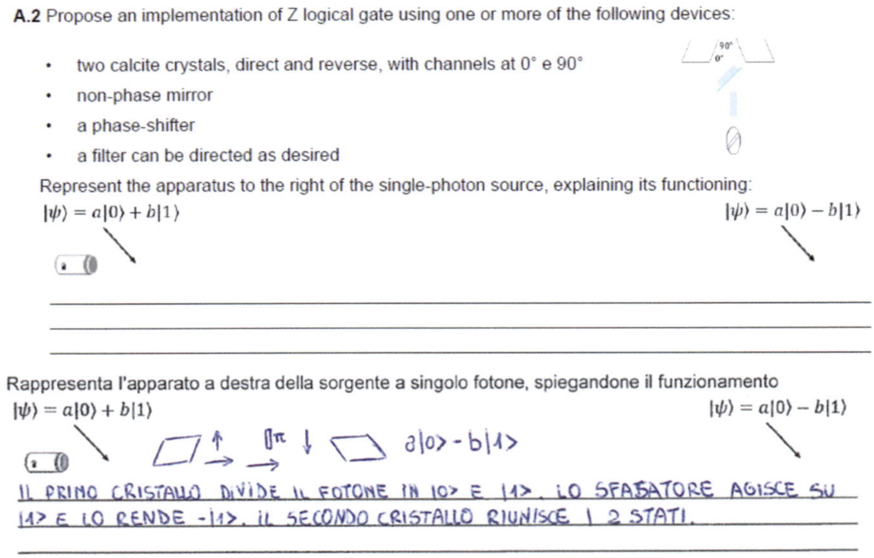

Fig. 2 Item on polarization encoding of qubits requiring students to design a Z gate. Example of a correct answer provided by a student. The written text can be translated as "*the first crystal divides the photon into* |0⟩ *and* |1⟩. *The phase shifter acts on* |1⟩ *and turns it into* −|1⟩. *The second crystal joins back the two states*"

Fig. 3 Item on which path encoding of qubits requiring students to design the circuit equivalent of a simple logic diagram. Example of a correct answer to sub-item A3.2 provided by a student

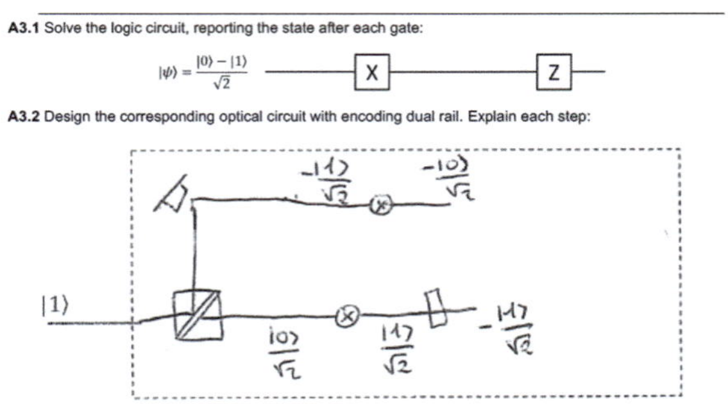

algorithm in three sequential processes, each connected to a property of quantum systems, seems productive in scaffolding students' learning.

6.4 Classical-Quantum Comparison

Several problems were identified in items asking to perform a comparison between the meaning of a superposition in the classical and quantum case (Fig. 5). A similar item, not shown here, was proposed also for the interpretation of spatial states. In

Interference: We know that the Hadamard gate on a superposition state creates interference phenomena such that $|0'\rangle \longmapsto |0\rangle$ and $|1'\rangle \longmapsto |1\rangle$.

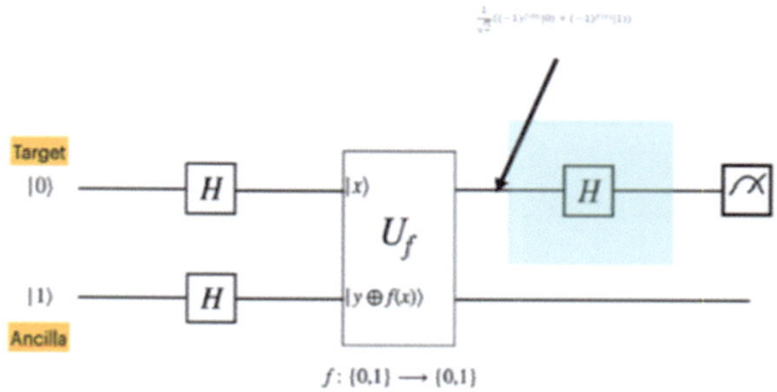

Question C1. Knowing that the final state before the last Hadamard is $|\psi\rangle = \frac{1}{\sqrt{2}}((-1)^{f(0)}|0\rangle + (-1)^{f(1)}|1\rangle)$, complete the table

Boolean function	State after the Oracle	State after the last Hadamard	Classic bit after measurement and probability
$f(0) = 0 \wedge f(1) = 0$	$\lvert\psi\rangle$ =_______ = _____	$\lvert\psi\rangle_{fin}$ =________	, p=
$f(0) = 0 \wedge f(1) = 1$	$\lvert\psi\rangle$ =_______ = _____	$\lvert\psi\rangle_{fin}$ =________	____, p= ____
$f(0) = 1 \wedge f(1) = 0$	$\lvert\psi\rangle$ =_______ = _____	$\lvert\psi\rangle_{fin}$ =________	____, p= ____
$f(0) = 1 \wedge f(1) = 1$	$\lvert\psi\rangle$ =_______ = _____	$\lvert\psi\rangle_{fin}$ =________	____, p= ____

Fig. 4 Worksheet item on the Deutsch algorithm

general, the number of students who fill this kind of sheet in an entirely satisfactory manner is very low, about 2–3/14 in the UPV sample, and none in either of the curricular samples. In FL, seeing students quite confused in the individual work, the teacher decided to perform the activity shown in Fig. 5 as a guided whole class discussion. In a future re-design of the TLS, these items will be profoundly revised, probably presenting them as multiple choice items based on the main difficulties emerging from this first trial.

Despite the poor educational outcomes, this kind of activities give us important indications on which aspects of the TLS may need to be revised in the future. We subsequently enumerate these indications synthetically.

For both the items on the polarization and spatial representation of classical and quantum states, a poor understanding of the classical representation of the E.M. field emerges. Students confuse, and sometimes reverse the meaning, of the terms amplitude and intensity. Student answers seldom contain any reference to energy in relation to the meaning of the coefficient squares.

	Polarisation of the classical plane electromagnetic wave $\mathbf{E} = a\mathbf{i} + b\mathbf{j}$	Photon polarization $\lvert\psi\rangle = a\lvert 0°\rangle + b\lvert 90°\rangle$
Physical interpretation and unit of measure of the vector in the left-hand side of the equation		
Space to which vector belongs		
Interpretation of coefficients and their square		
Physical interpretation of the superposition sign		

Fig. 5 Worksheet item on the comparison of classical and quantum representation of polarization states

In the analysis of the quantum case, most students correctly identify the squares of coefficients as probabilities, but struggle to provide an interpretation of the coefficients themselves. In many cases, especially in the curricular samples, students write explicitly that a change in the relative phase in a superposition is irrelevant because its effect is unmeasurable. This is a specific difficulty in student understanding of superposition states well documented in the literature (Passante et al. 2015) and one of the main open issues in our TLS, as will be discussed in the Conclusions.

7 Discussion and Conclusion

Globally, data from the discussed trials shows that the formalism is not the primary challenge for students when dealing with basic concepts of quantum information and computation. Students appear to master the vector representation of quantum states reasonably well, and are able to move with good confidence between different representations (mathematical-formal, graphical in space state, diagrammatic, circuital) of quantum processes. In this sense, we have evidence that the strategy of using multiple representations can be very productive in supporting students' reasoning.

Data on student understanding of quantum algorithms (the Deutsch algorithm in particular) shows that the tripartite subdivision of the quantum information processing phase is useful for scaffolding student thinking and imbue with meaning the internal workings of quantum algorithms. According to the teachers involved in the trials here presented, the outcomes in terms of school evaluations for the material treated in the TLS were in line with those obtained for other topics in the curriculum.

In this sense, the educational reconstruction of the topic of quantum information science can be said to have been at least partially successful.

However, it is also evident that many students still hold an underlying classical vision and do not perform a complete classical-quantum conceptual change, and this manifests itself when asked for verbal explanations, especially concerning the concepts of state and superposition. On one hand, these difficulties call for higher attention to these conceptual issues in the initial introduction to quantum theory. On the other hand, specific difficulties, such as the one highlighted in Sect. 6.4, could be more efficiently addressed at the time of the introduction to quantum computation. For example, for the purpose of highlighting an experimental consequence of the sign (relative phase) in a superposition, one could perform an activity in which they have to find the succession of gates (a combination of H and Z gates) necessary to transform the initial superposition state into one which is measured with certainty as $|0\rangle$ or $|1\rangle$, a combination which will be different depending on the sign of the initial superposition.

The double coding of qubits, using polarization and spatial states of the photon, is demanding for students in the initial stage of learning, when they tend to confuse optical devices producing transformation in one or the other coding. Proceeding with instruction, also with the help of visual aids such as representing devices and gates corresponding to either qubit coding in different colours, these issues are greatly reduced and students acquire, again using the strategy of switching and translating between representation, the capability to deal with moderately complex two-qubit computations. Still, we have at the moment no significant support for the design hypothesis (Sect. 2.3) that the double coding could help students identify the quantum state as a vector in an abstract space rather than one representing a physical quantity. More data from future trials will help clarify the issue.

Although we have not drawn strong conclusions from this study, which represents a step in an ongoing DBR project, it is still useful to recapitulate its main limitations, mentioned also elsewhere in the article. First of all, the sample studied is not very large (55 students in total) and partially (the UPV sample) composed of students with a special interest in physics and/or interested in choosing physics as a university career. All students attended a science-intensive secondary school (Liceo Scientifico), so no conclusion at all can be drawn for students with different backgrounds. The scarcity of previous research in this field can be also considered a limitation, in the sense that we have no data from other studies to compare our results with. The validity of the instruments used to probe students' understanding of quantum physics and quantum information so far relies only on the approval from experts in the field, and the creation and validation of more robust evaluation instruments is certainly a priority for later stages of this research.

References

Angara PP, Stege U, MacLean A (2020) Quantum computing for high-school students an experience report. In: Müller HA, Byrd G, Culhane C, DeBenedictis E, Humble T (eds) Proceedings of the 2020 IEEE international conference on quantum computing and engineering (IEEE), pp 323–329. https://doi.org/10.1109/QCE49297.2020.00047

Bakker A (2018) Research questions in design research. In: Bakker A (ed) Design research in education. Routledge, London, pp 73–86. https://doi.org/10.4324/9780203701010-7

Bennett CH (1988) Notes on the history of reversible computation. IBM J Res Devel 32(1):16–23. https://doi.org/10.1147/rd.321.0016

Deutsch D (1985) Quantum theory, the church-turing principle and the universal quantum computer. Proc Math Phys Eng Sci 400(1818):97–117. https://doi.org/10.1098/rspa.1985.0070

Duit R, Komorek M, Wilbers J (1997) Studies on educational reconstruction of chaos theory. Res Sci Educ 27:339–357. https://doi.org/10.1007/BF02461758

Englert BG, Kurtsiefer C, Weinfurter H (2001) Universal unitary gate for single-photon two-qubit states. Phys Rev A 63(3):032303. https://doi.org/10.1103/PhysRevA.63.032303

IBM (2023) Open source quantum development. https://qiskit.org/. Accessed 27 Jan 2024

Kaur M, Venegas-Gomez A (2022) Defining the quantum workforce landscape: a review of global quantum education initiatives. Opt Eng 61(8):081806. https://doi.org/10.1117/1.OE.61.8.081806

Landauer R (1996) The physical nature of information. Phys Lett A 217(4–5):188–193. https://doi.org/10.1016/0375-9601(96)00453-7

Malgieri M, Onorato P, De Ambrosis A (2017) Test on the effectiveness of the sum over paths approach in favoring the construction of an integrated knowledge of quantum physics in high school. Phys Rev Phys Educ Res 13(1):010101. https://doi.org/10.1103/PhysRevPhysEducRes.13.010101

Marshman E, Singh C (2015) Framework for understanding the patterns of student difficulties in quantum mechanics. Phys Rev Phys Educ Res 11(2):020119. https://doi.org/10.1103/PhysRevSTPER.11.020119

Munfaridah N, Avraamidou L, Goedhart M (2021) The use of multiple representations in undergraduate physics education: what do we know and where do we go from here? Eurasia J Math Sci Technol Educ 17(1):em1934. https://doi.org/10.29333/ejmste/9577

Passante G, Emigh PJ, Shaffer PS (2015) Student ability to distinguish between superposition states and mixed states in quantum mechanics. Phys Rev ST Phys Educ Res 11(2):020135. https://doi.org/10.1103/PhysRevSTPER.11.020135

Pospiech G (2021) Quantum cryptography as an approach for teaching quantum physics. In: Jarosievitz B, Sükösd C (eds) Teaching-learning contemporary physics: from research to practice. Springer, Cham, pp 19–31. https://doi.org/10.1007/978-3-030-78720-2_2

Riedel M, Kovacs M, Zoller P, Mlynek J, Calarco T (2019) Europe's quantum flagship initiative. Quantum Sci Technol 4(2):020501. https://doi.org/10.1088/2058-9565/ab042d

Satanassi S, Ercolessi E, Levrini O (2022) Designing and implementing materials on quantum computing for secondary school students: the case of teleportation. Phys Rev Phys Educ Res 18(1):010122. https://doi.org/10.1103/PhysRevPhysEducRes.18.010122

Sutrini C, Malgieri M, Macchiavello C (2022) Quantum technologies: a course for teacher professional development. J Phys Conf Ser 2297:012018. https://doi.org/10.1088/1742-6596/2297/1/012018

Sutrini C, Zuccarini G, Malgieri M, Bondani M, Macchiavello C (2023) An educational model of the Deutsch algorithm for secondary school. Eur J Phys 45(2):025302. https://doi.org/10.1088/1361-6404/ad1399

Walsh JA, Fenech M, Tucker DL, Riegle-Crumb C, La Cour BR (2021) Piloting a full-year, optics-based high school course on quantum computing. Phys Educ 57(2):025010. https://doi.org/10.1088/1361-6552/ac3dc2

Zuccarini G, Malgieri M (2022) Modeling and representing conceptual change in the learning of successive theories: the case of the classical-quantum transition. Sci Educ 1–45. https://doi.org/10.1007/s11191-022-00397-1

Zuccarini G, Michelini M (2022) Promoting the transition to quantum thinking: development of a secondary school course for addressing knowledge revision, organization, and epistemological challenges. https://arxiv.org/abs/2301.00239

Cultural Understanding of Physics: Instruments and Methods

Marco Giliberti, Grzegorz Karwasz, Luisa Lovisetti, Giovanni Organtini, Tommaso Rosi, Pasquale Onorato, Stefano Oss, Graziano Surace, and Eugenio Tufino

Abstract Although nowadays sciences are considered to be extremely important for society, their human, profound aspect—the one touching human soul—is still missing. Sciences are, in fact, considered mainly for their utilitarian aspects. If we want physics to be truly considered cultural heritage of humankind, we must bring a deep, cultural vision of physics to all educational levels, promoting synergies with other areas of human knowledge. Here, we report the results of a symposium with different contributions in the research field of cultural understanding of physics. We start by discussing the relevance of Virtual Reality and Augmented Reality for a cultural appropriation of physics. We proceed by suggesting experimental activities

---

M. Giliberti · L. Lovisetti (✉)
Department of Physics "Aldo Pontremoli", University of Milan, Milan, Italy
e-mail: luisa.lovisetti@unimi.it

M. Giliberti
e-mail: marco.giliberti@unimi.it

G. Karwasz
Institute of Physics, Nicolaus Copernicus University, Torun, Poland
e-mail: karwasz@fizyka.umk.pl

G. Organtini
Department of Physics, Sapienza Università Di Roma, Rome, Italy
e-mail: giovanni.organtini@uniroma1.it

T. Rosi · P. Onorato · S. Oss · E. Tufino
Department of Physics, University of Trento, Trento, Italy
e-mail: tommaso.rosi@leveluptrento.com

P. Onorato
e-mail: pasquale.onorato@unitn.it

S. Oss
e-mail: stefano.oss@unitn.it

E. Tufino
e-mail: eugenio.tufino@unitn.it

G. Surace
Department of Mathematics, Sapienza Università Di Roma, Rome, Italy
e-mail: graziano.surace@uniroma1.it

Z. Ješková and J. Hanč (eds.), *Advancing Physics Education*, Challenges in Physics Education, https://doi.org/10.1007/978-3-031-94207-5_11

with digital tools to involve students in the design and construction of apparatuses including aspects, such as beauty, traditionally reserved for the humanities. We then present a large-spectrum, inter-disciplinary proposal concerning the opportunities given by discussing very different aspects, such as science vs religion, or Big Bang and biological evolution. Finally, we show the tangled links between physics and culture through the example of a historical-pedagogical reconstruction of the birth of quantum mechanics aimed at providing an epistemological and educational cultural framework for this theory.

Keywords Cultural understanding of physics · Multi and interdisciplinary framework for physics · Virtual and augmented reality · Physics with Arduino · History of quantum physics · Cultural framework for quantum mechanics

1 Introduction

The relationship between physics, arts and humanities—painting, sculpture, poetry, theatre—has become increasingly intense and fruitful in recent centuries, so much so that stimulating connections continue to be found that do not only concern the technical aspects related to science—before the invention of photography, cinematographic art could certainly not exist!—but also involve hitherto unsuspected similarities between the ways of discovering reality that are common to research in the arts and basic research in physics (Giliberti 2023; Rustichelli and Stefanon 2012). However, in major theatres, it is still difficult to see science plays put on the bill; and, in general, not even important paintings inspired by physics are exhibited in the main museums of the world, and no poet has received the Nobel Prize for literature with poems concerning physics.

Needless to blame art gallery directors or theatre operators; in fact, few people think that science is able to touch people's hearts as well as their minds, as art is expected to do instead. "Mona Lisa", not "Conic sections", is a painting by Leonardo, and "Broken Symmetries" is not the title of a bestseller on the relationship between men and women. Who among the *habitués* would go to the theatre on a winter's night for a rerun of "The Hilbert space"? Indeed, such a title would make us smile. The situation is symmetrical on the scientific side; indeed, there is still little theatre, painting or poetry in university physics courses; or perhaps, there is none at all. This fact is so ingrained in our minds that it may even seem right and inevitable to us, may be that it is difficult even to think that things could be different, and that a true interdisciplinary construction of knowledge could be useful and productive.

Since we, however, do really think that an interdisciplinary construction of knowledge is essential both to acquire a mature vision of the world and to be able to better understand the disciplinary aspects of physics itself, we feel the urge to change the current situation. We believe it essential to improve the cultural perception of physics at all school levels—starting from what is proposed in university courses—and to make explicit the links, often profound, between physics with other disciplines, including those of a humanistic and artistic nature to generate that large network of knowledge which we call culture. To foster the production of meanings, necessary for this to happen, different interpretative tools are necessary; they can, in fact, help the construction/explication of these links in the active perspective of creating disciplinary knowledge.

At the turn of this century, a cultural superiority of the humanities over scientific disciplines was perceived (Giliberti and Lovisetti 2023) and led to a decline in vocations to scientific faculties (Brandi et al. 2005). Now times have changed, but it is the technological or social aspects of science that are often put in the foreground (Tolstrup Holmegaard et al. 2014).

The present situation has deep historical roots and, as a problem, dates back to the 1950s and 1960s of the last century, when the enormous volume of funding coming from the US government—mainly originating from war defense, and arising from the "success" of the Manhattan project—began to give rise to concerns of an economic and social nature that have begun to undermine the image and perception of science (and especially of physics) in society (David 2012; Kragh 1999). Problems linked to the safety of nuclear plants for peaceful purposes, pollution and the environment (think for example of radioactive fall-out) have contributed to building an image of science as an instrument of exploitation rather than knowledge. The pacifism of the 1960s and 1970s and the philosopher-theorists of revolutionary movements of students (for example Marcuse) as well as the thought of some philosophers of science such as Feyerabend had the effect of further undermining the image of science as a human work linked to the arts and spirituality; an image already so often more or less ruined over the entire human history for millennia (just think of the myth of Prometheus who is punished by the Gods—i.e., the spirituality—for having brought knowledge to man (Carpineti et al. 2011)). Back to the present day, the image of science we are referring to probably arises from both an erroneous conception of science (particularly of physics) and a too restricted idea of art within our culture. Indeed, when topics are meaningful to people (because they are perceived as useful, beautiful, or fascinating (Giliberti 2021), interest, attention and the desire to understand are strong. In fact, it is completely natural to aspire to a happier life. It is precisely in this sense that, perhaps, the most significant aspect of culture manifests itself. On the contrary, culture is often spoken of as a citizen's duty, or as a social elevator, while the most important and characterizing element of culture is, in general, left apart: culture, in fact, is "culture" when it speaks to us in depth, modifying our vision of the world and of life, and giving us the possibility to experience the pleasure of research and of understanding. Culture is the pleasure of never-ending research.

Therefore, if we truly believe that physics should be perceived in a cultural way not only by a small group of insiders, we must ask ourselves how the image and vision of the world that emerges from modern physical science are useful for enriching our lives; and, therefore, how science, and physics in particular, can really play a role that helps us live better.

In this paper we present the contributions to a Symposium organized by the GIREP Thematic Group "Cultural understanding of physics: instruments and methods" addressing precisely the issues briefly stated above, by discussing concrete research examples of disciplinary methods and tools that can help the cultural learning of physics and improve its cultural image in society.

We begin with the presentation of the instruments of Virtual Reality (VR) and Augmented Reality (AR); in fact, these are some of the main new technological tools that science and physics education researchers have been focusing on in the past decade. By exploring new interactions between computer simulations and the real world, what will be referred to as "augmented simulations" may offer a new way of culturally understanding what reality is through the simultaneous presentation of physical conceptual interpretation and sensory perceptions.

We continue by discussing how modern digital tools can be used in physics teaching in order to involve students in the design and construction of experimental apparatuses to set up engaging activities connecting physics to students' own experiences, leveraging more on the cultural aspects of the laboratory experience, rather than on achieving accurate and precise values from the measurements. In particular, exploiting "beauty". The proposed approach fosters the integration of physics with other disciplines, even non-scientific ones, allows teachers to engage low achievers in science learning, and can be exploited to experiment with alternative assessment methods.

We then present inter-disciplinary aspects of physics education and teaching both, in schools and universities, with examples of actions about how to expand the teaching of physics to intercultural subjects, i.e., an interdisciplinary teaching of science and religion and the origin of the universe.

Finally, a new approach to physics—based on the simultaneous presentation of poems, and literary and philosophical passages is discussed with various examples (second level Master courses for teacher training, physics orientation courses for high school students and theatrical performances)—but referring above all to a new research work which aims to explore the importance of educational research for the history of physics (while the inverse relationship is usually underlined) by bringing together educational acquisitions and historical rigor based on hundreds of primary sources, and concerning quantum physics education.

2 Coding Augmented Reality Ad Hoc Software for the Laboratory of Physics and Communication Events (Based on the Presentation of Tommaso Rosi at the Symposium)

Inherently, physics demands a multimedia approach for comprehensive understanding. From its inception, physics has utilized words, numbers, symbols, formulas, sketches, images, tables, diagrams, and graphs for understanding and communication. Recently, the introduction of new media like pictures, videos, slides, simulations, visualizations, and electronic equipment has further enriched physics education. Throughout its history, the use of multimedia in physics education has aimed to make learning more accessible, interactive, and engaging, transforming the traditional lecture model into a more dynamic and student-centred learning experience. As technology continues to advance, multimedia tools are expected to play an even greater role in shaping physics education.

Over the past two decades, the growing availability and user-friendliness of VR and AR technologies have positioned them at the forefront of science and physics education research (Ibáñez and Delgado-Kloos 2018).

Various AR setups, based on smartphones, tablets, 3D cameras, projectors, glasses, and headsets, have become more accessible and user-friendly over time, thanks to decreasing costs and the advent of new, user-friendly software. The Physics Education Research (PER) group of the University of Trento has actively engaged in diverse projects, leveraging AR in laboratory settings and communication events, which we will discuss here below.

2.1 AR in the Laboratory

Over time, the PER group at the University of Trento has developed several different hardware/software setups for the use of AR in the physics laboratory: in this subsection, we will briefly present two setups. The first, known as "The Augmented Laboratory" (https://augmentedlaboratory.wordpress.com/), leverages a 3D camera and a projector. This system can create 3D reconstructions of the physical world, captured by the 3D camera. These reconstructions can be visualized in real-time as data is acquired or explored later through interactive playbacks. The setup is proficient in tracking multiple objects in 3D (Rosi et al. 2017) and projecting pertinent data directly onto and around the objects used in real physics experiments—what we term an "augmented experiment." In Fig. 1, on the left, you can observe two frames from a live video captured during an augmented experiment using this setup.

This setup has also unlocked the potential for new interactions between computer simulations and the real world, which we call "augmented simulations." An example of augmented simulation is the augmented ripple tank (see Fig. 1, right). This setup

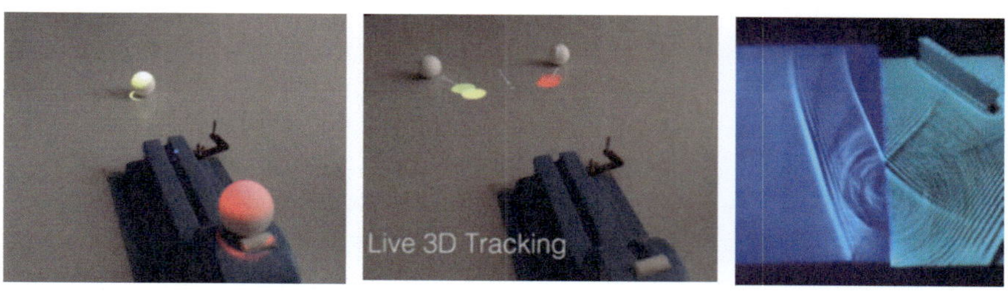

Fig. 1 Left: the live tracking of the position of two colliding spheres is shown; the results of the tracking can be recorded and seen again, also adding additional information such as vectors and graphs. Right: An "augmented simulation" of an AR ripple tank

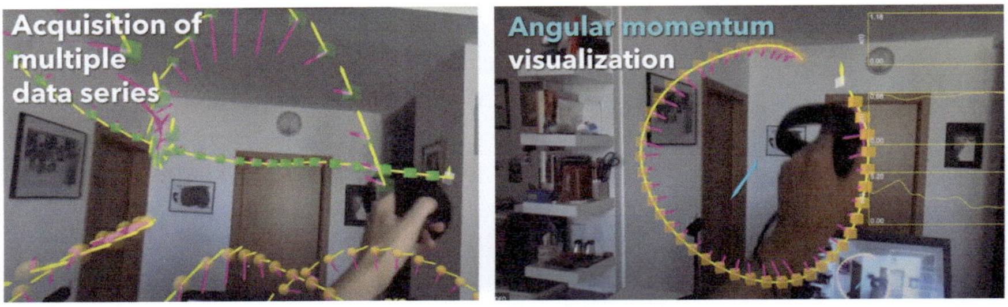

Fig. 2 Realtime 6 degrees of freedom tracking of a controller, as seen through the AR headset and on the computer monitor

enables the execution of various wave propagation experiments using physical objects to represent wave sources, barriers, lenses, and zones with varying wave speeds. The software can recognize and differentiate these elements. Consequently, it facilitates an intuitive and clear discussion of phenomena such as reflection, refraction, diffraction, the Doppler effect, and the sonic boom. We have extensively tested this setup in various educational and outreach contexts, ranging from formal settings like university and high school courses to informal environments such as public exhibitions.

The second main setup on which we have been working has been based on VR and AR headsets (Rosi et al. 2021). The idea behind their use is similar to the previous setup: being able to get quantitative measurements using the setup and visualizing the results of these measurements directly in the physical world (e.g., tracking the controllers and visualizing their 3D motion in real time using the headset, see Fig. 2) and being able to mix computer simulations and real experiments.

Again, we have been testing this setup both in formal and informal contexts, proving the high quality of quantitative measurements of these devices and the increase in students' engagement (Rosi et al. 2021).

2.2 AR for Cultural Understanding of Physics

Images have been pivotal in elucidating physics concepts, evolving from simple sketches to complex computer-generated simulations and visualizations. They serve not only as tools for researchers, educators, and communicators but also as a bridge between physics and art, inspiring novel explorations at the intersection of these fields, as exemplified by initiatives like the Arts at CERN Project (https://arts.cern/).

AR, with its unique ability to intuitively represent abstract concepts and meld real-time data acquisition with captivating visualizations, paves the way for new cultural expressions of physics. These expressions are inherently interdisciplinary, potentially intertwining physics with visual arts, theatre, literature, music, and dance.

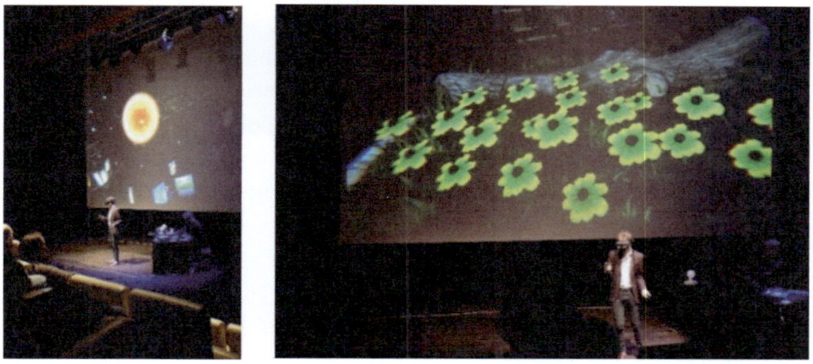

Fig. 3 Images from the augmented lecture "HyperVision"

The PER group at the University of Trento has ventured into this interdisciplinary domain, designing and implementing a performance titled "HyperVision" (https://leveluptrento.com/hypervision/). This performance falls into a new category of theatre known as "augmented lecture" (Brunello and Oss 2022), a format where scientific topics are conveyed through an interactive dialogue between a scientist and an artist. HyperVision integrates the physics of light and colours with digital visual arts experienced in AR, complemented by live electronic music tailored for the performance (Fig. 3). In the augmented lecture, topics are highlighted such as: How does vision work? How does our colour perception work? What and how do animals see? What distinguishes transparency from opacity? (https://www.arditodesio.org/teatro-scienza-hypervision.html). This blend between art and physics exemplifies the interdisciplinary potential of AR in contributing to the cultural understanding of physics.

3 Physics with Smartphones and Arduino (Based on the Presentation of Giovanni Organtini at the Symposium)

The study of physics is riddled with pitfalls, often due to misinterpretations of terms that physicists often borrow from ordinary language, but which, unlike the latter, take on a very special meaning in the discipline, as pointed out, e.g., by Brookes and Etkina (2009). An example of this difficulty is the use of the term "apparent force", to define a force which is not such, since it is measured. Perhaps there are no more concrete forces than apparent ones, from the point of view of our perception. Experiments involving non-inertial systems may be extremely effective in clarifying the meaning of such a definition.

Another source of confusion is the excessive levity with which mathematical models of physical phenomena are constructed, neglecting numerous details.

Although such details may be irrelevant to many physicists, it is important to explain to students how the model formulation was arrived at, and why. For example, the role of the string to which a mass is attached to a pendulum is that of keeping constant the distance of the mass from the axis of rotation. Suspending a real pendulum with two or more strings makes it clear that the "mathematical string" does not translate into the necessity of having just one point of attachment to the mass, thus clarifying the role of the model and how it is obtained.

Finally, physicists are accustomed to reformulating equations into such forms as to make them "beautiful." This is often done unknowingly, following a habit acquired during the study of physics. Beauty is not an end in itself but serves to make it easier to extract meaning from it. Just compare the equations written by James Clerk Maxwell in his "Treatise" (Clerk Maxwell 2011) with those used today, the result of Oliver Heaviside's work: from the former it is almost impossible to draw the idea that electromagnetic phenomena lead to the production of waves.

We believe, supported by Deslauriers et al. (2019), Crouch et al. (2004) and our experience, that a good way to overcome the difficulties outlined above is to awaken students' interest in STEM topics by means of their greater involvement in laboratory practice, which helps them better understand disciplinary content as well as stimulate critical thinking. By experimenting with phenomena themselves, students gain a better understanding of the meaning of terms; comparing a model with a self-constructed apparatus highlights the limitations of these models and provides clues as to what can actually be neglected; building one's own apparatus and analysing large amounts of data allows one to appreciate the role of "beauty" in science-based activities.

On the other hand, the direct involvement of students in non-scientific disciplines is often part of their education and serves to confront them with the difficulties of rendering complex concepts (emotions, moods, intricate plots) which is, in fact, the work of writers, painters, sculptors, and composers. In this way, students can appreciate a poem even more for how effectively the verses bring to mind certain moods; a painting for how it conveys the painter's gaze; a piece of music for how masterfully it describes a scene or situation. Similarly, the student should have the opportunity to experience how complex (and at the same time interesting and rich in insights) the task of extracting meaning (physics laws) from a mass of data is. After all, the work of a physicist is quite similar to that of a sculptor who extracts, from a shapeless block of marble, the forms that he can perceive to be hidden within. Indeed, the interest in science among young students is declining since time (Carpineti et al. 2011), as also reported in a publication of the EU dating back to 2012 (Forsthuber et al 2012). The executive summary of the document highlights the facts that integrated science teaching occurs mostly at the lower level of education, splitting it into separate subjects with age, that there are no specific support measures for low achievers in science, and that traditional assessment methods still prevail.

The current availability of easy-to-use, affordable digital tools, makes it possible to set up engaging activities connecting physics to students' own experiences, leveraging more on the cultural aspects of the laboratory experience, rather than to achieving accurate and precise values from the measurements.

3.1 Methods

Technology pervades every aspect of our daily lives, yet teaching often favours the use of traditional instruments and methods in the mistaken belief that they are easier to understand. In this way, physics seems to celebrate the past rather than strive for the future. But even with modern equipment, which is specifically designed to eliminate systematic effects as much as possible, students' perception of science is that of a disjointed, decontextualized collection of facts, unrelated to their own experience (Osborne et al. 2003).

Students cannot see the relevance of what they learn for their lives and interests, also because physics is often taught in terms of mathematical models that are necessarily crude approximations of the real world, thus avulsed from everyday experience. Often, there is no clear difference between physics and mathematics (Pospiech et al. 2019), and, indeed, physics is perceived as a kind of "dirty" mathematics.

Phyphox (Staacks et al. 2018) is an App that turns a phone into a real instrument for the measurement of acceleration, angular velocity, sound and light intensity, magnetic fields and pressure. With Phyphox and readily available materials, it is possible to design a number of experiments in which a lot of precise data can be collected in a short time (Kuhn and Vogt 2022). The same aim can be achieved using Arduino boards (Organtini 2018).

For the sake of brevity, we concentrate, here, on smartphones, considering Arduino boards as effective substitutes in experiments in which the use of smartphones is impossible or inappropriate (though often there are more specific reasons to adopt them).

In a Phyphox-driven teaching-learning process, beauty, often considered a topic meaningful only for fine arts, plays a role. Extracting meaning from data to create plots is similar to sculpting a sculpture from a block of marble, as outlined above. Similarly, creating a painting to express emotions felt during a sunset is comparable to a scientist writing an equation that concisely and precisely represents experimental observations.

Beauty, in physics, is not pursued only as an end in itself, but is functional to bring out more clearly the meaning of equations. We do compare the reconstruction of physics laws to poetry and literature. Moreover, beauty is a fundamental component of experiments' design: beautiful experiments are often more reliable. In other words, beauty, even in artworks, is not an end, but a means to understanding.

Many of the activities that can be set up using Phyphox and Arduino, as well as with any other readily available material, can be linked to the territory, strengthening the sense of community. Moreover, they easily engage low achievers, who can exploit and showcase their talents even to teachers who traditionally would have no way to appreciate them.

3.2 Considerations About the Evaluation of the Method

Several activities of the type outlined above are already taking place in Italy, as school-university collaborative projects, and are the basis of an experimental laboratory course, in which we do not just teach the technicalities of statistical data analysis, but deeply address the cultural aspects of doing experiments. It is still too early to draw conclusions on the effectiveness of this approach. Yet, our impression is that, while low achievers gain interest in learning science at school, even the best university students benefit: the answers they provide to the questions on the exam are more complete and deeper than the classical ones. We are currently defining metrics and methods to assess the validity of this method, which will be reported in a future publication.

4 Bringing a Contribution from Physics to the Global (Human) Culture (Based on the Presentation of Grzegorz Karwasz at the Symposium)

Physics is quite a general science: for the ancient Greeks it meant *nature* (φύσις). Aristotle in his work *Metaphysics* (Aristotle 1999) classified it as "one of three maximal theoretical sciences", together with mathematics and theology. According to him, physics deals with objects that are extended and exist in matter, mathematics deals with objects that are extended but need not exist in matter, and metaphysics (theology) deals with objects that are extended but cannot exist in matter.

The metaphysics is still present in modern physics: questions such as materialism, objectivity, determinism, causation, time, space, human free will, have been discussed by major physicists (Einstein 1934; Planck 1934; Heisenberg 1969) and are still under discussion (Crease 2017; Rovelli 2017). What about theology? The separation between questions about nature (astronomy, geography, physics) and Christian theology dates back to Bacon (1928) and Aquinas (1965). Present relations between science and religions were classified by Peters (2018) into ten categories, ranging from open conflict to some convergence and concordance. These polarizations are particularly acute among young people, who are in search of their personal identity: therefore, the discussion in school cannot be avoided.

Our brief discussion aims at giving a few suggestions and examples about how improving the social perception of physics, how making physics more attractive to young people, and how making an interdisciplinary interesting teaching. Always keeping in mind that cognitive didactics suggests starting from the thinking of the addressee, not of the physicists.

4.1 Constructing a Mutually Fruitful Dialog

As noted by an English experimental physicist, McLeish (2014), the conflict between religion and science harms both sides, and science even more. Can we extend the methodology of contemporary physics—experimental evidence, well-defined limits of knowledge, extrapolation procedures, etc.—to validate the dialogue between science and faith? Can we test this approach in education? Numerous authors have indicated that the search for subjects covering both sciences and cultural aspects rises the overall efficiency of teaching: "*The best reason therefore of including issues of religion in the science classroom would be, if, by so doing, learners gain a better understanding of science*" writes Reiss (2013). But the positive influence is for both sectors, as we discuss here.

We proved in practical actions how expanding the teaching of physics to intercultural subjects (Big Bang, genesis of mankind, physics in medicine and archaeology, mass spectroscopy and arts, etc.) allows to trigger both students' cultural interest and their specific notions interest in sciences.

As examples of our multi and interdisciplinary approach, we propose the subject of Big Bang/Creation and the origin of Homo sapiens as discussed in popular lessons/lectures on Science and Faith, by one of the authors (GK). Although addressed also to a broad audience, lectures were scientifically documented with original papers featuring breakthrough discoveries. Together with science, also notions from the Bible—to compare with—and some examples of extraordinary pieces of arts—to enlarge imagination—were presented.

While speaking about remote (and hypothetical) events such as the beginning of the Universe, the imagination of a scientist or an artist are similar. In the cycle of creation of the Solar System (about 1170–1179) in the Cathedral of Monreale (Sicily), we note a young God, keeping a technical drawing in his left hand, who, with his right arm outstretched, creates the light of the sun for day and the light of the moon for night. As a complementary picture to the previous one, we can sketch how experimental physics answers similar questions: authentic experimental results with discoveries of "exotic" particles, i.e., strange, charming, bottom quarks, with pictures and drawings taken from scientific journals (often from papers that led to Nobel Prizes), see Karwasz (2010).

Discussion with people is extremely fruitful, for instance concerning how in the history of man, questions about the origins of the universe have found answers from different viewpoints: this is the reason why science, philosophy and theology were born. These forms of knowledge do not conflict with each other, but only if they are respected in their methods of investigation and if they engage in dialogue to seek the "truth" and the good for humankind; if they are seen as complementary aspects to help man find the direction to give to the power offered by knowledge and progress.

Another example is given by the discussion concerning the birth of humankind. Always of key importance is the discussion of authentic scientific results in the didactical message: from the presentation of the detailed parentship between human mitochondrial (i.e., female) DNA in Europe, Asia and Africa as given in Olivieri et al. (2006). Similar research was also done for male DNA, so that we can show the common "Eve" and "Adam" overlap in time and space, some 120 thousand years ago.

4.2 Interdisciplinary Teaching on Science and Faith

As other examples, lessons that we describe have been developed over some twenty years, starting from extra-curricular lectures for secondary-school students in Trento (Italy) and local inhabitants (via different cultural associations) of the Trentino region. In Italy, these lessons have been recently (2022/23) offered as online courses for active teachers of primary and secondary schools in Sardinia, Puglia and Sicily and evolved into lessons in lower secondary schools. In Poland we implemented these subjects as in- and extra-curricular lessons in secondary schools, as extra-scholastic lectures (paid on voluntary basis) held at premises of the University, and, from 2014, as university courses, open to students of all faculties.

One of the requirements for this type of didactics, in times of TikTok and bla-bla, is that, in order to convince young people, original objects must be presented–hands-on experiments, ancient manuscripts or copies of scientific papers with significant discoveries. Another requirement nowadays is to be illustrative. We bring students to imaginative and comparative thinking. When you ask a student to draw an electron, she/he will sketch a point; when you ask about proton–it will be another small point. This is when we try to expand students' imaginations by making them aware that all the things that concern man can seem small from afar. "*Also the Earth seen from the space is a point, and here, we have the soccer club Inter quarrelling with Milan. Use your imagination! What is a point, and who are we?*". We then discuss the role of representations and modelling: we show pictures of quarks (i.e., particles that intrinsically cannot be extracted from either proton or neutron), as three-coloured (according to quantum chromodynamics), small "dogs", with their mass being proportional to the fourth power of their linear dimensions. We bring also real, touchable objects: in fact, the present, "virtual" world, interacting with real objects becomes a constructivist adventure (Karwasz and Wyborska 2023).

In our lessons (see: https://scire.umk.pl/workshops-in-the-year-of-copernicus/), students have also "lateral"-instructive experiences: they can touch an electron (a Greek one-cent, 2.7 g coin, called lepton), a heavy proton (almost ten pounds cube of iron) made of three quarks confined inside, a 3D postcard with the Big Bang, a beer can in a form of Volta's pile, a quark cheese from Alto Adige, a Belarusian folk artwork in wicker to illustrate the precession of Mercury's orbit in the General Relativity Theory, etc. The concepts of science are everyday words and objects, but what a difference in meaning!

The scope of such interdisciplinary teaching is two-fold: the main reason is to compare alternative points of view on subjects like the beginning of the Universe, chemical variety of elements, origin of life (in the interstellar space?). However, by-products of such lessons are even more important. Attracting the attentions of students from non-science faculties (and keeping their inter-disciplinary attention vivid, by including arts, economy, geo-sciences, history, literature into the contents of lessons), allow us to present important issues of modern science, like genetics, elementary particles, cosmology, and so on.

4.3 The Impact

As already mentioned, essentially the same contents are presented at different levels, with different discussion methodologies, and in different environments: most of the times triggered by invitations from outside the university. The large number of lessons delivered (and the variety of the didactic environments) leaves little time for testing the efficiency but in special circumstances such as university course. Evaluations given by students at the university course "Science and faith" in Toruń are very high. In 2022/23, out of 17 answers given (usually less than half of the students fill out the internet questionnaires, which are not obligatory), only single students gave "1", "3", "4" scores, and remaining evaluated the course with the higher mark, i.e., "5").

Comparing the didactical efficiency is difficult. However, for example, issues concerning the age of the Universe, of the Solar System and about a hypothetical mitochondrial Eve should constitute a cultural background already in the secondary school. Students of the first year in Toruń who followed the course, faced these contents and, from a disciplinary point of view, they were able to learn the contents (3 wrong answers out of 60 in total) much better than in a control group given by other "similar" similar first-year students who followed another approach and gave proportions pretty stochastic: 15 wrong answers over 45 in total.

The contents of lessons, especially those at the universe level yielded scientific papers (Karwasz 2015), and also a book (Karwasz 2019), being within 50 bestsellers in its category in Italy. The book is currently under translation to Polish and English, within two Templeton Foundation grants that our activity has been awarded.

The cultural context of teaching physics in an interdisciplinary way is important not only for education, but first of all for understanding our human, philosophical condition; as Richard Cheetham wrote: "*Both science and religion have a major, albeit variegated, influence across the world on how humankind understands the nature of reality. This in turn has a substantial effect on how we live—so the engagement between science and religion is a crucial conversation for the twenty-first century.*" (Cheetham 2019).

5 Quantum Physics and Culture—An (En)tangled Tale (Based on the Presentation of Luisa Lovisetti at the Symposium)

What do we gain from studying and teaching physics? Well, the ancient Greek scientist Euclid would have smiled at such an apparently silly question, simply dismissing us with a coin. However, even today, twenty-three hundred years later, this aspect is instead extremely topical and important, especially in the case of scientific disciplines, such as physics. Why, then, is studying and teaching physics so important? What culture (understood as the formation of man as a person, that is Greek *paidéia* and Latin *humanitas*) derives from it? In order to be able to answer, it is clear that, first of all, we need to grasp what conceptual, philosophical, historical, and general change in thinking lie behind the acquisition of new knowledge. We also need to understand that new knowledge changes the way we relate to the world around us, giving us new eyes to look at the universe, thus making us, in Feynman's words, *"atoms with consciousness…matter with curiosity."* (Feynman 1955, p. 14).

For what concerns our topic, physics is defined through the study of phenomena, the construction of theories, and the formulation of laws, written and described through the specific language of the discipline. Understanding laws and phenomena requires knowledge of their epistemological foundations and also of their historical development, in order to establish, through our language, the meaning of a phenomenon and its role within the theory in which it is interpreted.

Therefore language, at all levels of teaching and learning, should be an expression of a deep understanding and an acquired awareness that allows us to marvel at a physical law, to understand its cultural value, and to grasp its intrinsic greatness and beauty, giving physics a strong cultural meaning. In fact, physics and culture are strongly connected and interlaced, but this bond—which is concretely very close, but often somehow confused, intricate, and tangled in the eyes of most—needs to be unravelled and brought to light.

5.1 Physics and Culture

One of the main cultural aspects of physics lies in its drive to transcend common thought, to transgress the usual interpretation of reality, and also to understand that questions often considered significant for centuries are, instead, erroneous or even meaningless, then asking new ones.

If we truly believe that physics should be widely perceived as culture, we have to ask ourselves whether the images and worldviews that emerge from physics are useful for enriching our lives. It is therefore necessary for society to invest in the value of knowledge before being able to reap its benefits, which, in general, will not be only technological. In fact, the greatest cultural aspect of physics for society lies more in its ability to ask questions about nature than in the answers we get. And when the topics covered are significant for people (being perceived as useful, beautiful or fascinating), then interest, attention and desire to understand become greater and greater.

Therefore, one of the most important challenges for PER consists precisely in showing how physics is culture and how it has great cultural significance, by exposing how physical knowledge is created and used (Stadermann and Goedhart 2021). On this direction, the PER group at the University of Milan designed the two initiatives "Principles and Equations of Physics", for high school students and teachers (Giliberti et al. 2024) (about 150 and 650 participants respectively), where physics was the central pivot of the discussion but where variations on the theme (taken from philosophy, poetry, art, etc.) were addressed, to give a wide cultural framework.

Both courses were really appreciated: in fact, the number of participants has greatly increased from the first to the second edition of the laboratory. Moreover, all interviewed participants said that they greatly liked the approach used, and teachers unanimously requested other laboratories to be carried out in the same way and concerning other aspects of physics. We therefore took advantage of this suggestion, and we extended the same approach to other courses, concerning Quantum Physics.

5.2 A Cultural Framework for Quantum Physics

In fact, a cultural understanding of quantum physics needs even greater and more urgent effort (Michelini and Stefanel 2021), as is required by a society increasingly based on scientific knowledge and quantum technologies, in which the application of physical knowledge risks overtaking the true meaning of physics itself.

The cultural relevance of quantum physics is inherent in the fact that it has generated a deep revolution, having radically changed our way of looking at and understanding reality. Therefore, it is configured as an indispensable cultural heritage for every citizen. To this purpose, the PER group at the University of Milan has written and performed an augmented lecture with a cultural perspective of quantum mechanics (Giliberti 2019), and organised three courses for high school students and teachers:

a. The IDIFO21 master course "Principia: from Quantum Mechanics to Quantum Field Theory", in which the links between physics and other disciplines (art, literature, philosophy, etc.) were shown.
b. The course "Old (but Gold) Quantum Theory", held in two different editions, aimed at discussing the great cultural impact generated by the gradual birth of the old quantum theory, with a multidisciplinary vision, to help building a foundation for critical thinking. This course has its roots in a broader research work, developed within a PhD research program. In fact, we are exploring a further perspective for quantum physics, based on an accurate historical reconstruction (Lovisetti 2024).

c. The course “Principles and Equations of Physics III: Quantum Mechanics”, in which the old quantum theory served as an introduction to the formal construction of quantum mechanics, with a presentation that merged historical and physical aspects, building step-by-step a comprehensive conceptual framework (Giliberti and Lovisetti 2025; Lovisetti 2024).

The use of history in physics teaching is (or could be) quite common; on the contrary, the use of didactics within a predominantly historical work is quite unusual. Therefore, we made a critical analysis of the historical path that led to the birth of old quantum theory and early quantum mechanics, based almost entirely on primary sources (more than 800 documents, including papers, books, letters, etc.), and enriched with several pedagogical explanatory comments and notes, specifically designed for “inquiring minds”. The presence of didactic considerations within an historical account allowed to highlight widespread historical-conceptual misconceptions (Kragh 1992), while also providing a broad cultural framework, and thus facilitating the contextualization and the understanding of the physics discussed. In fact, the commented path is aimed at enhancing understanding and engaging the natural curiosity and the desire to make sense of the quantum issues, as well as at providing a concrete insight into the cultural significance of the first quantum revolution.

This project was not a mere historical tale, but rather aimed at exploiting the historical narrative to show how a physical theory (in our case, quantum mechanics) was born, and came to light, showing in which historical, social, philosophical context it has its roots, and trying to understand from which physics problems and from which experimental and conceptual aspects it originated (Giliberti and Lovisetti 2024). History alone cannot be enough, since our primary goal is to teach physics. To this purpose, quantum physics must be carefully unravelled and explained in its physical aspects, both experimental and theoretical, to highlight the peculiarities of this theory (Lovisetti et al. 2023), and to avoid giving it only a “mysterious” or oversimplified image.

The usual practice of reducing quantum physics to a simplified quantum tale (for example, in popularisation, or even in school textbooks) has several negative implications. First and foremost, isolating experimental and theoretical aspects from their context and disregarding their origins and scientific roots does not contribute to a profound comprehension of physics (Lovisetti 2024). Secondly, reducing quantum physics to that physics born from an alleged “crisis of classical physics” is a factual distortion that misrepresents the nature of science (Kragh 1992; Leone et al. 2018). Lastly, but certainly not least, simplifying the intricate, gradual, and challenging birth of quantum physics into a mere pseudo-historical narrative, focused on isolated and disconnected ad hoc models, not integrated in a comprehensive framework (as is unfortunately often done in high school curricula (Krijtenburg-Lewerissa et al. 2017; Stadermann et al. 2021) and textbooks, on an international scale) is undeniably a missed opportunity for recognizing that the evolution of physics is frequently far from straightforward, being an extremely tangled path, even with its failures and its critical moments (Kragh 1992; Leone et al. 2018; Lovisetti 2024). This is a deep cultural issue, because a simplified quantum tale gives us a distorted view of the physical world and also of quantum physics in the world.

Here the fundamental importance of "doing" good quantum physics, also shows its deep cultural value (in addition to its technological relevance): definitely an opportunity that physics education cannot afford to lose.

6 Discussion and Conclusions

The discussion following the previous proposals led us to some considerations and observations.

In the past, research in physics education has often focused on the most effective ways to obtain an adequate appropriation of disciplinary knowledge (see for example (Levrini et al. 2020)). It is a well-known example that the first principle of dynamics is difficult to understand in its meaning, so that, when asked what forces act on a body launched upwards, many students (including university students, who have taken physics courses) answer that, when the body is going up the force is directed upwards, when the body stops the force is zero, and when the body is going down the force is directed downwards. Research in physics education certainly knows useful methods and tools, for improving the learning of the principles of dynamics, which allow students to answer the previous question in a convinced "correct" manner. In general, however, these methods are not able to improve student's understanding of the cultural aspects of the principles of dynamics, and do not foster the student to frame them—historically, philosophically, and artistically—in a context that can fill him/her with wonder and make his/her change, at least in part, his/her *Weltanschauung* (Galili 2022; Giliberti et al. 2024). Only recently has the question arisen of how this PER know-how can help integrate disciplinary knowledge into a global framework and a global vision that is culturally significant for the student. But now the awareness of the necessity of this cultural integration is well present, so much so that the title of the GIREP-EPEC 2023 Conference, which this symposium refers to, is "Physics learning promoting culture and addressing societal issues"; however, much work still remains to be done.

The proposals of this symposium provide important research and action ideas to help build a personal network centred on the physical discipline but also involving many other areas of knowledge.

The real world of usual experiments is enriched through AR so that those who have these augmented experiences find themselves virtually living in the world of mechanics, looking at moving objects with the physicist's eye: the trajectories of objects become as "concrete" as the objects themselves, and so do the velocity and acceleration vectors. Abstract mathematical objects become a common experience, they become personal appropriation, part of everyone's way of looking.

In the introduction of this paper, we complained that poets no longer talk about science, but what do scientists do to make a new poem perceived? We have seen that new virtual tools, such as AR, can generate artistically meaningful situations and environments. Indeed, we probably can no longer talk about new things (like new physics) with old tools, and we need to learn how to talk to artists by showing them new opportunities. From the opposite perspective, we must also seize the purely disciplinary opportunities provided by art. How beautiful are the "old" scientific instruments in wood and brass, well-crafted by the hands of artisan artists! So why not propose a new pact between craftsmanship and science for schools too? Technology can successfully combine with design for an artistically and scientifically attractive work. This is what is proposed, for example, by the Design for Physics project (partnership between the European Institute of Design of Rome and the Physics department of Sapienza University of Rome) which proposes designing teaching tools intended to improve the learning of physics in synergy with the help of smartphones and Phyphox (https://www.ied.it/news/design-for-physics).

Culture is also a lack of fear, it is a set of tools to face the world, so why not also try to draw parallels with other interpretative systems other than physics? Knowledge also deepens through awareness of what is not science, and at the same time, through the awareness that science is not the only perspective worthy of attention. The notable proposal we have discussed regarding "Science and faith" is of this type. Alternative points of view foster scaffolding physics knowledge into cultural networks.

For a cultural presentation of physics, we need to go and get people where they are and not where we would like them to be. This is the guiding spirit behind the proposal concerning the old quantum theory, a proposal inserted in a cultural environment rich (we underline rich, not enriched) with poems, philosophical and even musical considerations. When the inevitable specialization of knowledge is recomposed into a significant network of relationships and suggestions, the disciplinary foundations are strengthened, and joy literally enters the hearts. This is what a teacher who participated in the IDIFO21 master course anonymously wrote, concerning the step-by-step construction of quantum mechanics (from its historical foundations) proposed during the course. Here is an example: *"I can't help but write to tell you that your lessons are truly spectacular. In reality, it is difficult to find a suitable adjective to describe them because, in addition to having an excellent preparation, you have an uncommon ability to disseminate. I have a degree in physics, and I have always looked for a wider scope than the presentation of the mere topics: that attraction for art, music, philosophy, poetry which, in the eyes of others, seems to be contradictory for those who should only deal with laws and formulas, [...] has been amply repaid by your way of explaining topics, even the most complicated ones."*

In conclusion, the contributions presented here show that closer attention should be devoted to cultural understanding of physics and new instruments and methods are needed to explore possible new ways to gain an "active cultural learning".

References

Aquinas T (1965) Summa Theologiae. Eyre & Spotiswoode, London
Arditodesìo website. https://www.arditodesio.org/teatro-scienza-hypervision.html. Accessed 20 Jan 2024
Aristotle (1999) The Metaphysics. Penguin Classics, London, Book VI, 1025, 20
Art at CERN website. https://arts.cern/. Accessed 20 Jan 2024
Bacon R (1928) Opus Majus. Volumes 1 and 2. Translated by R B Burke. University of Pennsylvania Press, Philadelphia
Brandi MC, Cerbara L, Misiti M, Valente A (2005) Giovani e scienza in Italia tra attrazione e distacco. JCOM 4(2):A01–A015. https://doi.org/10.22323/2.04020201
Brookes DT, Etkina E (2009) "Force" ontology, and language. Phys Rev ST Phys Educ Res 5:010110. https://doi.org/10.1103/PhysRevSTPER.5.010110
Brunello A, Oss S (2022) Augmented lectures: a unique resource to inspire, outreach and understand science science & theatre: communicating science and technology with performing arts. In: Weitkamp E, Almeida C (eds) Leeds: Emerald Publishing Limited, pp 143–150
Carpineti M, Cavinato M, Giliberti M, Ludwig N, Perini L (2011) Theatre to motivate the study of physics. JCOM 10(01):1–10. https://doi.org/10.22323/2.10010201
Cheetham R (2019) Lies, damned lies, science, and theology: why everyone needs to know the truth about science and religion. In: Billinsgley B et al (eds) Science and religion in education. Springer, Cham, pp 227–238
Clerk Maxwell J (2011) A treatise on electricity and magnetism. Cambridge University Press, Cambridge
Crease RP (2017) Philosophy of physics. Institute of Physics Publishing, Bristol
Crouch C, Fagen AP, Callan JP, Mazur E (2004) Classroom demonstrations: learning tools or entertainment? Am J Phys 72(6):835–838. https://doi.org/10.1119/1.1707018
David K (2012) How the hippies saved physics—science, counterculture, and the quantum revival. W.W. Norton & Company, New York—London
Deslauriers L et al (2019) Measuring actual learning versus feeling of learning in response to being actively engaged in the classroom. PNAS 116(39):19251–19257. https://doi.org/ https://doi.org/10.1073/pnas.1821936116
Einstein A (1934) Mein weltbild. Ullstein, Amsterdam
Feynman RP (1955) The value of Science. Eng Sci 19:13–15
Forsthuber B, Horvath A, Almeida Coutinho AS, Motiejūnaitė A, Baïdak N (2012) Science education in Europe: national policies, practices and research. Eurydice, Brussels
Galili I (2022) Scientific knowledge as a culture: the pleasure of understanding. Springer, Cham
Giliberti M (2021) Stories and theatre for teaching physics at primary school. Educ Sci 11(11):696–709. https://doi.org/10.3390/educsci11110696
Giliberti M, Lovisetti L (2023) Quousque tandem? G Fis 64(3):191–211. https://doi.org/10.1393/gdf/i2023-10527-2
Giliberti M, Lovisetti L (2024) Old quantum theory and early quantum mechanics: a historical perspective commented for the inquiring reader. Springer, Cham
Giliberti M, Lovisetti L (2025) Teaching quantum mechanics within a nature-of-science framework. Am J Phys 93(2):172–179. https://doi.org/10.1119/5.0210636
Giliberti M (2023) Theatre: the other side of physics new challenges and opportunities in physics education. In: Streit-Bianchi M et al (eds) New challenges and opportunities in physics education. Springer, Cham, pp 175–189
Giliberti M, Albanese F, Lovisetti L (2024) Principles and equations of physics: a multidisciplinary laboratory. JPCS 2750:012028
Giliberti M (2019) There are no things inside things. In: McLoughlin E, van Kampen P (eds) Concepts, strategies and models to enhance physics teaching and learning. Springer, Cham, pp 203–213. https://doi.org/10.1007/978-3-030-18137-6_18

Heisenberg W (1969) Der Teil und das Ganze. Gespräche im Umkreis der Atomphysik. Piper, München
HyperVision webpage. https://leveluptrento.com/hypervision/. Accessed 20 Jan 2024
Ibáñez MB, Delgado-Kloos C (2018) Augmented reality for STEM learning: a systematic review. Comp Educ 123(6):109–123. https://doi.org/10.1016/j.compedu.2018.05.002
IED Design for Physics website. https://www.ied.it/news/design-for-physics. Accessed 08 Feb 2024
Karwasz GP, Wyborska K (2023) How constructivist environment changes perception of learning: physics is fun. Educ Sci 13(2):195. https://doi.org/10.3390/educsci13020195
Karwasz G (2010) Physics is Fun—come si costruisce la Fisica Moderna. In: Michelini M (ed) Fisica Moderna per la Scuola. LithoStampa, Udine, pp 193–201
Karwasz GP (2015) Il costante progredire della frontiera tra teologia e scienza: "Fisica". Sci et Fides 3(1):61–86. https://doi.org/10.12775/SetF.2016.012
Karwasz GP (2019) Scienza e fede. Un breve manuale. Aracne Editrice, Roma
Kragh H (1992) A sense of history: history of science and the teaching of introductory quantum theory. Sci Educ 1:349–363. https://doi.org/10.1007/BF00430962
Kragh H (1999) Quantum generations: a history of physics in the twentieth century. Princeton University Press, Princeton, NJ
Krijtenburg-Lewerissa K, Pol HJ, Brinkman A, van Joolingen WR (2017) Insights into teaching quantum mechanics in secondary and lower undergraduate education. Phys Rev Phys Educ Res 13:010109. https://doi.org/10.1103/PhysRevPhysEducRes.13.010109
Kuhn J, Vogt P (eds) (2022) Smartphones as mobile minilabs in physics. Springer, Cham
Leone M, Robotti N, Verna G (2018) "Rutherford's experiment" on alpha particles scattering: the experiment that never was. Phys Educ 53(3):035003. https://doi.org/10.1088/1361-6552/aaa353
Levrini O, Levin M, Fantini P (2020) Fostering appropriation through designing for multiple access points to a multidimensional understanding of physics. Phys Rev Phys Educ Res 16:020154. https://doi.org/10.1103/PhysRevPhysEducRes.16.020154
Lovisetti L (2024) "Can old-quantum-theoretical description of physics be rendered coherent?": A pilot experimentation for high schools. Il Nuovo Cimento C 47(5):277. https://doi.org/10.1393/ncc/i2024-24277-6
Lovisetti L, Organtini G, Giliberti M (2023) Inducing the construction of formal axioms of quantum mechanics and fostering their comprehension by high school students: the effectiveness of a conceptual approach. Il Nuovo Cimento C 46(6):200. https://doi.org/10.1393/ncc/i2023-23200-1
McLeish T (2014) Faith and wisdom in science. University Oxford Press, Oxford
Michelini M, Stefanel A (2021) Approaches on T/L quantum physics from PER literature. In: Jarosievitz B, Sükösd C (eds) Teaching-learning contemporary physics from research to practice. Springer, Cham, pp 3–17. https://doi.org/10.1007/978-3-030-78720-2_1
Olivieri A et al (2006) The mtDNA legacy of the levantine early upper palaeolithic in Africa. Science 314(5806):1767–1770. https://doi.org/10.1126/science.1135566
Organtini G (2018) Arduino as a tool for physics experiments. J Phys Conf Ser 1076(1):120–126. https://doi.org/10.1088/1742-6596/1076/1/012026
Osborne J, Simon S, Collins S (2003) Attitudes towards science: a review of the literature and its implications. Int J Sci Educ 25(9):1049–1079. https://doi.org/10.1080/0950069032000032199
Peters T (2018) Science and religion: ten models of war, truce and partnership. Theol Sci 16(1):11–53
Planck M (1934) Where is science going? Norton, New York
Pospiech G, Michelini M, Eylon BS (eds) (2019) Mathematics in physics education. Springer, Cham
Reiss MJ (2013) Religion in science education. In: Mansour N, Wegerif R (eds) Science education for diversity. Theory and Practice. Springer, Dordrecht. https://doi.org/10.1007/978-94-007-4563-6_15
Rosi T, Onorato P, Oss S (2017) Multiple object, three-dimensional motion tracking using the Xbox Kinect sensor. Eur J Phys 38(6):065003. https://doi.org/10.1088/1361-6404/aa8183

Rosi T, Perini M, Onorato P, Oss S (2021) Commercial virtual reality headsets for developing augmented reality setups to track three-dimensional motion in real time. Phys Ed 56(2):025016. https://doi.org/10.1088/1361-6552/abd5a2

Rovelli C (2017) L'ordine del tempo. Adelphi, Milan

Rustichelli F, Stefanon M (2012) The European project "Immersion in the science worlds through the Arts" Il nuovo Saggiatore 5–6:46–49

Staacks S, Hützl S, Heinkel H, Stampfer C (2018) Advanced tools for smartphone-based experiments: phyphox. Phys Educ 53(4):450–459. https://doi.org/10.1088/1361-6552/aac05e

Stadermann HKE, Goedhart MJ (2021) Why and how teachers use nature of science in teaching quantum physics. Phys Rev Phys Educ Res 17:020132. https://doi.org/10.1103/PhysRevPhysEducRes.17.020132

Stadermann HKE, van den Berg E, Goedart MJ (2021) How high schools teach quantum physics—a cross-national analysis of curricula in secondary education. J Phys Conf Ser 1929:012045. https://doi.org/10.1088/1742-6596/1929/1/012045

The Augmented Laboratory website. https://augmentedlaboratory.wordpress.com/. Accessed 20 Jan 2024

Tolstrup Holmegaard H, Møller Madsen L, Ulriksen L (2014) To choose or not to choose science: constructions of desirable identities among young people considering a STEM higher education programme. Int J Sci Ed 36(2):186–215. https://doi.org/10.1080/09500693.2012.749362

Workshops in the year of Copernicus website. https://scire.umk.pl/workshops-in-the-year-of-copernicus/. Accessed 16 Feb 2024

Pre-service Teacher Education and University Education

School-University Collaborations: An Exploration of Issues and Possible Solutions

David Sands, Sarah Zoechling, C. Damián Rodríguez-Fernández, Mattia Ostinato, and Carmen Martin Valderrama

Abstract This EPS-sponsored symposium followed on from a Round Table conducted at the EPS forum in Paris in 2022. We consider physics teaching in schools from a Europe-wide perspective and examine issues in teacher professional development. We present views from the Physics Education Division of the European Physical Society, the Young Minds initiative of the European Physical Society, various sections of which actively work with schools around Europe, and the Teacher and Student Programmes section at CERN, which has been conducting many different Physics Education Research projects over the last ten years. We discuss in particular the origin of the research-practice gap in schools and consider ways in which schools might better utilise the outcomes of Physics Education Research. We explore briefly the possibility of a European initiative in school-university collaboration.

Keywords In-service teachers · Professional development · Research-practice gap · Professional learning communities · Outreach · Young minds

---

D. Sands (✉)
Independent Physics Education Consultant, East Yorkshire, UK
e-mail: david.sands@physicseducationinternational.com

S. Zoechling
European Organization for Nuclear Research (CERN), Meyrin, Switzerland
e-mail: sarah.zoechling@cern.ch

C. Damián Rodríguez-Fernández
Departamento de Física E Ciencias da Terra, Universidade da Coruña, Coruña, Spain
e-mail: Damian.rodriguez@usc.es

C. Damián Rodríguez-Fernández · M. Ostinato · C. M. Valderrama
European Physical Society, EPS Young Minds Action Committee, Mulhouse, France
e-mail: mattia.ostinato@ub.edu

C. M. Valderrama
e-mail: c.martin@nanogune.eu

Z. Ješková and J. Hanč (eds.), *Advancing Physics Education*, Challenges in Physics Education, https://doi.org/10.1007/978-3-031-94207-5_12

1 Introduction

The immediate impetus for this EPS symposium grew out of a Round Table at the EPS Forum, Paris 2022, in which a panel of distinguished educational physicists discussed the issue with interested members of the physics community, but the idea has its roots in the extensive literature on physics education research, particularly as it relates teacher education.

Two webinars hosted by the University of Malta in November of 2020 and 2021 dealt with this last issue. Organised under the aegis of GIREP, they were originally intended as a single seminar to be run in situ, but the outbreak of the COVID pandemic and the restrictions placed on international travel during 2020 led to the event in November of that year being reduced in scope and held entirely on-line. The intention was to run the seminar in situ the following year, but that was also held online. The output of both seminars now comprises an extensive record of the most pressing issues in teacher education and development (Borg Marks et al. 2022, 2023), which might be expected to feature in some form in collaborations between schools and universities. Some of the issues dealt with in Malta include the use of multimedia and technology in teaching, laboratory practice, initial teacher education and continuing professional development, all of which could feature in different collaborations.

Some of the outcomes of the Malta webinars are unsurprising, but others give pause for serious reflection. The leaders of working group six,[1] on in-service teacher professional learning strategies, ran an informal survey among delegates to the webinar to get a sense from teachers themselves of their professional development needs. There were 38 respondents from all over Europe and some from the United States and Chile. The vast majority (over 80%) taught physics as a single subject and over 90% taught in secondary schools. The length of service was broken down into 4 crude intervals from 0–10 years, 10–20 years, 20–30 years and over 30 years and all four groups were well represented. Therefore, we had a view of professional developments needs from across the entire spectrum of physics teachers, from relatively new to very well-established, with over 50% in the combined mid-career group spanning 10–30 years. We asked delegates whether they felt they needed professional development in three specific areas: content/science topics, pedagogy and classroom practice, with teachers left to define these terms for themselves. Over 70% wanted development in content, nearly 53% wanted development in pedagogy and over 81% wanted to develop their classroom practice. With only 26% of respondents having been teaching for 10 years or less, it is clear that large numbers of well-established teachers feel the need for in-service professional development in a range of areas, but especially classroom practice. We also asked about development opportunities in these three areas and it is clear that for a significant number of teachers, such opportunities are lacking.

The results of this rather informal survey have been supported to some extent by a more formal study of teacher professional development by Horvat et al. (2022) in which enhancement of both the curriculum and concepts and models were considered

[1] Eilish McLoughlin, Wim Peeters, David Sands and Jenaro Guisasola.

as the two most important needs among a total of 10 items considered in the study. Enhancement of the curriculum related specifically to particle physics in Horvat's study and although less definitive, our rather informal survey of delegates to the Malta webinar is perhaps more general and therefore more informative in as much as it provides an insight into possible areas of school-university collaboration. If we take content, for example, there was a clear need across a large number of teachers with presumably very different experiences. Whilst there is a clear role for disciplinary expertise in helping teachers to understand areas of so-called "modern physics", such as quantum mechanics (Merzel et al. 2022; Michelini and Stefanel 2022), Einsteinian relativity (Kersting 2022; Postiglione and de Angelis 2022), and, as discussed, particle physics (Horvat et al. 2022; Horvat 2022), attention should not be restricted to just these. One question that was not asked in this survey concerned the educational background of the respondents. We do not know what fraction, if any, of the respondents have a first degree in physics or a qualification in an unrelated science, such as biology. For non-specialists, development of content could relate to any aspect of physics.

Content cannot always be neatly separated from pedagogy and the two come together in the concept of PCK, or pedagogical content knowledge. This was emphasized by de Winter and Airy (2019) in an article entitled, "What makes a good physics teacher?" Quoting Shulman, they describe PCK as going "beyond knowledge of subject matter per se to the dimension of subject matter for teaching". The relevance of expertise within universities for non-specialist teachers is obvious. Take Newton's laws of motion, for example. Students' difficulties with these are well known and there are too many papers to cite, but it might not be obvious to a non-specialist that Newton's first law is simply a consequence of the second. Quite why Newton established it as a law on its own is not a matter for discussion here, but it is clear that a nett force is needed to change the velocity, that is, the speed and/or the direction of motion. Two things follow immediately from this. First, if there is no nett force the velocity will not change, which is the first law. Secondly, as long as the nett force is not zero, the velocity will continue to change, which is at odds with the very common misconception among students that a nett force is required for a constant speed. It would seem sensible at first sight to teach Newtonian mechanics with reference to the second law first, but whether this is in fact appropriate can only be decided after research into student learning (as discussed for mechanics, e.g., by (Spatz and Wilhelm 2021). This raises the question as to whether teachers should become researchers themselves or rely principally on the research literature.

One aspect of physics teaching in which sequencing is crucial is in experimental work. Working group 2 in the November 2020 webinar in Malta explicitly recommended collaboration between schools and universities as a way to improve practical instruction at the secondary school level. Experiment and observation are the bedrocks upon which science has been built and teaching practical classes is not just a way to cement knowledge of concepts, but of teaching the scientific method, that is, the Nature of Science (Lederman 2007). Again, though, the best way to teach practical work should be based firmly on research.

We come to the crux of school-university collaboration. The preceding has argued that there is much that can be transferred from physics education research conducted at universities into schools. The question naturally arises as to why the findings of such research are not currently being transferred into much of school practice. Is it that they are they not being taught in either pre-service teacher education or in-service teacher education? Or is there some other reason connected with the need for on-going support and development? A genuine collaboration between schools and universities would offer a long-term relationship that is mutually beneficial to both organisations and ways of establishing such connections will be explored in this symposium.

2 The Symposium

There were 3 talks in this symposium aimed at addressing the issue of school-university collaborations and trying to identify a possible solution that goes beyond individual arrangements, offers a long-term, mutually beneficial relationship and may be susceptible to support from trans-national bodies like the EPS or CERN. David Sands, at that time chair of the Physics Education Division of the European Physical Society (EPS), introduced the issue by expanding on the themes introduced above. Sarah Zoechling, a physics education researcher based at CERN, described physics education research from the perspective of a huge trans-national organisation with an extensive network of teachers developed through its teacher training programmes and its educational offers for high school students. C. Damián Rodríguez-Fernández presented an overview of the activities of the EPS Young Minds initiative, which aims to give young physicists, which includes everyone from the undergraduate student to the postdoctoral researcher, experience of outreach, research and educational leadership among other things. The symposium then opened up a discussion to allow attendees the opportunity to contribute with a view to helping define possible initiatives. The outcome of these discussions is summarised at the end of the paper.

2.1 The Research-Practice Gap in Secondary Education

It has long been recognised that a gap exists between physics education research in universities and the incorporation of the outcomes into teaching practice within schools. As reported by Vanderlinde and van Braak (2010), there is limited empirical data available, but widespread recognition that there should be more cooperation between researchers and practitioners. In this paper, we are interested specifically in physics education research (PER) and the barriers to enabling the implementation of its outcomes in teaching practice.

PER means different things to different people and there are different aspects. First, there are the students, who might be the subjects of such research. Of primary interest then are the cognitive and affective factors that are associated with learning, though often the affective is neglected in favour of the cognitive (Broekkamp and van Hout-Wolters). In short, the focus is on what students have learned and what are the barriers to learning, such as misconceptions. The students might be in schools or in higher education. We are, of course, interested in school children in this paper. Not only are the methods of teaching different between schools and universities, but the cognitive demands of the physics and associated mathematics are also different. Secondly, there are the universities themselves. The majority of physics education researchers, regardless of whether their research is focused at the school or university level, will be academic researchers. Academic researchers might take as their focus university students, school students, or even teachers, either in-service or pre-service.

Schools and universities naturally lend themselves to a mutually beneficial relationship. Teachers do not necessarily have the time or the knowledge to conduct extended research projects, whereas academic researchers will not have access to young learners without the cooperation of teachers. Universities also educate teachers and develop research-based teaching approaches and materials for schools. It is perhaps surprising, then, that research findings are not taken up by schools more widely and that there is perceived to be a gap between educational research and practice.

PER can be "understood as research on education and research for education" (Biesta 2007; Bauer and Fischer 2007). Biesta (2007) explores what is meant by the term "research" and what it means for the outcomes to influence educational practice. He makes the point that there is a common assumption that research should produce knowledge which in turn can be used in educational practice, but, summarizing work carried out in the Netherlands in the 1980s, Biesta makes the point that there are two ways of looking at the possible impact of research on educational practice. The first is through technical, or instrumental knowledge, which Biesta defines as, "knowledge that indicates what one should do in order to achieve a particular result or outcome", such as effective teaching practices or effective assessment practices. The second way that research can inform educational practice is through a cultural role, which Biesta defines as, "the provision of different interpretations and understanding of educational practice". Rather than providing a direct route to changes in practice through the implementation of definite outcomes, the cultural role helps practitioners acquire a different understanding of their practice. Biesta cites the widespread uptake of constructivist ideas, but equally one can think of the widespread adoption of inquiry-based learning as an example of the change of culture that has come about through the influence of research on educational practice.

Biesta concluded that, from this perspective, research has had "a massive impact on educational practice". Nonetheless, citing the literature reviewed by Broekkamp and Van Hout-Wolters (2007), Biesta also concludes that there is a widespread perception that educational research should be instrumental and have a direct impact on classroom practice and discussed in detail what this means. We shall consider this later. In some sense, the instrumental view is entirely natural. As Biesta put it: "Educational research is, after all, never simply research on education but always in some sense also research for education", but the question of what research can do for education is complex. Biesta argues that learning is a complex process in which learners interpret information to make sense of what they have been taught and that in consequence there is not, and perhaps never can be, a causal relationship between teaching and learning. That is to say, education cannot be reduced to a technology in which theoretical knowledge and skills can be applied to a given situation to produce a defined outcome. Effective teaching requires professional judgement, not the "mechanical application of research-based rules for action".

All this impacts on the nature of the research-practice gap. Discussing the work of Bauer and Fischer (2007) Biesta considers three different kinds of gap: a communications gap, a "loop" gap, and an interaction gap. A communications gap arises when there is a perceived lack of communication from research to practice, which is to say that practitioners may be unaware of the kinds of research being conducted, its outcomes and potential impact on their practice. A loop gap is so-called because it describes the situation in which the "loop" between researchers and practitioners has not been closed and researchers fail to address the issues of concern to practitioners. The interaction gap describes the lack of practitioner involvement "throughout all the stages of the research process". In the context of this symposium, different kinds of school-university collaborations might be envisaged, depending on the type of problem to be solved.

A similarly complex view of the research-practice gap was put forward by Broekkamp and van Hout-Wolters (2007), who argued that the gap is often seen from a mono-causal perspective. A single cause implies a single solution, with other possibilities then being excluded. In order to arrive at a more rounded view, Broekkamp and van Hout-Wolters (2007) undertook a literature review, held a symposium and distributed a questionnaire. From a literature review, four basic problems were identified, but each problem is not in itself characterised by a single cause. The reader is referred to Broekkamp and van Hout-Wolters (2007) and the references cited therein for details of the four problems, which can be characterised as:

1. "Educational research yields only few conclusive results". Among the causes cited under this problem are: neglect of the affective in favour of cognitive aspects, which ignores the complexities of the learning process; a lack of quality among researchers, which might be due to a lack of expertise or control over the educational environment; researchers neglecting to make use of prior knowledge and theories; reward systems that encourage competition rather than collaboration.
2. "Educational research yields only few practical results". This follows on largely from the above. If results are not conclusive, they are unlikely to be considered practical. This also relates to the "loop" gap described by Biesta. If the issues investigated by researchers seem remote and of little interest to practitioners, they are hardly likely to seem practical, even if they are conclusive. The list of causes under this heading is quite extensive and rather than attempt to summarise them, we refer instead to the words of Broekkamp and van Hout-Wolters (2007): "In other words, there are too few easily accessible results available which, when applied, could be expected to make a substantial contribution to the enhancement of education".

3. "Practitioners believe that educational research is not conclusive or practical". Citing Gore and Gitlin (2004), the claim here is that "teachers considered research to be inaccessible, irrelevant, and unreliable". A similar claim is made in respect of politicians.
4. "Practitioners make only little (appropriate) use of educational research". The negative perceptions alluded to above play a significant role in this, but there are other factors at work. Practitioners are neither trained nor supported to access the research literature. This can lead not only to a failure to access research, but when research is accessed it may be used inappropriately.

Having discussed at length the issues revealed through a survey of the literature, Broekkamp and Van Hout-Wolters (2007) then considered possible solutions contained in the literature and grouped the strategies into four models.

1. The Research Development Diffusion (RDD) model. Practitioners conduct research based on fundamental research in order to assess whether the findings from the academic research can be applied to "realistic educational situations". This model requires mediators to identify appropriate research for practitioners (the development phase) and also select the results for diffusion more broadly to the profession.
2. The Evidence-Based Practice (EBP) model. This is similar in some respects to the RDD model, but whereas the RDD model might encompass broad research results, including theoretical studies, the EBP model is concerned exclusively with empirical evidence as the basis for identifying effective methods. Practitioners can either conduct research themselves within their own educational context or rely on methods which have been shown by other practitioners to be effective.
3. The Boundary-Crossing Practices (BCP) model. Tasks from different professional domains are combined. The example is given of teachers working with policy makers and researchers to design innovative practices and the conducted research contributes to professional development activities.
4. The Knowledge Communities (KC) model. A group of people sharing the same interest can benefit from each other's expertise in the generation of new knowledge. There is no set structure or membership for such communities, nor is there a preferred method of interaction, but some models work better than others.

These four models also relate to Biesta's different conceptions. The RDD and EBP models involve a uni-directional flow of information from research to practice and clearly depend on having effective communication channels, whereas both the BCP and KC models are interactive. Whichever model might be chosen in any given context, however, there are a host of other issues to resolve and Broekkamp and Van Hout-Wolters (2007) discuss these in some depth.

Following on from their survey of the literature, Broekkamp and Van Hout-Wolters issued a questionnaire prior to running a symposium. The questionnaire was distributed widely in the Netherlands and the respondents included educational researchers, teachers, teacher trainers, policy makers and students. Table 1 shows a selection of statements from the survey together with the number of respondents answering the particular question (N), the mean score (M), on a scale of 1–7, with 7 indicating strong agreement, and the standard deviation on that score (SD). These statements are selected because they support the perception reported in the literature, and summarised by Biesta (2007), of a wide gap between research and practice.

Biesta (2007) concluded that the instrumental view of research, namely that it should directly transform practice by providing evidence-based methods of teaching and assessment, is probably unattainable, but that the cultural view, in which research has changed the educational landscape, has brought about a significant change. More recent studies on the research-practice gap support this view. As with the study by Broekkamp and Van Hout-Wolters (2007), these have been conducted within particular national or educational contexts, but the findings are remarkably consistent both with the findings in Broekkamp and Van Hout-Wolters (2007) and across the different studies. For example, in Finland educational research has featured heavily in the education of teachers since the 1970s. As reported by Koski et al. (2023), "during the initial teacher education, Finnish student teachers familiarise themselves with educational research, practise research skills and become acquainted with different research methodologies. Thus, they should have the necessary skills to make use of educational research when they start working as teachers." Nonetheless, Koski et al. found that teachers "reported engaging with research literature rarely, and even less with actual research activities" and that they perceive educational research to have little practical value. Echoing the findings by Broekkamp and Van Hout-Wolters (2007), it is "not relevant to them, and they lack the skills to utilise it in their daily teaching tasks" (Koski et al. 2023).

Table 1 Selected statements from the survey by Broekkamp and Van Hout-Wolters (2007) with the mean score (M) and standard deviation (SD). These statements support the wide-spread perception of a research-practice gap

Statement	N	M	SD
Educational practitioners and policy-makers have a low opinion on educational research	151	4.76	1.51
Reports on education research are inaccessible	150	4.55	1.55
National educational policy is not based on research	144	5.13	1.55
Teaching materials (textbooks, tests, etc.) are not based on research	142	4.53	1.56
Educational practitioners and policy-makers do not get the time and the means to use the results of educational research	144	5.40	1.28
Educational practitioners and policy-makers very rarely cooperate with researchers	141	3.45	1.58
Current educational research could contribute much more to the field than is generally assumed	148	5.68	1.43

Singapore is another country with high level initiatives to encourage the use of research in teaching, with research "specifically funded to produce rigorous knowledge relevant to practice" (Baildon and Ong 2022). Nonetheless, most teachers in the Singaporean study also perceived research as having a low utility. In fact, the situation described by Baildon and Ong (2022) is complex. Access to research among teachers is encouraged but this mainly occurs within professional learning teams (PLT) because time set aside for work in PLTs also provides a time and a place to carry out such an activity. However, the research is generally accessed for specific purposes and is generally regarded as not providing something useful that can immediately affect practice. Even so, many teachers expressed the aspiration that "research could inform practice and support education system goals in instrumentalist terms".

These two studies, by Koski et al. (2023) and by Baildon and Ong (2022), reveal a tension among teachers between what they hope educational research can provide and a belief that in practice it has little immediate impact in the classroom. Koski et al. reported, for example, that teachers generally fell into one of three broad categories. The first comprised those who were broadly familiar with the notion, and broadly accepting of the aims, of educational science, research and theory (referred to by Koski et al. as ESRT) in the cultural sense defined by Biesta in (2007). The second category comprised those who regarded ESRT as primarily instrumental, and the third category consisted of those who considered ESRT as insignificant. However, even those teachers in the first category who Koski et al. found had an optimistic view of ESRT, like those teachers in Baildon and Ong's study who hoped that educational research would change practice, regarded educational research having a low utility.

It seems that teachers in general find it challenging to engage with educational research through a lack of time and opportunities, regardless of their national context. An extensive study in Australia as part of a project to develop a reliable instrument for measuring teachers' engagement with research (Gleeson et al. 2023) specifically mentions teachers' "enduring low engagement with research in both Australian … and international … contexts, despite increasing efforts of jurisdictions worldwide that are focused on developing national and/or federal infrastructures for research use in education". They attributed this to personal factors as much as organizational, the latter including "…the leadership and organizational features of "research-engaged schools" …, or system contextual factors, such as "mediation" mechanisms that include funded dissemination of research. Personal factors centred around "individuals' research use beliefs, motivations, and confidence", which have been found to be "very relevant influences".

If, as reported by Baildon and Ong (2022), there is now "near-universal agreement that research generated insights are an insufficient basis for practice" and teachers around the world doubt the utility of educational research, the question obviously arises as to how best to use the outcomes of educational research in the classroom and close the research-practice gap. McKenney and Schunn (2018) suggest that an effective way to achieve large-scale impact is to think about educational designers as the logical recipients of the research so that the results will diffuse down to teachers via the RDD model. Biesta (2021), writing in an introduction to a 2021 special issue of the British Educational Research Journal on this topic describes the question of the relation between theory and practice as "one of the enduring questions of our field". He makes an important distinction between theory and practice and considers that education is itself a largely experimental practice, but "never a matter of blind trial and error". Describing teachers as "thoughtful practitioners" who draw on a wide range of knowledge, including theories, to inform their practise, he suggests that the research-practice gap should not be regarded as a frustrating and continuing problem to be solved but a feature of the field that properly belongs to it. He argues again, echoing the views expressed in his 2007 paper (Biesta 2007), that "a degree of critical distance" may be beneficial as teaching should be informed by theory but not subsumed by it. He concludes by saying that, "discussion about research, theory and practice is complicated and … far from settled".

In view of what Biesta calls, "the contested relationships between educational research, theory and practice" and the persistent difficulties in providing an immediate impact in the classroom, initiatives that translate the outcomes of research into practical, usable resources are likely to prove beneficial in overcoming the perceived lack of utility. In this final section on the research-practice gap we would like to consider a coherent programme of influencing teaching practice underway in Austria that goes some way to transcending Biesta's division between cultural and instrumental impacts by embracing elements of both. These are combined in four different aspects.

1. *Pre- and in-service teacher education.* Within Universities and University Colleges of Teacher Education in Austria, pre-service teachers have to participate in obligatory courses on physics education (research) and can write their Bachelor and Master theses on PER (e.g. they conduct their own research projects). In addition, in-service teachers can participate in teacher training programmes on physics education (research).
2. *Online offers.* The PER group at the University of Vienna, the Austrian Educational Competence Centre (AECC) Physics has a website in the German language (https://aeccp.univie.ac.at/) with a tab for "Teachers" and also a tab for "Teaching concepts". These link to different materials as well as research-based teaching materials.
3. *Book.* There is an openly accessible book in German language "Teaching concepts for physics education" (Wilhelm et al. 2021) aimed at pre- and in-service teachers as well as physics education researchers. There is a chapter for each physics content area presenting different materials, such as quantum physics, basic ideas, goals, and physics education research supporting the material, including where to access it.
4. *Curriculum.* There is a new physics curriculum for grades 5 to 8 in Austria. A corresponding commentary on the physics curriculum highlights already existing research-based teaching concepts and links to the website (under the tab for "Teaching concepts": https://aeccp.univie.ac.at/lehrer-innen/unterrichtskonzeptionen/). Teachers are encouraged to access the book and the online material. In addition, a MOOC is currently being developed to train the teachers on the new physics curriculum.

In summary, although there are multiple reasons for the research-practice-gap in physics education, and we have considered only a few of them, coherent solutions such as the above might offer a way to bring research and practice closer together.

2.2 *The EPS Young Minds Programme*

Started in 2010 by the European Physical Society (EPS), the EPS Young Minds (YM) programme is intended to foster connections between young people active in physics across Europe. Members of the YM programme are predominantly young physicists who may be either undergraduate or postgraduate, and are involved in various activities, such as outreach, informal education, and professional development. YM Members are grouped together in YM Sections, which are structured groups formally established in universities or research centres, with the objective of organizing and performing activities beneficial to their institution and local communities. YM Sections form the smallest building block of the YM Programme and there are currently more than 60 sections distributed all over Europe. Each section has formally appointed officers with fixed duties (President, Vice President, Treasurer and Secretary), and a Faculty Advisor, i.e. a member of the academic staff serving as an official link between the Section and its parent institution. The YM Programme also has a leading body, the Action Committee, presided over by a Chair, with roles of coordinating the Sections, managing the Programme, and serving as a link between the YM Members and the EPS.

The organization of YM is such that the Action Committee is not heavily involved in the activities organized by the sections. Each section is autonomous and is encouraged to plan and undertake whatever activities are appropriate for its members within its local community and possibly more broadly. Representatives of the sections gather regularly, fostering connections between different groups and a sense of a single organization, despite the geographical separation and independence of the different sections.

Not all of the activities undertaken by the different sections involve schools, but many do and it is this aspect what is of particular interest in this symposium. Despite the autonomy that characterizes the activities of each section, Young Minds is still an EPS-programme and if there is a role for EPS in fostering school-university collaborations across Europe, it is natural to think about how an organization like Young Minds, with sections across Europe, can contribute.

In this respect, the YM Sections act as educational agents. They design, schedule and perform activities along three major lines of action: networking, professional development and science dissemination or outreach. However, since each Section is unique and independent from the others, these strategic lines materialize in very different initiatives according to the geographical and cultural contexts, such as the cultural characteristics of the country of origin as well as the local features of the hosting city. The composition of the Sections is often unique, but always heterogeneous, with each section having different percentages of undergraduates, masters and PhD students, and post-doctoral researchers, depending on the kind of hosting institution. For example, it is unlikely for a Section established within a research centre to have undergraduates among its members.

The most common kind of intervention carried out by YM Sections in collaboration with schools comprise interactive dissemination activities. The performance of any such activity in a classroom or auditorium is the culmination of a complex learning process involving both YM Members and school or high school students. The members of the YM Section work together to create the contents of an event which must be fully suited to the target public. Members with more experience mentor those with less, and all members work together to fulfil the requirements of the activity. Therefore, members of the YM Sections learn by mentoring (Colvin and Ashman 2010), by teaching (Duran 2017) and by doing (Bot et al. 2005). The last aspect augments and consolidates knowledge and skills gained within a degree programme. Following the design of the activity, a schedule of events must be developed in collaboration with schools, high schools, other institutions such as museums and possibly public events like science fairs. The performance of the activity in a public arena can be considered as a form of learning: a so-called "learning-service intervention" (Speck 2001), where physics students teach about physics to students at early educational stages or to other kinds of public.

From the perspective of the schools and high schools, children and teenagers benefit from carefully planned teaching interventions. These interactive experiences are dynamically adapted to the level of physics of the class while the contents can be either complementary to the material studied in their courses or totally new. In both cases, the teaching strategy, based on experiments and on the dialogic interaction with the children, offers a unique opportunity for the school students to learn about physics in a different way. In this regard, it is important to highlight that the chance of performing novel physics experiments with the students is greatly appreciated by the science teachers in schools and high schools as well as by the students themselves (Holubova 2019). In fact, beyond the specific contents of physics addressed in the interventions, these kinds of activities are precious for raising awareness about the relevance of science in our current world and awakening future passions for science.

In summary, the Young Minds Programme and its constituent YM Sections represent an effective mechanism for disseminating physics and constitute a powerful resource at the service of schools and high schools in their respective local communities. Their intrinsic local character combined with their high degree of internationalization are a great example of how international scientific organizations, such as the European Physical Society, can establish strategies to achieve a direct interaction within the societies of different countries. Furthermore, it also is a valuable example of how universities can approach schools and high schools to initiate a sincere dialogue in which students and teachers from both institutions collaborate and learn together.

2.2.1 Young Minds Collaboration with Schools (Primary Education)

About 40% of the YM Sections perform activities in collaboration with local schools around their area. The target public are usually children at the final stage of primary education, around 10–12 years old. In addition, YM Sections scheduling activities with schools visit on average 5 different educational centres per year, and, once there, they often perform their activity in at least two different classrooms. The selection criteria for deciding the centres to visit are highly dependent on the YM Section, but proximity or direct contact from the centre's side are common factors. Furthermore, a large number of YM Sections have special predilection for centres with the greatest need: perhaps located in rural areas, or with scarce resources, or working with disadvantaged children. There are even cases of activities performed in special education centres or hospitals.

The most common activity revolves around interactive experiments on different topics on physics, which sometimes are gamified (Forndran and Zacharias 2019) to enhance engagement. These hands-on experiences are usually enthusiastically received by the children, who have fun whilst learning something new. They also discover that science is more than what is written in a textbook: it is experimentation and practical knowledge about the world around us. The feedback from the teachers is positive. The activities are often scheduled for the future incoming years and notes are taken to enable replication of the experiments with other students. Indeed, one of the most common comments of primary school teachers is that they lack the materials and hands-on experience to prepare by themselves for these types of experiences.

An especially good example of an activity directly addressing these needs of primary school teachers is the activity "Children are the future" of the Bethlehem YM Section. What, on the face of it, seems to be an interactive workshop about physics performed with everyday materials also aims to impact teachers and the whole school by preparing kits with the required materials which are gifted to the schools after the workshop. In addition, specific training is offered to the teachers. In this way, everything necessary to replicate the workshop in other classrooms or in the following years is provided in a genuine form of university-school collaboration in the informal framework of dissemination. In addition, as these kits are prepared with everyday materials such that their contents are easy to replace, the teachers can use the ideas for many years.

2.2.2 Young Minds Collaboration with High Schools

Another important part of the activities performed by YM Sections is done in collaboration with high schools. About 60% of the YM Sections perform activities in and/or in conjunction with local high schools every year. On average, these sections schedule at least 5 activities per year, visiting at least two different classrooms each time. The target audience in this case are 16–18 year-old students, who are very likely thinking about what to do after high school. Therefore, these activities can go beyond a simple learning experience and provide students with inspirational scientific models. As with schools, the selection criteria for high schools is highly dependent on the individual YM Section.

There are three different types of activities: experiments, scientific or motivational talks and visits. The experiment-related activities are similar to those described in relation to schools, with a prevalence of hands-on experiences and gamification-based approaches adapted to the level of the students. The scientific or motivational talks try to help students know what it is like to pursue a bachelor's degree in sciences and what the job of a scientist entails. Scientists comprise a variety of different kinds of people, so that pupils should be able to feel that they can become a scientist themselves. Lastly, visits to laboratories are done in order to show the actual work of a scientist and the work environment as well as to perform a wider variety of experiments to the high school students.

A very interesting activity to highlight among the Young Minds collaborations with high schools is the work done by the Catania YM Section in the framework of the Italian "Percorsi per le Competenze Trasversali e per l'Orientamento" (Pathway for Transversal Skills and Orientation), better known as PCTO. The PCTO is a compulsory period during which Italian high school students are given real work experience by joining a company or a public service. In this case, the Department of Physics and Astronomy 'E. Majorana' of the Catania University and the Italian National Plan for Scientific Degrees in collaboration with the Catania YM Section, offer several PCTO positions for developing a supervised scientific project. During the project, YM members serve as lecturers of specific scientific topics and soft skill workshops, and provide mentorship to the high school students, helping to explain the ins and outs of the scientific world. At the last stage of the project, the students have an opportunity to share their own experience with the rest of the classmates in the form of a final talk on their research in their high school. Therefore, the collaboration between the Catania YM and the high schools through the PCTO positions contributes to the training of high school students in technical and practical skills, such as teamwork, communication, problem solving and time management, while working on real research.

2.2.3 Young Minds Assists Learning Outside of School Related Environments

Complementing the activities carried out directly in primary or high schools, YM Sections also extend their connection with educational institutions and students by taking part in learning activities performed outside school-related environments. Our data show that a high percentage of YM Sections engage in educational activities in scientific museums (26%), scientific summer camps (26%), theatres (4%) or, predominantly, in auto-organized (4%) or pre-existent (67%) science fairs, showing a propensity for informal learning activities (Schugurensky 2000). Such contexts, although external to learning institutions, are highly targeted to, and attended by, students, both during school time (as scheduled activities) and during their free time. Moreover, the majority of such activities take place in venues or events that are managed by other entities, which also allows YM members to learn how to engage in coordinating and communicating with other entities in middle to large scale organizational efforts.

The opportunity for YM Sections to take part in such activities depends also on the degree of contact with the organizers, as well as the frequency with which these events are organized in their local community, with many sections participating in periodic events which can be locally organized, like scientific summer camps, yearly science fairs or local cultural events (e.g. Nuremberg "Blue Night"), or part of larger efforts, such as the European Researchers' Night or the World Quantum Day. Generally, these events attract large and varied audiences, allowing for a higher reach and challenging the YM members with the choice of their target. Some sections focus on given age ranges, with 5–10 and 16–18 being common choices, while others design flexible activities suitable for diverse audiences. Such activities are usually layered with different levels of complexity so that they can be adapted to the composition of the audience.

The range of activities depends on both the YM Section and the context. Common choices are interactive experiments at science fairs, workshops in science museums and hands-on activities in science camps, but also scientific games, talks and other kinds of experiences have been reported. Some examples of this include exhibitions of art inspired by Physics (YM Niš), robotics projects designed by collaborating schools (YM Tbilisi), the design and implementation of physics based Escape Rooms (YM Naples) or even touring Science Shows, where experiments and demonstrations are integrated into a coherent narrative frame (YM Valladolid).

An especially good example of an out of school learning event which encompasses several different activities is the "STEM Summer Camp" organized in Armenia in 2022 by the Yerevan YM Section with the objective of encouraging enrolments in STEM related courses. The event involved 50 students from 14 to 17 years old, selected from different high schools and was carried out in collaboration with other entities, like universities and NGOs. Featuring a dense agenda with a highly varied range of activities, it included visits to universities, museums and science centres, meet-ups and seminars involving local academics and hands-on workshops and projects, tackling a variety of scientific disciplines in a way to reinforce and extend the knowledge acquired by students in formal learning contexts. Following the event, the organizers monitor the participants' university career choice in order to evaluate the effectiveness of the process.

Activities like those described in this section show a diversity of ways in which YM Sections can contribute and assist in learning STEM subjects also in informal contexts, providing a bridge between schools, cultural institutions and universities and opening the possibility for further and more long term collaborations between these entities.

3 Discussion and Conclusion

The discussion in the symposium was opened up to the audience, but a large portion of the time was given over to Wim Peeters, well known to the GIREP community, to describe his work in the development of Professional Learning Communities (PLCs) as a possible solution to the research-practice gap and the development of teachers. Wim was an attendee at the symposium but his experience in this area was invaluable in bringing the ideas to a fruitful conclusion.

The essential idea behind the development of PLCs is that professional development by delivery rarely changes teachers' practices, but by engaging in practitioner inquiry (PI) teachers can investigate issues of concern within their own practice or within their school. The aim of the PI is not to turn teachers into researchers, but to give them the tools to act within their own environment to bring about evidence-informed change. In the context of school-university collaborations, having an academic educational researcher connected to the PLC would enhance the level and scope of PI and enable the transfer of important knowledge to the school whilst benefiting the academic by providing enhanced contact with schools.

Biesta has written (2007): "Whereas attempts to bridge gaps between research and practice are … generally laudable, it is also important to remain aware of differences in expertise and responsibility between the stakeholders. This, in sum, reveals that it is as important to try to bridge gaps between research and practice as it is to keep a critical distance between the two, both from the side of educational research and from the side of educational practice". The development of a PLC with academic input would conform to this view.

Owing largely to limitations on the time available, the role of the EPS Young Minds programme in more formal school-university initiatives did not form part of the discussion, but it remains an interesting and very effective vehicle for universities and research institutes to work with schools. As a Europe-wide activity involving collaborations between schools and universities, it already addresses some of the issues addressed in this symposium, but the possibility of further development of this trans-national network remains something for future consideration.

Acknowledgements The Young Minds Action Committee (CDR, MO CM) would like to thank the Bethlehem, Catania, Niš, Tbilisi, Naples and Yerevan Young Minds Sections for providing detailed information about the activities mentioned in this text. The participants would like to thank Wim Peeters for his valuable contribution to the discussion. David Sands would like to acknowledge the support and cooperation of the Physics Education Division of the European Physical Society (EPS) and also Anne Pawsey, David Lee and Ophelia Fornari of the EPS.

References

Baildon MC, Ong MWL (2022) Singapore teachers' perceptions of research in practice: contexts, constraints, and possibilities. Teach Teach Educ 119(2):103850. https://doi.org/10.1016/j.tate.2022.103850

Bauer K, Fischer F (2007) The educational research-practice interface revisited: a scripting perspective. Educ Res Eval 13(3):221–236. https://doi.org/10.1080/13803610701626150

Biesta G (2007) Bridging the gap between educational research and educational practice: the need for critical distance. Educ Res Eval 13(3):295–301. https://doi.org/10.1080/13803610701640227

Biesta G, Aldridge D (2021) The contested relationships between educational research, theory and practice: introduction to a special section, Br Educ Res J 47(6):1447–1450. https://doi.org/10.1002/berj.3772

Borg Marks J, Galea P (eds) (2023) Physics teacher education: more about what matters. Springer Series, Challenges in Physics Education, Springer, Cham. https://doi.org/10.1007/978-3-031-44312-1

Borg Marks J, Galea P, Gatt S, Sands D (eds) (2022) Physics teacher education: what matters?, Springer Series, Challenges in Physics Education, Springer, Cham. https://doi.org/10.1007/978-3-031-06193-6

Bot L, Gossiaux PB, Rauch CP, Tabiou S (2005) Learning by doing: a teaching method for active learning in scientific graduate education. Eur J Eng Educ 30(1):105–119. https://doi.org/10.1080/03043790512331313868

Broekkamp H, van Hout-Wolters B (2007) The gap between educational research and practice: a literature review, symposium, and questionnaire. Educ Res Eval 13(3):203–220. https://doi.org/10.1080/13803610701626127

Colvin JW, Ashman M (2010) Roles, risks, and benefits of peer mentoring relationships. Mentor Tutoring Partnersh Learn 18(2):121–134. https://doi.org/10.1080/13611261003678879

de Winter J, Airey J (2019) What makes a good physics teacher? Views from the English stakeholder community. Phys Educ 55(1):015017. https://doi.org/10.1088/1361-6552/ab5215

Duran D (2017) Learning-by-teaching. Evidence and implications as a pedagogical mechanism. Innov Educ Teach Int 54(5):476–484. https://doi.org/10.1080/14703297.2016.1156011

Forndran F, Zacharias CR (2019) Gamified experimental physics classes: a promising active learning methodology for higher education. Eur J Phys 40(4):045702. https://doi.org/10.1088/1361-6404/ab215e

Gleeson J, Cutler B, Rickinson M, Walsh L, Ehrich J, Cirkony C, Salisbury M (2023) School educators' engagement with research: an Australian Rasch validation study. Educ Assess Eval Account 35:281–307. https://doi.org/10.1007/s11092-023-09404-7

Gore JM, Gitlin AD (2004) [Re]Visioning the academic-teacher divide: Power and knowledge in the educational community. Teach Teach Theory Pract 10(1):35–58. https://doi.org/10.1080/13540600320000170918

Holubova R (2019) The impact of experiments in physics lessons—"Why, when, how often?" AIP Conf Proc 2152:030007. https://doi.org/10.1063/1.5124751

Horvat AK, Wiener J, Schmeling S, Borowski A (2022) Learning goals of professional development programs at science research institutions: a Delphi study with different stakeholder groups. J Sci Teacher Educ 33(1):32–54. https://doi.org/10.1080/1046560X.2021.1905330

Horvat AK (2022) Particle physics in high-school education, Doctoral thesis. University of Potsdam. https://publishup.uni-potsdam.de/frontdoor/index/index/docId/56026

Kersting M (2022) A case for conceptual approaches in general relativity education. Astron Educ J 2(1). https://doi.org/10.32374/AEJ.2022.2.1.040op

Koski A, Puustinen M, Säntti J, Salminen J (2023) Exploring the research-based approach of academic classroom teachers in Finland. Teach Teach 29(7–8):833–849. https://doi.org/10.1080/13540602.2023.2208034

Lederman NG (2007) Nature of science: past, present, and future. In: Abell SK, Lederman NG (eds) Handbook of research on science education, 1 st edn. Lawrence Erlbaum Associates, Mahwah, NJ, pp 831–879. e-book: https://doi.org/10.4324/9780203824696

McKenney S, Schunn CD (2018) How can educational research support practice at scale? Attending to educational designer needs. Br Edu Res J 44(6):1084–1100. https://doi.org/10.1002/berj.3480

Merzel A, Weissman EY, Katz N, Galili I (2022) Toward teacher training for teaching quantum physics in high school. In: Borg Marks J, Galea P, Gatt S, Sands D (eds) Physics teacher education: what matters. Challenges in physics education series. Springer, Cham, pp 161–172. https://doi.org/10.1007/978-3-031-06193-6_12

Michelini M, Stefanel A (2022) Implementing research-based intervention modules for teachers of quantum mechanics. In: Borg Marks J, Galea P, Gatt S, Sands D (eds) Physics teacher education: what matters. Challenges in physics education series. Springer, Cham, pp 93–109. https://doi.org/10.1007/978-3-031-06193-6_7

Postiglione A, de Angelis I (2022) Introducing general relativity in high school: a guide for teachers. In: Borg Marks J, Galea P, Gatt S, Sands D (eds) Physics teacher education: what matters. Challenges in physics education series. Springer, Cham, pp 205–213. https://doi.org/10.1007/978-3-031-06193-6_15

Schugurensky D (2000) The forms of informal learning: towards a conceptualization of the field. Centre for the Study of Education and Work, OISE/UT, NALL Working Paper No.19. https://utoronto.scholaris.ca/server/api/core/bitstreams/2d91687e-a45f-4e36-9666-cd78356ee49c/content

Spatz V, Wilhelm T (2021) Zweidimensionaler Mechanikansatz. Eine Chance zum Überwinden von Lernschwierigkeiten, Naturwissenschaften im Unterricht Physik 32(181):10–13. http://www.thomas-wilhelm.net/veroeffentlichung/Mechanikansatz.pdf

Speck BW (2001) Why service-learning? N Dir High Educ 114:3–13. https://doi.org/10.1002/he.8

Vanderlinde R, van Braak J (2010) The gap between educational research and practice: views of teachers, school leaders, intermediaries and researchers. Br Edu Res J 36(2):299–316

Enhancing University First-Year Physics Education: Impact of Interactive-Engagement Over Traditional Approaches Visualised via Open Modern Data Science

Zuzana Ješková, Jozef Hanč, and Dominik Borovský

Abstract In our research, a model for teaching first-year university physics courses within the European university context has been developed, incorporating interactive engagement (IE) methods into the currently existing course format. During the lectures, interactive lecture demonstrations and conceptual questions with online e-voting were implemented using the peer-instruction method. The problem-solving classes were enhanced by more interactive student-centred activities using digital tools for data collection, mathematical modelling, data analysis and interpretation. Finally, students in groups also work on a long-term project, solving a research problem with the help of experimenting and modelling. The model has been employed for the last ten years within the study program for future physicists and physics teachers, demonstrating a higher and statistically significant impact on the conceptual understanding of mechanics concepts compared to traditional methods. Furthermore, by utilising open modern data science tools for analysis, the data processing was significantly simplified and the interpretation of student results was improved, further enhancing the evaluation of the effectiveness of IE methods in European educational settings.

Keywords Introductory physics course · Interactive-engagement methods · Interactive lecture demonstration · Mathematical modelling · Project-based methods · Open modern data science

Z. Ješková (✉) · J. Hanč · D. Borovský
Faculty of Science, P. J. Safarik University, Košice, Slovakia
e-mail: zuzana.jeskova@upjs.sk

J. Hanč
e-mail: jozef.hanc@upjs.sk

D. Borovský
e-mail: dominik.borovsky@student.upjs.sk

Z. Ješková and J. Hanč (eds.), *Advancing Physics Education*, Challenges in Physics Education, https://doi.org/10.1007/978-3-031-94207-5_13

1 Introduction

The traditional way of teaching at universities is predominantly based on lectures and problem-solving methods. Researches in the field of higher education show that traditional way of instruction based on lecture and problem-solving methods impart little conceptual understanding of physics concepts. In recent years, this traditional approach has shifted in many countries towards implementing interactive-engagement methods, which promote students' active learning. This change has emerged from extensive research in the field of physics education at the university level conducted over more than 20 years. Over this period, a large amount of data was collected that analysed the effectiveness of traditional vs. interactive-engagement methods.

Interactive engagement methods (IE methods), originally defined by Hake (1998) are those designed, at least in part, to foster conceptual understanding through the interactive engagement of students in heads-on (always) and hands-on (usually) activities. In addition, these activities yield immediate feedback through discussions with peers and/or instructors. Numerous IE methods have been developed in line with Hake's definition, such as Peer Instruction (Mazur 1997), lectures with interactive demonstrations (Sokoloff and Thornton 2004), Just-in-time learning (Novak 1999), interactive engagement models, simulations and animations (Christian and Belloni 2004), flipped learning methods (Talbert 2017), project-based learning (Holbrook and Rannikmäe 2014) and many others.

The results of extensive research demonstrate that using IE methods significantly increases the effectiveness of university courses compared to traditional methods (Antwi et al. 2011; Bašista and Ješková, 2017; Blum and Leiß 2007; Crouch and Mazur 2001; Drinkwater et al. 2014; Freeman et al. 2014; Hake 1998; Katramatou 2017; Lombardi et al. 2021; Schiltz et al. 2019; Thornton et al. 2009). However, although implementing interactive methods in higher education has become common in many U.S. universities, it has not been adopted so widely in European universities (Schiltz et al. 2019).

In Slovakia, traditional methods are still dominantly used at the University level. Based on positive research results, we have adopted and implemented selected interactive methods into the first introductory physics course that students, future physicists and future physics teachers, enroll in after entering university. Two-decades-long research on evaluating the impact on students' conceptual understanding was conducted in order to answer the following main research question: *Is the designed introductory physics course effective with regard to conceptual understanding of mechanics?*

2 Introductory Physics Course and Its Innovative Design

The structure of the first introductory physics course for future physicists and physics teachers is based on two traditional forms of instruction: lectures (four hours per week) running for all students at the same time and problem-solving classes (two hours per week) running separately for smaller groups of maximum 12 students. In addition to these classes, students can join an optional two-hour class. Students participate in this extra problem-solving class regularly each year, and as a result, the course finally involves four hours of lectures and 4 h of problem-solving classes per week.

The course content covers the topics of mechanics as the main part and also the basics of molecular physics and thermodynamics. The innovated course design keeps the original content structure while interactive-engagement methods were selected and adapted to be implemented consistently and systematically within the course in both forms of instruction. The course has online support via the LMS Moodle e-learning platform.

The basic elements of the course design related to implemented interactive-engagement methods are presented in Fig. 1.

2.1 *Lectures and IE Methods*

Within lectures, interactive lecture demonstrations (ILD) are regularly conducted. Demonstrations follow a procedure designed by Thornton and Sokoloff (2004). They involve simple conceptual experiments in which students are asked for predictions while engaged in group discussion, and their predictions are followed by a

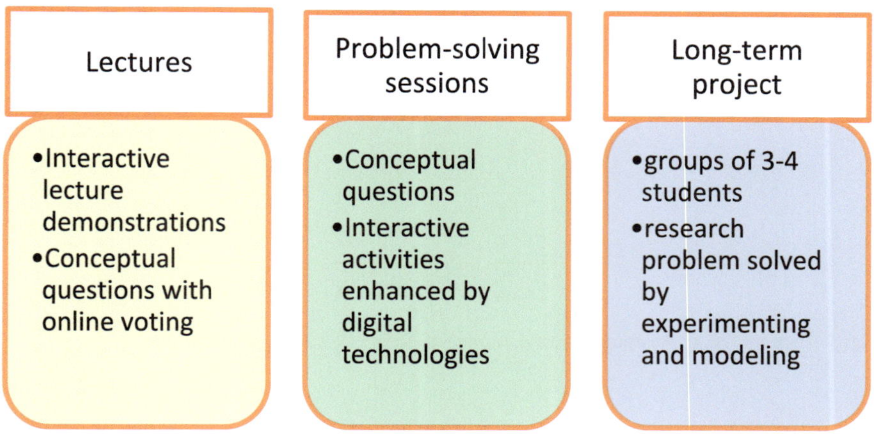

Fig. 1 IE methods implemented within the introductory physics course

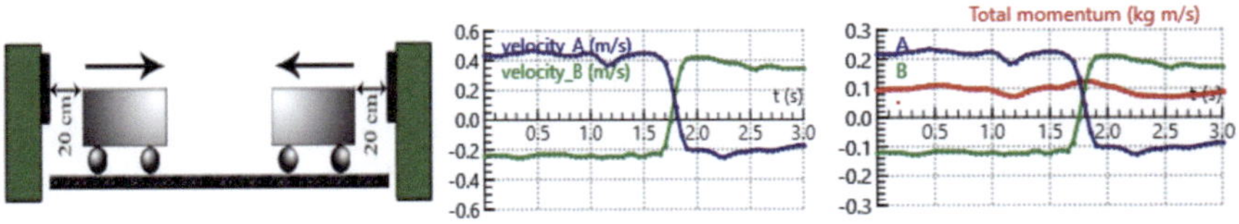

Fig. 2 Example of interactive lecture demonstration aimed at the momentum conservation

real demonstration, usually enhanced by digital tools to compare with their prior knowledge. Figure 2 presents an interactive lecture demonstration on the concept of momentum and its conservation and the real-time measurement results of two equal carts bouncing off each other after the collision.

During the lecture, a teacher engages students by posing prepared conceptual questions, usually in the form of multiple-choice questions. In this case, the procedure of the Peer instruction method (Mazur 1997) is followed when students are first expected to answer individually, then discuss the problem with their peers and select the answer using the online voting tool. This way, the teacher and also students get immediate feedback about their level of understanding. In case of a low gained score, the teacher can immediately return back to the identified misconception and discuss the concept again. In Fig. 3 a conceptual question about fluid dynamics and the understanding of Toricelli's law is presented.

Two bottles with the bottom cut off are filled with water to the same height. One bottle has an opening connected to a long, thin tube and the other to a short tube. From which bottle will the water come out sooner?

a) It will take the same time
b) Bottle with long tube
c) Bottle with short tube

Fig. 3 Example of a conceptual question aimed at Torricelli's law understanding

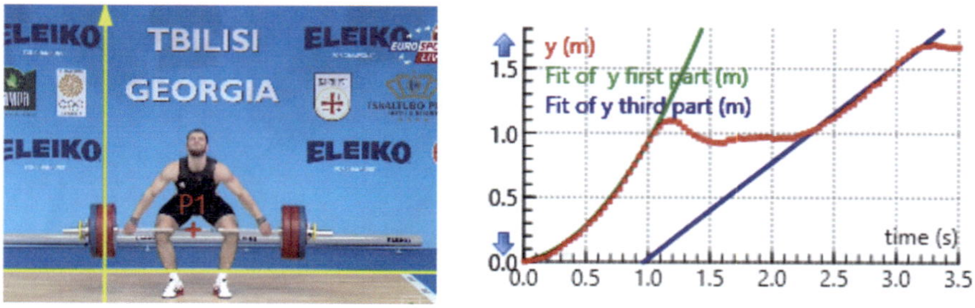

Fig. 4 Videomeasurement of the weightlifter lifting the barbell in the snatch. The barbell position-time data are fitted with appropriate functions

2.2 *Problem-Solving Classes and IE Methods*

The original structure of problem-solving classes is enhanced by implementing more interactive student-centered activities. Students solve more realistic problems by applying the physical principles that they have learned. When solving problems, they use computer-based tools for data collection, mathematical modelling, data analysis, and interpretation. In Fig. 4, a videomeasurement of a weightlifter can be seen. Students are expected, based on measuring the position of a barbell, to determine how much work is done by the weightlifter and what power he exerted during the lift.

Besides that, students regularly answer a conceptual question, usually related to the last lecture. They submit their detailed explanations, which are discussed in the next session. We have drawn many ideas for this type of questions from Paul Hewitt's Conceptual Physics (Hewit 2006).

2.3 *Long-Term Project*

Students in groups of 3–4 work on a long-term project. The project assignment is formulated by the teacher at the beginning of the course. In this case, students solve a more complex problem compared to the activities that they conduct in regular classes. In order to solve the assigned research problem, students apply the scientific approach through experimenting and modelling, which should lead to experiment-model correspondence. They follow the steps described in Fig. 5. In the modelling part itself, students follow the modelling cycle (Blum and Leiß 2007), going through different phases by structuring and simplifying the problem context (e.g., drawing a rough picture of the situation), then creating and manipulating algebraic formulas in the form of a computer model. Finally, the mathematical results are interpreted and validated in the context of the original problem. If the results are inadequate for the real situation, the modelling process starts again.

Students work in groups on a problem assigned by the teacher following these steps:

1. Design and conduct an experiment
2. Analyze and interpret gained data
3. Design a computer model
4. Compare experimental with modelling results in order to find the best correspondence
5. Draw conclusions
6. Present and defend the project outcomes in front of the whole class

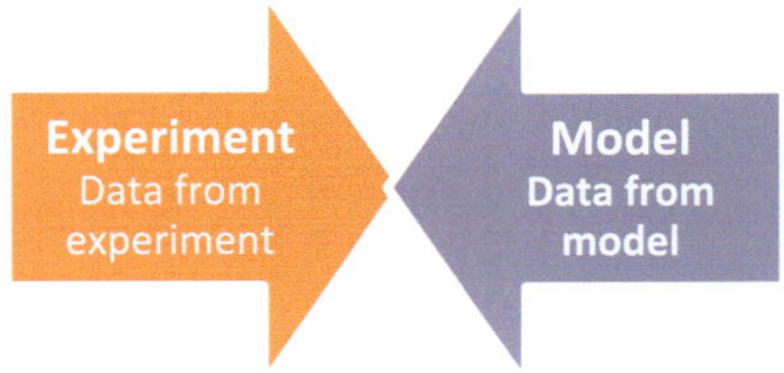

Fig. 5 Procedure that students follow when conducting a long-term project

Examples of project assignments involve:

- Throw a hammer or an axe so it spins during its motion. How is the centre of mass moving?
- Investigate the motion of the centre of mass of two carts of different masses before and after collision (carts are moving in opposite or in the same direction).
- A ball is thrown vertically upward. What is the time of ascent and descent? Compare it for a heavy ball and a light Styrofoam ball.
- How does a small sphere fall through two kinds of oil of different viscosity?
- Serve the shuttlecock. How does it move?
- Two identical spheres start moving at the same time, one moves vertically with zero initial speed, the other one is thrown horizontally. Analyse the motion and the time of flight of both spheres.
- How do water and oil heat up? Conduct an experiment, design a model and compare experimental and modelling results.
- Investigate the sliding and rolling motion of an object moving down an inclined plane.
- Analyze the uniform and decelerated motion of a bicycle wheel.
- Move a bucket with water hanging on a thread in a vertical circle so the water does not flow out. Analyse its motion and the forces acting on the bucket bottom at different positions.

On the basis of the project results students elaborate a project report that they present at the end of the term in groups in front of the whole class. The presentation is followed by a discussion between the project authors, other students and the teacher. Students are expected to reason and justify the design, results and interpretation of their experimental and modelling results.

In Figs. 6 and 7 students' results of the shuttlecock problem is presented. Students recorded a video, then they analysed it with the help of videoanalysing tools. Students designed a model based on numerical solution of differential equation describing the shuttlecock motion. Finally they compared the results simulating the model for the best correspondence.

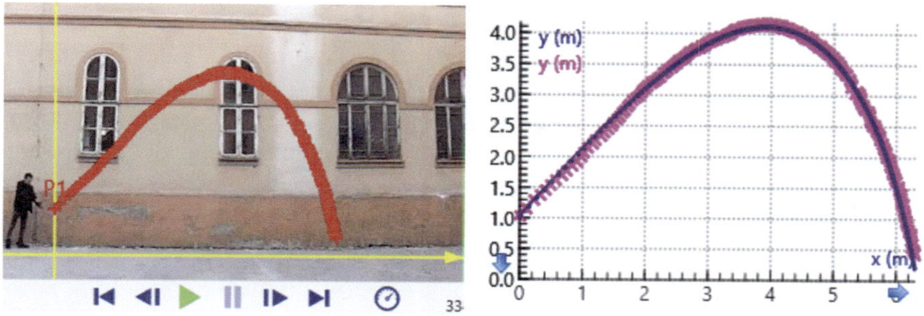

Fig. 6 Videomeasurement of a shuttlecock motion and its trajectory. Compariosn of experimental (crosses) and modelling results (line)

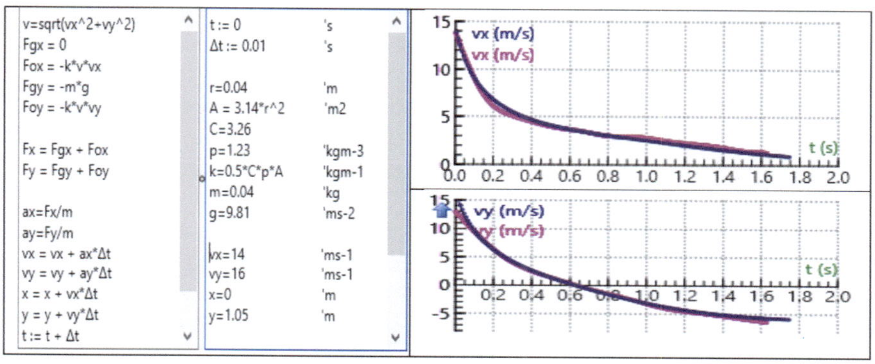

Fig. 7 Model of a shuttlecock motion and corresponding horizontal and vertical velocity components (experimental vs. modelling results)

3 Research and Methods

3.1 *Research Design, Research Instrument and Research Sample*

In order to evaluate the effectiveness of the introductory physics course design in terms of its impact on conceptual understanding, we have conducted a research addressing the following research question: *What is the effect of interactive-engagement methods on students' conceptual understanding in an introductory physics course compared to traditional pedagogical approaches?*

The effectiveness of the introductory physics course model we are developing is evaluated by the most commonly used content survey in mechanics, the well-known Force Concept Inventory (FCI) (Hestenes et al. 1992). We have prepared the FCI Slovak version, which is also available on the PhysPort platform (Hanč and Tóth 2006).

Our research documents the development of student results over 20 years, from 2002 to 2022. The course design has undergone several milestones and significant changes in applied methods, forms, and tools. These milestones naturally delineated corresponding specific research periods. In the first period (denoted as TRAD1) up to 2006, the course was led by a renowned physicist without formal pedagogical training. The course applied traditional approaches (rigorous lectures and recitation), albeit with high-quality textbooks, but pedagogical interventions were based only on personal empirical experiences.

Based on the initial FCI measurements, since 2007, the course has been taken over by one of the authors of this paper (ZJ), a Physics Education Research (PER) researcher, with the aim of applying best research-based practices. Until 2011 (period TRAD2), the course remained in the original format. However, best practices and interactive elements were implemented within the framework of traditional lectures and problem-solving classes (Redish 2003). These include setting the context, chunking material, focusing on effective speaking techniques, maintaining clear handwriting and layout on the blackboard, and employing just-in-time authentic experimental demonstrations, illustrative problems supplemented with authentic questions and heuristic discussions. Equally important from PER was focusing on evaluating the impact of instruction and basing the course on students' prior learning (preconceptions).

In the following three years (2012–2014, denoted as ILD1), the course incorporated and adapted the mentioned interactive demonstrations and more student-centred classes. From 2015 to 2019 (denoted as ILD2), the course fully integrated and adapted the peer instruction method with e-voting as the second interactive key element to ILD. The last research period (COVID, 2020–2022) corresponds to the teaching and generation of students fully impacted and influenced by the COVID-19 pandemic.

The distribution and properties of our research sample are presented in Table 1. The table captures the overall numbers, as well as the distribution according to three categorical variables (factors): research period (or instruction), study program (single-subject physics or physics teacher education), and gender of students. Since only a portion of the students were able to complete both the pre and post-FCI tests, the numbers in parentheses indicate the percentage of those who participated in at least one measurement, either at the pre or post-FCI test.

3.2 Data Collection and Analysis

Data collection for our research study began in 2006, utilising paper versions of the assessments. In this initial year, we also retrospectively tested students from all cohorts with a posttest, i.e., those who had completed the introductory course from 2002–2006. This paper-based collection method continued until 2012, when we transitioned to using Google Forms administrated via computers or smartphones.

Regarding the analysis of the continuously collected data, we applied standard descriptive and inferential statistics, including the calculation of recommended effect

Table 1 Distribution and properties of our research sample

	Study programme	*N* %	Gender	*N* %
Research sample	Physics	149 (75)	Male	134 (62)
	Physics teacher	124 (53)	Female	139 (64)
	All	273 (63)	All	273 (63)
Traditional 1 (TRAD1)	Physics	14 (61)	Female	13 (41)
2002–2006	Physics teacher	15 (31)	Male	16 (41)
Traditional 2 (TRAD2)	Physics	47 (76)	Female	50 (63)
2007–2011	Physics teacher	43 (54)	Male	40 (66)
IE methods 1 (ILD1)	Physics	32 (86)	Female	19 (68)
2012–2014	Physics teacher	13 (48)	Male	26 (72)
IE methods 2 (ILD2)	Physics	33 (72)	Female	26 (57)
2015–2019	Physics teacher	28 (56)	Male	35 (70)
Covid (COVID)	Physics	23 (74)	Female	26 (87)
2020–2023	Physics teacher	25 (86)	Male	22 (73)

size measures: Cohen's d and average normalized gain $\langle g \rangle$ (Cumming 2012; Marx and Cummings 2007; Madsen et al. 2016). Additionally, we supplemented our analysis with the machine learning method of Decision Regression Trees, used in open data science (Aiken et al. 2021; Lowndes et al. 2017).

Our data analysis was conducted using modern open data science tools based on the programming languages Python and R (Scavetta and Angelov 2021), which have become fundamental in the last decade due to the enormous advances in open data science. Particularly, we performed very effective data preprocessing, wrangling, and descriptive statistics using Python packages Pandas (The Pandas Development Team 2024) and NumPy (Harris et al. 2020). Inferential statistics (95% CI's) were based on the R packages Effectsize (Ben-Shachar et al. 2020). Decision regression trees were generated and plotted using the R package rpart (Therneau and Atiknson 2023). Powerful data visualisation, allowing us to create tailored plots for our specific data analysis, was done using the versatile Python package Matplotlib (The Matplotlib Development Team 2024). All manipulations, calculations, and coding were conducted in Jupyter Notebooks within the JupyterLab environment (https://jupyter.org/), where the rpy2 interface (https://rpy2.github.io/) allows for the simultaneous use and combination of Python and R commands and packages in one data analysis.

4 Results/Discussion

A numerical summary of descriptive and inferential statistics is presented in Table 2. Graphical summaries of the analysis are presented in Fig. 8, depicting Cohen's d on the left, with levels (small, medium, large, very large and huge) based on Cumming (2012) and Sawilowsky (2009).

The average normalized gain in Fig. 8 also includes empirical levels (spans) typical for traditional and interactive approaches (Hake 1998). The impact of factors

Table 2 Numerical summary of FCI test results during given research periods

Instruction	TRAD1 (5y)		TRAD2 (5y)		ILD1 (3y)		ILD2 (5y)		COVID (3y)	
Years	2002–2006		2007–2011		2012–2014		2015–2019		2020–2022	
N	29		90		45		61		48	
Test	pre %	post %	pre %	post %	pre %	post %	pre %	post %	pre %	post %
Mean	44	49	45	59	56	73	54	70	46	62
Std	23	26	24	24	22	22	21	21	25	22
Min	10	10	10	10	17	17	20	27	13	30
25%	30	30	27	40	37	60	40	53	26	43
50%	43	43	40	62	53	77	53	73	37	60
75%	50	57	57	82	80	90	67	87	64	81
Max	100	100	100	100	97	100	97	100	97	100
$t(N\text{-}1)$	2.20		9.59		8.17		8.82		8.11	
p value	0.036		2.3E-15		2.3E-10		2E-12		1.8E-10	
d	0.41		1.01		1.22		1.13		1.17	
95% CI *d*	(0.03, 0.79)		(0.75, 1.26)		(0.83, 1.6)		(0.8, 1.45)		(0.8, 1.53)	
$\langle g \rangle$	0.13		0.28		0.44		0.38		0.31	
95% CI $\langle g \rangle$	(−0.01, 0.26)		(0.22, 0.33)		(0.34, 0.54)		(0.3, 0.45)		(0.24, 0.39)	

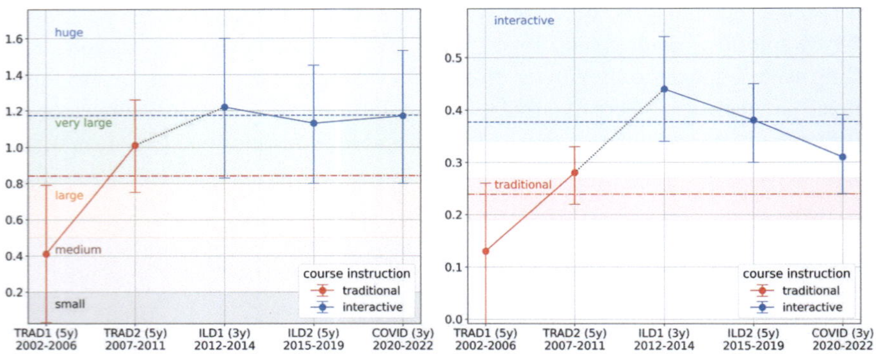

Fig. 8 Effect size measures: Cohen's *d* and normalized gain $\langle g \rangle$ with 95% confidence intervals

on normalized gain is shown in Fig. 9. A paired t-test (Table 2) demonstrated that in each period, there was a statistically significant improvement in effectiveness, at least at the 0.05 level (for all $p \leq 0.036$). We can draw the following conclusions about the effectiveness based on the numerical and graphical summarization of the results.

Traditional instruction (periods TRAD1, TRAD2)

- Initial instruction (TRAD1) achieved a medium effect size ($d = 0.41$). After changing the instructor, the effect size jumped to a very large level ($d = 1.01$). This increase aligns with findings in (Hattie 2015), emphasising the importance of evaluating instruction impact and student prior learning ($d > 0.9$).

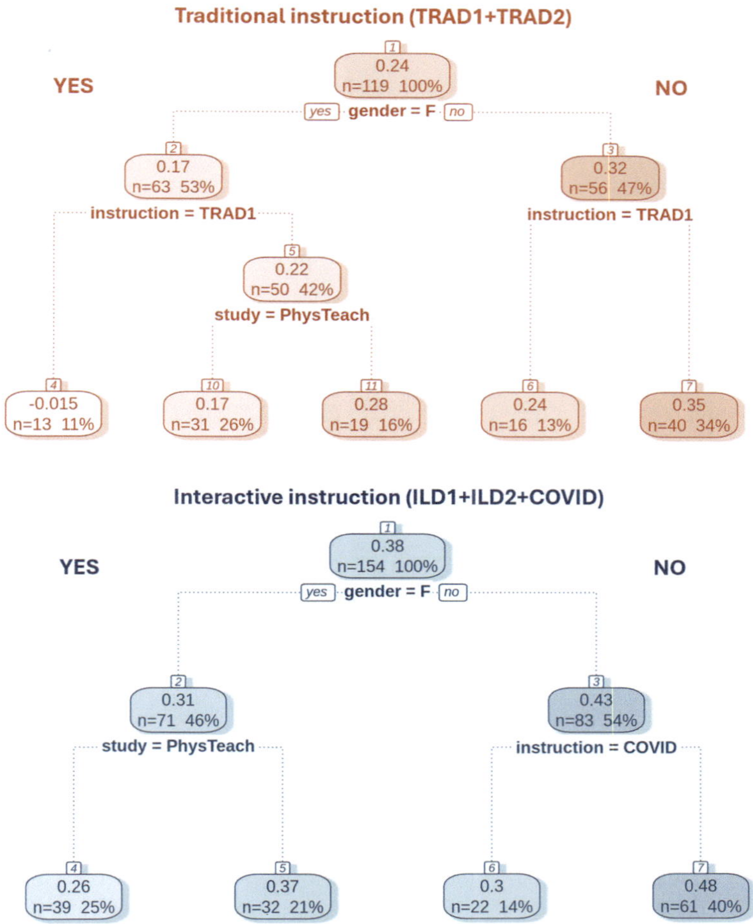

Fig. 9 Modern decision regression tree models for the normalized gain during traditional and interactive instructions with respect to three factors: instruction period, gender and study program

- The normalized gain ($\langle g \rangle = 0.13$) for TRAD1 instruction is below the typical $\langle g \rangle$ for traditional courses (red span $\langle g \rangle = 0.23 \pm 0.04$, see Hake (1998) and Redish (2003)).
- As shown by the regression tree model (branch 1-2-4), the reason is that TRAD1 approach did not suit female students ($\langle g \rangle < 0$). Further analysis revealed that most of our female students in physics teacher education programs were not primarily motivated to study physics but the second subject.
- The TRAD2 approach increased the normalized gain to $\langle g \rangle = 0.28$, above the typical range for traditional courses. This shift was due to the improvement in $\langle g \rangle$ from negative to 0.17 for female teacher students, for whom the instructor change was crucial. According to the regression tree model, male students, enrolled in the physics program, also benefited. Their $\langle g \rangle$ improved from 0.24 to 0.35, reaching the lower bound of the effectiveness for interactive methods (blue span $\langle g \rangle = 0.48 \pm 0.14$).

Interactive Instruction (periods ILD1, ILD2, COVID)

- Incorporating interactive methods (ILD1, ILD2, COVID) led to achieving the lower bound of a huge effect size (d $\approx$ 1.2), clearly attributable to this pedagogical innovation (Hattie 2015).
- The normalized gain for ILD1 and ILD2 periods reached values within the effectiveness range for interactive approaches (blue span $\langle g \rangle = 0.48 \pm 0.14$) (Hake, 1998; Redish 2003).
- Interestingly, enhancing interactivity in ILD2 led to a slight decrease in effectiveness concerning $\langle g \rangle$. However, the regression tree model (branch 1–2-4) shows that this is again due to female teacher education students, who are generally less active and confident in peer instruction discussions (Bruff 2009).
- In the COVID period, the effectiveness of interactive instruction ($\langle g \rangle = 0.31$) fell below the lower bound of the interactive methods range. According to the regression tree model (branch 1-3-6), this decline was caused predominantly by the impact of COVID restrictions on male students, where $\langle g \rangle$ decreased from 0.48 to 0.30.

5 Conclusions

The results of our research strongly affirm the crucial role of the instructor, particularly one who understands and implements PER findings, a conclusion from the first two research periods under traditional course settings (TRAD1, TRAD2). Even with only minor enhancements through interactive teaching elements, changing the instructor significantly increased the effect size (from $d = 0.41$ to 1.01). This underscores the pivotal impact of instructors on student outcomes. Institutions should carefully evaluate and revise instructional strategies and instructors to maximise educational effectiveness.

In subsequent research periods (ILD1, ILD2, COVID), the first students experiencing the curriculum reform that drastically reduced science lessons in secondary schools entered the university. Despite this, the systematic, consistent, and thoughtful implementation of interactive methods led to an even more significant effect size ($d \approx 1.2$) in these periods, with average normalized gains reaching levels typical for successful interactive approaches. These results indicate that interactive methods are highly effective and should be prioritised in educational programs to enhance student engagement and learning.

However, during the COVID period, the average normalized gain fell below the typical range for interactive methods. This decline illustrates the challenges that external disruptions pose to interactive instruction and emphasises the benefits of in-person learning. Moreover, the decrease in effectiveness for specific subgroups—such as female teacher education students during the ILD2 period and male students during the COVID period—suggests that interactive methods do not uniformly benefit all students. This finding highlights the need to consider diverse needs and engagement levels when implementing interactive methods. Tailoring these approaches to address different student demographics could enhance their overall effectiveness.

In this research, modern data analysis and visualising methods, such as decision regression trees, played a crucial role in uncovering the underlying causes behind some of these quantitative results. The integration of more sophisticated data analysis techniques and user-friendly open tools not only simplified the research process but also provided deeper insights into the effectiveness of instructional methods. These advancements underscore the importance of leveraging modern technologies to enhance educational research and practice.

Acknowledgements This work was supported by the Slovak Research and Development Agency under the Contract no. APVV-21-0369 and no. APVV-21-0216.

References

Aiken JM et al (2021) Framework for evaluating statistical models in physics education research. Phys Rev Phys Educ Res 17:020104. https://doi.org/10.1103/PhysRevPhysEducRes.17.020104

Antwi V et al (2011) The impact of Interactive-engagement (IE) teaching on students understanding of concepts in mechanics: the use of force concept inventory (FCI) and mechanics baseline test (MBT). Int J Educ Plan Adm 1(1):81–90

Bašista J, Ješková Z (2017) Interactive methods in introductory physics course In: Proceedings of the 9th international conference PTEE. University of Žilina, Žilina, pp 95–102

Ben-Shachar MS et al (2020) Effectsize: estimation of effect size indices and standardized parameters. J Open Source Softw 5(56):2815. https://doi.org/10.21105/jpss.02815

Blum W, Leiß D (2007) How do students and teachers deal with modelling problems? In: Haines C et al (eds) Mathematical modelling. Woodhead Publishing, Cambridge, pp 222–231

Bruff D (2009) Teaching with classroom response systems: creating active learning environments. Jossey-Bass, San Francisco

Christian W, Belloni M (2004) Physlet physics: interactive illustrations, explorations, and problems for introductory physics. Prentice Hall, Upper Saddle River

Crouch CH, Mazur E (2001) Peer instruction: ten years of experience and results. Am J Phys 69:970–977. https://doi.org/10.1119/1.1374249

Cumming G (2012) Understanding the new statistics: effect sizes, confidence intervals, and meta-analysis. Taylor & Francis Group, New York

Drinkwater MJ et al (2014) Managing active learning processes in large first year physics classes: the advantages of an integrated approach. Teach Learn Inq 2(2):75–90. https://doi.org/10.20343/teachlearninqu.2.2.75

Freeman S et al (2014) Active learning increases student performance in science, engineering, and mathematics. PNAS 111(23):8410–8415. https://doi.org/10.1073/pnas.131903011

Hake RR (1998) Interactive-engagement versus traditional methods: a six-thousand-student survey of mechanics test data for introductory physics courses. Am J Phys 66(1):64–74

Hanč J, Tóth J (2006) Force concept inventory—Slovak version. https://www.physport.org/assessments/assessment.cfm?A=FCI

Harris CR et al (2020) Array programming with NumPy. Nature 585:357–362. https://doi.org/10.1038/s41586-020-2649-2

Hattie J (2015) The applicability of visible learning to higher education. Scholarsh Teach Learn Psychol 1(1):79–91. https://doi.org/10.1037/stl0000021

Hestenes D, Wells M, Swackhamer G (1992) Force concept inventory. Phys Teach 30:141–58. https://doi.org/10.1119/1.2343497

Hewitt PG (2006) Conceptual physics, 10th edn. Benjamin-Cummings, San Francisco

Holbrook J, Rannikmäe M (2014) The philosophy and approach on which the PROFILES project is based. Cent Educ Policy Stud J 4(1):9–29

Katramatou M (2017) Active learning teaching for introductory physics. Paper presented at the 6th International conference on new frontiers in physics education ICNFP2017, CERN, 17–29 August 2017

Lombardi D et al (2021) The curious construct of active learning. Psychol Sci Public Interest 22(1):8–43. https://doi.org/10.1177/152910062097397

Lowndes JSS et al (2017) Our path to better science in less time using open data science tools. Nat Ecol Evol 1:0160. https://doi.org/10.1038/s41559-017-0160

Madsen A, Sayre E, McKagan S (2016) Effect size: what is it and when and how should I use it? PhysPort, https://www.physport.org/recommendations/Entry.cfm?ID=93385. Accessed 1 Feb 2024

Marx JD, Cummings K (2007) Normalized change. Am J Phys 75:87–91

The Matplotlib Development Team (2024) Matplotlib: visualization with python. https://zenodo.org/records/11201097. Assessed 1 Feb 2024

Mazur E (1997) Peer instruction: a user's manual. Prentice Hall, Upper Saddle River

Novak G et al (1999) Just-in-time teaching: blending active learning with web technology. Addison-Wesley, Boston

The Pandas Development Team (2024) Pandas. https://zenodo.org/records/10957263. Assessed 1 Feb 2024

Redish EF (2003) Teaching physics with the physics suite CD. Wiley, New York

Sawilowsky SS (2009) New effect size rules of thumb. J Mod App Stat Meth 8(2):597–599

Scavetta RJ, Angelov B (2021) Python and R for the modern data scientist. O'Reilly Media Inc., Sebastopol, CA

Schiltz G, Feldman G, Vaterlaus A (2019) Active-learning settings and physics lectures: a performance analysis. J Phys Conf Ser 1286:012019. https://doi.org/10.1088/1742-6596/1286/1/012019

Sokoloff DR, Thornton RK (2004) Interactive lecture demonstrations: active learning in introductory physics. John Wiley & Sons, New York

Talbert R, Bergmann J (2017) Flipped learning: a guide for higher education faculty. Stylus Publishing, Virginia Sterling

Therneau T, Atkinson B (2023) Rpart: recursive partitioning and regression trees. https://CRAN.R-project.org/package=rpart. Assessed 1 Feb 2024

Thornton RK et al (2009) Comparing the force and motion conceptual evaluation and the force concept inventory. Phys Rev ST Phys Educ Res 5:010105. https://doi.org/10.1103/PhysRevSTPER.5.010105

A Scaffolded Curriculum to Foster Experimental Skills Acquisition in an Introductory Physics Lab Course

Micol Alemani

Abstract This article outlines a comprehensive redesign of an introductory physics laboratory course aimed at promoting active student engagement, enhancing scientific experimental skills and the development of expert-like attitudes. The redesign process involved redefining specific and measurable learning goals and the development of tailored activities and assessment rubrics to reach and assess those goals. In order to implement course scaffolding, we made significant modifications to the previously existing course format, space, physics topics contexts, and assessment. Instructors' roles evolved from imparting knowledge to promoting student decision-making and fostering a collaborative learning environment. To guide the laboratory transformation we used the results of a research-based survey. This work offers practical insights for educators seeking to redesign laboratory courses, emphasizing authentic scientific practices and student-driven learning experiences in physics education.

Keywords Active learning · Scientific reasoning · Inquiry-based instruction · Curriculum transformation

1 Introduction

Teaching methodologies that actively involve students in the collaborative construction of knowledge have demonstrated greater efficacy compared to passive information reception. A comprehensive review of the literature on this subject is available in Meltzer and Thornton (2012). Physics laboratory courses (PLCs) could be considered inherently active learning environments, as students are active when working on lab activities. Recent research has however raised doubts regarding the efficacy of traditional prescriptive PLCs in enhancing physics content knowledge (Wieman and Holmes 2015), developing a deeper understanding of measurement uncertainties

M. Alemani (✉)
Physics and Astronomy Institute, University of Potsdam, Potsdam, Germany
e-mail: alemani@uni-potsdam.de

Z. Ješková and J. Hanč (eds.), *Advancing Physics Education*, Challenges in Physics Education, https://doi.org/10.1007/978-3-031-94207-5_14

(Volkwyn et al. 2008), and fostering positive attitudes towards experimental physics (Wilcox and Lewandowski 2018).

Reflecting on these results, one could argue that *hands-on* activities do not necessarily mean *minds-on* activities (Redish 2004) since just "interpreting" a prescriptive lab manual is not engaging students with higher cognitive levels, which eventually involve (i) the creation of new outputs by students beyond the information received by the instructors as well as (ii) collaborative aspects of learning (Chi and Wylie 2014). The comparison of the cognitive tasks used by professional experimental physicists with those performed by students in a traditional confirmatory introductory laboratory course shows that the majority of the cognitive tasks are already done by instructors (Wieman 2015).

A recent historical review on instruction in PLCs in the United States highlights that there have been recurrent and consistent criticisms of traditional prescriptive instruction in PLCs in the past 200 years (May 2023). However, only in recent years a shift in focus from laboratory course structures to their learning goals had happened.

Consequently, novel active learning pedagogies have been proposed and their effectiveness has been systematically studied using rigorous evidence-based methods (Wieman and Holmes 2015; Volkwyn et al. 2008; Etkina et al. 2006; Brookes et al. 2020; Zwickl et al. 2015).

Nevertheless, despite studies highlighting the deficiencies of the traditional laboratory courses, traditional prescriptive and confirmatory labs endure in numerous physics programs. On the other hand, the working group of the GIREP community with focus on laboratory instruction recently wrote a position paper (Bearden et al. 2022) stating that the main goal of lab courses should be to (i) help students to learn how to think like a physicist/scientist, (ii) help students to learn how to plan, design and conduct experiments and experimental procedures, and (iii) help students to learn how to collect, analyze, present data and report the findings.

In light of the literature discussed above and aligned with the reflections emphasized by the GIREP work group on laboratory courses, we have undertaken a redesign of our PLC for physics major students at the University of Potsdam (UP). Our redesign entails a transition from a conventional "teacher-focused" and "concept-based" approach to a more "student-centered" and "skill-based" paradigm. This shift aims to promote cognitive development and foster collaborative engagement among students during learning activities as well as students understanding of the nature of experimental physics. While the assessment of our course transformation was discussed in detail elsewhere (Teichmann et al. 2022, 2024; Alemani 2023c; Tufino et al. 2024), we want to focus here on the discussion of the development of a scaffolded curriculum to foster experimental skills and expert-like attitudes acquisition.

2 Existing Course Description

The introductory laboratory course here described addresses physics major students. It is composed of a sequence of four courses which take place during the first four semesters of the physics curriculum. Typically there are between 40 and 60 students enrolled in each instance of the course. Each part of the course is obligatory and except the last part of the sequence, it is not graded.

Before the transformation students were doing overall 30 different experiments, one per lab appointment. Students were working on different physics topics encountered in the physics lecture for the allocated time of about three hours. All experiments had the same structure: Students were given a setup already built and previously optimized, received a detailed list of instructions to interpret and follow for the data taking and the data analysis with the aim of reproducing well known results from theory. They were confronted with many different measurement and data analysis techniques but were not asked or allowed to engage in any of the steps involved in designing an experiment or choosing a data analysis strategy. There was also no space for iteration of experimental steps, the experimental process was linear. The sessions ended with the lab report homework either concluding that the theory was confirmed by the experiment or in case of disagreement by proposing reasons for it. The success of the experiment was biased towards the end result. Experimental problems were usually experienced by students as a source of frustration.

As evidenced by existing literature, our experience revealed insufficient levels of student engagement (Wieman 2015; Holmes and Wieman 2018), identified a gap in the development of crucial experimental skills (Volkwyn et al. 2008), and expert-like attitudes (Wilcox and Lewandowski 2018). To address these shortcomings, we initiated a comprehensive transformation of our course. In the subsequent sections, we provide a detailed account of the steps involved in this transformative process.

3 Model for the Course Transformation

The process of redesigning a laboratory course is inherently complex, involving a multitude of steps and considerations. To effectively navigate this complexity, we employed a backward design methodology, as proposed by Zwickl et al. (2013).

Figure 1 provides a simple visualization of the process we undertook. Our approach began with delineating our rationale, which involved reviewing the literature on labs, setting a transformation strategy, and defining new learning goals, grading methods, and course scaffolding. This was followed by implementation, which included adjusting the course structure, physics topics, lab equipment, redesigning the lab space, and developing tailored activities to achieve the learning goals. The final step involved assessing students' learning. It is important to note that these three aspects of the lab course redesign are interconnected (as shown by the arrows

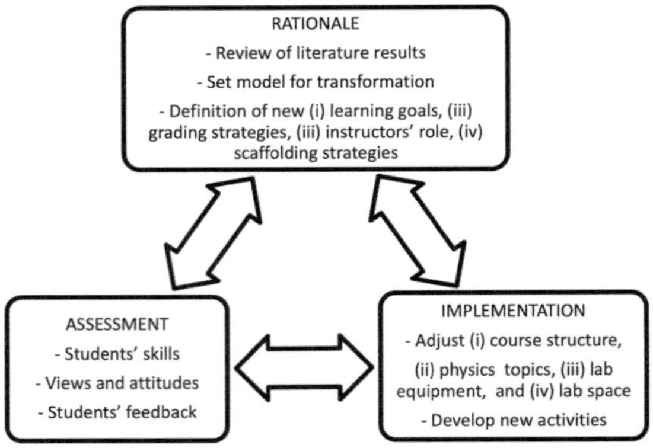

Fig. 1 Visualization of the lab redesign process with its steps and components

in Fig. 1) and were iteratively revised based on the assessment results. The following sections will detail the different aspects related to the first two steps of the lab redesign, while the assessment process will be briefly outlined.

4 Course Learning Goals, Objectives and Outcomes

Our goal is to foster students' acquisition of scientific experimental skills. To classify those skills we used the framework developed by Zwickl et al. (2013) and divided them into the categories *design*, *modeling*, *technical-computational*[1] and *communication*. As a further goal we seek to promote the development of experts-like views and attitudes towards experimental physics. We do this by offering students authentic laboratory experiences in which students engage in scientific practices. In fact, we consider this second aspect crucial for studentsâŁ™ achievement of the first learning goal.

Based on these general goals, we defined the following course-wide objectives phrased here as learning outcomes: *By the end of the four courses of the introductory laboratory sequence, students should be able to*:

- Design measurement setups, methods, procedures and measurement strategies
- Use and refine models of physical and measurement systems to make sense of experimental results both qualitatively and quantitatively based on evidence
- Use available instrumentation and programming environments to collect and analyze data or manipulate the real world
- Communicate clearly processes and outcomes of an experiment supporting arguments by evidence and using formats utilized by professionals

Finally for each category of skills, we have meticulously outlined a comprehensive set of specific and measurable learning outcomes (LOs). Each LO is accompanied

[1] Note that we add the computational aspect to the technical skills category of Zwickl et al. (2013).

by designated activities designed to facilitate student practice and assessment. These LOs represent finely detailed educational objectives. Several of these LOs could be grouped into broader categories to streamline the learning process. We intentionally defined very specific LOs to be able to scaffold practice activities. The list of our specific LOs for the category *communication* is shown as an example in Fig. 2. The complete list including the categories *design* and *modeling* and *technical and computational* can be found in Open Science Framework (2025). To formulate these LO we made use of the revised Bloom taxonomy (Anderson et al. 2001). This way, we checked that all levels of the cognitive domain are involved. Note that these LOs are the results of various iterations. For example we validated those LO by re-examining our activities and grading rubrics and monitoring how often each LOs was practiced by students. After such a check, one might find out that some of the defined LOs are not actually requested for the successful achievement of the lab activities.

Communication (C)

1. Understand why physicists use laboratory notebooks (LNs) daily:
    - State why LNs are useful
    - State what type of information is recorded in laboratory notebooks (LNs), what are the criteria for a good LNs and what are the difficulties encountered when writing a LN
2. Describe in a clear and complete way the experimental processes in a LN using the knowledge of 1C (see also list "Labbook Best Practices")
3. Use LNs to present preliminary results in discussions
4. Facilitate discussions between groups and make sure that the necessary information about experimental processes is exchanged
5. Present an experimental project using formal authentic formats such as (i) a proposal (including write a time plan for the project and a budget plan)
6. Present an experimental project using formal authentic formats such as (ii) a final report
7. Present an experimental project using formal authentic formats such as (iii) an oral presentation
8. Present outcomes, claims, and conclusions as arguments that are supported by and follow from evidence
9. Distinguish between good and bad graphs
10. Create clear, informative and complete figures or graphs (including using figure captions)
11. Make Python/Arduino IDE codes for data analysis understandable by others using comments and the capabilities of Jupyter Notebooks
12. Motivate in a convinging way why a research question is interesting

Fig. 2 Specific learning outcomes for the laboratory course sequence in the skills category communication

5 Modifications Imparted to the Existing Laboratory Course

To allow the achievement of the newly established learning goals, we made use of scaffolding techniques. Consequently significant modifications have been implemented in the existing laboratory course. These alterations encompass various facets, such as the redesign of the laboratory space and the course structure. In the subsequent sections, we provide an overview of each of these modifications.

5.1 Scaffolding

Conducting experiments necessitates the utilization of a diverse array of skills. We counted up to about 30–40 different of our "fine-grained" skills per experimental activity. To ensure that students engage in targeted practice to master each of these skills, we structured our curriculum and activities using scaffolding techniques.

Rather than overburdening students with a multitude of new skills from the beginning, we strategically staggered the introduction of new LOs in each lab activity (up to 11), thereby ensuring that students encounter a manageable number of LOs during each session. In the new curriculum, students initially practice fundamental skills during specialized active learning tutorials conducted in the first and second semester. These focus specifically on essential aspects of experimentation like estimation of measurement uncertainties, (visual) data representation and analysis and authentic use of laboratory notebooks (Stanley and Lewandowski 2016, 2018; Hoehn and Lewandowski 2020; Alemani 2023a, b). In other tutorials we introduce computational aspects of data handling and analysis in Python using Jupyter Notebook (Tufino et al. 2024). In a tutorial at the beginning of the second semester, students are introduced to the "Modeling Framework for Experimental Physics" developed by Lewandowski and colleagues (Zwickl et al. 2014, 2015). This framework was developed to guide students during experiments. It clarifies the importance of modeling the experimental measurement system in addition to the physical system to be studied, and highlights the iterative aspect in modeling.

Furthermore, our curriculum is sequenced such that more complex cognitive tasks are introduced after students have gained proficiency in foundational skills. For instance, students first learn to comprehend and apply data analysis methods before progressing to the development of their own data analysis strategies within experiments. Similarly, students initially practice documenting experiments using instructional videos before transitioning to documenting their own experiments independently. Note that, the difference to traditional concepts is the focus on the LOs, not mixing them with all other goals in pre-defined experimental work.

For the lab activities, this type of scaffolding is realized also by gradually increasing the complexity of experimental demands. However, it is important to note that one should not confuse the complexity of experimental setups with the level of the cognitive tasks that students engage with.

An integral aspect of our scaffolding approach involves also the gradual reduction of explicit instructional guidance over time. While students receive substantial guidance through the use of probing questions at the onset of the course, this guidance gradually diminishes as students become proficient in several scientific skills and progress through the courses. Towards the end of the courses time, we installed students' projects, which foster greater independence and self-directed learning (an aspect discussed in the literature as *agency* (Holmes et al. 2020)). In Fig. 3 some examples of the setups designed and built by students during their projects are shown.

5.2 *Laboratory Course Structure*

As a result of the outlined scaffolding, we reorganized several components of the course structure.

Prior to the transformation, various experiments with different setups were provided and students were randomly assigned to an order in which to process the experiments during the course of the semester. While this approach allows for a reduced number of each setup and a large variety of experiments, it limits scaffolding for skill acquisition. Consequently, we eliminated the rotating structure and transitioned to a uniform activity structure for the majority of activities, ensuring all students engage in the same scaffolded tasks in the same order.

The scaffolding strategy outlined in the preceding Sect. 5.1 has led to a decrease in the variety of experiments conducted by students and the range of physics topics

Fig. 3 Example of setups designed and built by students during their projects

covered (as detailed in the subsequent Sect. 5.3). Presently, students do not switch experiments at every lab session but may continue working on the same experiment across multiple appointments.

5.3 Physics Context

Before the transformation, the laboratory course covered many of the physics topics treated during the physics introductory lectures, spanning from mechanics and fluid dynamics to optics, thermodynamics, atomic physics and radioactivity. Since the reconstructured lab course is not aimed to reinforce physics concepts studied in lecture, we identify three physics areas as particularly suitable for the acquisition of experimental skills: Mechanics, Electricity/Electronic and Optics. Mechanics was chosen for its simplicity in apparatus, electricity/electronics and optics were chosen because they allow students to practice troubleshooting and iteration (American Association of Physics Teachers 1998; Zwickl et al. 2014).

5.4 Laboratory Space Redesign

We redesigned the laboratory space to foster a culture of collaboration and innovation among students. The introduction of open discussion areas facilitates group interactions during tutorials and experimental sessions, promoting peer-to-peer learning and knowledge exchange. Each room has been equipped with whiteboards, flip chart boards, and smart boards to encourage interactive learning. The integration of versatile furniture arrangements enables flexible use of space, accommodating varying group sizes and project requirements. Furthermore, the provision of fabrication tools, such as soldering stations, 3D printing facilities and very recently, a laser cutter, empowers students to engage in rapid prototyping and to materialize their design concepts effectively.

The removal of pre-existing setups optimized space utilization, creating ample room for students' projects and promoting a conducive environment for creative exploration. We established a reading corner at the lab entrance, featuring journals and magazines to inspire experimental endeavors. We attached the posters of the "Modeling Framework for Experimental Physics" (Zwickl et al. 2015, 2014; Dounas-Frazer and Lewandowski 2018) in each lab room. These posters are used by students and instructors while practicing modeling. In Fig. 4, one can see two views of an area for students discussion. One can notice a whiteboard, a blackboard, movable tables, and on the back the poster of the "Modeling Framework for Experimental Physics" and a small sketchnote poster about lab notebooks. On the side (see Fig. 4 on the left) there are soldering stations and a 3D printer. Not visible are on the other side a smartboard, a movable flipchart and benches for doing experimental work. Students in Fig. 4 are working together on solving data analysis problems.

Fig. 4 Pictures of a lab area redesigned to allow students' discussions and foster communication

5.5 *Laboratory Equipment*

Regarding the laboratory equipment, we prioritized leveraging existing resources while acquiring essential items aligned with our educational objectives. Notably, our learning outcomes did not hinge on the use of specialized or costly laboratory apparatus, as evidenced by the adaptability demonstrated during the COVID-19 pandemic, where students engaged in experimentation at home despite restricted access to expensive equipment. Nonetheless, our transformation necessitated the acquisition of both fundamental and sophisticated instruments. Notably, we transitioned predominantly to open-source software for data acquisition and analysis, minimizing the need for software investments.

For the first semester, students utilize basic instruments from our inventory. For the second semester, we acquired electrical equipment and established four soldering stations to facilitate setting up customized electrical measurements (see Fig. 4). Optical equipment such as breadboards, optics holders and different optical components was obtained for the third semester, enabling students to construct setups independently.

For the fourth semester, we introduced Arduinos and associated materials, providing each student with a comprehensive kit including LEDs, resistors, cables, and sensors for in-class and at-home experimentation. This adaptation, initially prompted by the COVID-19 pandemic, has since become integrated into regular coursework with modifications for in-person instruction. The use of Arduinos reduces investment costs. Additional sensor or material purchases are made as needed for project work, averaging 100–200 Euros per semester. Students are encouraged to explore available resources and coordinate with instructors for equipment borrowing, ensuring safe and efficient project execution.

5.6 Use of Authentic Forms of Laboratory Notebooks

In the traditional class format prior to the redesign, students wrote comprehensive final reports detailing their experimental setups, procedures, data analysis, and conclusions following each lab session. This process demanded substantial time investment from students, overshadowing the documentation of in-lab activities within their reports.

To foster deeper engagement in scientific practices during laboratory sessions and to emphasize the importance of laboratory notebooks for professional experimental physicists, we opted to transition our assessment focus from laboratory reports to laboratory notebooks (LNs). As noted by Edward H. Carlson, "learning to keep a good lab notebook is a difficult skill and it is tightly coupled to good work habits" (Carlson 1986). We now require final reports only for students' projects, not for all the other activities in the course. Student performance is primarily evaluated based on the quality of their LNs for each lab activity. The pedagogical advantages of utilizing LNs have been previously underscored in the literature (Stanley and Lewandowski 2016, 2018; Hoehn and Lewandowski 2020). They allow instructors to redirect attention from mere end-results to the entire experimental process. Instructors have the opportunity to recognize and commend students for acknowledging and reflecting on their mistakes within their lab notebooks. Additionally, LNs facilitate the documentation of iterative processes, such as model refinement and experimental apparatus adjustments, fostering the development of expert-like attitudes among students. Furthermore, LNs serve as a platform to instill crucial habits, such as immediate data plotting during or directly after measurements, enabling students to closely track experimental progress and make informed decisions based on their findings. Encouraging students to reflect on their work within the lab setting empowers them to refine their experimental techniques and enhance their overall learning experience.

5.7 Rubrics

As recommended in the literature (Etkina et al. 2006), we initiated the implementation of rubrics for student assessment within our laboratory courses, recognizing immediately their benefits. Primarily, rubrics serve to align student assessment with established learning objectives, thus mitigating the risk of divergent assessment among different instructors. Additionally, the process of crafting rubrics enables curriculum designers to scrutinize the validity of predetermined learning objectives, providing a tangible means to evaluate what is effectively practiced and measured within the curriculum. Furthermore, rubrics empower students to focus their efforts on the key facets for which each activity was designed, offering them a structured guide to orient their work effectively.

5.8 *Students' Coaching by Instructors and Teaching Assistants*

To foster students' decision-making and self-confidence in the laboratory we reconsidered the role and the teaching practices of the laboratory instructors. They slipped into the role of facilitators that support students in taking decisions in a non-invasive way. They now use their experience while authentically discussing students' ideas in a critical way. Enabling students to make decisions and attempt to resolve issues independently, as well as allowing them to practice troubleshooting, necessitates effective and coordinated interaction between instructors and students.

To calibrate instructors interventions based on students' self-reflection about their experimental process status, we adopted a system from colleagues at the University of Hannover (Barth et al. 2015). Each group of students is provided with three cones: green, yellow, and red (see detail on the table of Fig. 4). The green cone is exposed by students to indicate that they are working smoothly without encountering difficulties. When students encounter problems or troubleshooting scenarios, they display the yellow cone, and when they have persistently attempted to resolve a problem without success, they display the red cone. This visual cue allows instructors to tailor their interventions based on the students' expressed state, for example avoiding seeking interaction with students too often when they feel they are working smoothly on a process. There are occasions when students prefer to troubleshoot independently, even when encountering difficulties, demonstrating their desire to persevere and resolve issues.

5.9 *Group Work*

Before the laboratory transformation, students typically worked in pairs, a practice initially retained under the assumption that smaller groups would afford more hands-on practice and deeper learning experiences. However, insights gleaned from the Colorado Learning Attitudes about Science Survey for Experimental Physics (E-CLASS) survey (Wilcox and Lewandowski 2018) regarding students' attitudes towards communication prompted a reassessment of this approach (Teichmann et al. 2022). The findings revealed a lack of perceived importance regarding peer communication in scientific practices, prompting a strategic pivot in our instructional methods. To address this concern, we reconsidered how we foster communication in the entire lab course and added tasks or activities specifically focused on letting students brainstorm, discuss, present their ideas and results to other peers. For example, we introduced new activities wherein students collaborated in larger groups with shared goals (Alemani 2023c). Within these activities, students subdivided themselves into smaller teams, each tasked with specific sub-goals, fostering intergroup communication to exchange insights, solicit feedback, and resolve challenges collectively. An example of such a collective task is to build and test a radio receiver in the middle waves.

Moreover, tutorial sessions were restructured to accommodate larger group sizes, ranging from 3 to 12 students depending on the nature of the tasks. Notably, we adjusted the typical group size for experiments to three individuals and up to four for projects. This modification facilitates real-time experimentation, data analysis (utilizing Python), and documentation through shared responsibilities within the enlarged groups. Emphasizing the importance of structured group dynamics, we implemented mechanisms to ensure equitable participation, such as rotating roles during sessions or assigning a student as a communication facilitator to streamline intra-group exchanges or we reserved time for students' presentations to peers.

5.10 Description of the Curriculum

The curriculum resulting from the above described transformation process is now coarsely described. A description of the scaffolded activities can be found in Open Science Framework (2025).

In the first semester, students first learn the fundamentals of measurement uncertainties, data analysis, Python programming and scientific documentation during active learning tutorials. Thereafter, they practice those skills further during laboratory activities. In the second semester we introduce *modeling* using the *Modeling Framework for Experimental Physics* (Zwickl et al. 2015, 2014; Dounas-Frazer and Lewandowski 2018) and tailored activities. Students practice modeling electrical circuits and apparatus, identifying the assumptions and limits of those models and refining them. For example students have to figure out electrical components inside a closed box. At the end of the semester students participate in a large group project in which they practice more independently working with electrical circuits.

Students continue practicing on *modeling* and the skills in the other categories in the third semester by designing, building, and characterizing various optical measurement apparatus.

In the fourth semester, students extend their technical skills learning how to use and programm Arduino microprocessors and to measure or modify the real world by means of sensors and actuators. In different experiments students continue practicing modeling. At the end of the semester, during projects, they work on designing experiments to answer a self-determined research question (see Fig. 3).

6 Assessment

We used the E-CLASS survey to assess and guide our course transformation. The results until 2022 can be found in Teichmann et al. (2022; 2024). Here we just summarize the main findings. Instruction had a positive impact on students views about experimental physics. In particular, students gain confidence in doing data analysis, in overcoming difficulties and in using new lab equipment. On the other hand, instruction had a negative impact regarding the importance of communicating scientific results to peers. As mentioned previously in Sect. 5.9, this result motivated us to focus on communication aspects in the years after the assessment. To monitor the effects of our changes regarding this aspect we plan to collect further E-CLASS data.

7 Conclusions

Our article has outlined the comprehensive redesign and reconstruction of an introductory laboratory course for physics major students. The goal is to engage students actively and provide them with genuine experiences in scientific practices while cultivating their acquisition of experimental skills and expert-like attitudes. We have delineated the newly defined learning objectives and the changes imparted to the existing course and the activities crafted to activate students and immerse them in high-level cognitive tasks.

This work serves as a valuable resource for educators embarking on similar endeavors, offering practical guidance and innovative strategies to enhance student engagement and promote active learning in laboratory settings.

Acknowledgements The author would like to thank all who contributed or supported the laboratory reconstruction process. Great gratitude goes to all students who provided their precious and continuous feedback and consistently encourage the author in her efforts. I acknowledge support from the Stifterverband and the Baden-WÃ¼rttemberg Stiftung (Project No. H1205228500830925).

References

Alemani M (2023a) Eigene laborbuecher führen. unterrichtsaktivitaeten zur einuebung der wissenschaftlichen dokumentation in der sekundarstufe ii. Unterr Phys 195+196:25–27

Alemani M (2023b) Laborbuecher. eine authentische form der wissenschaftlichen dokumentation mit didaktischem potenzial. Unterr Phys 195+196:14–16

Alemani M (2023c) The redesign of an introductory physics laboratory course. Il Nuovo Cimento C 46:181. https://doi.org/10.1393/ncc/i2023-23181-y

American Association of Physics Teachers (1998) Goals of the introductory physics laboratory. Am J Phys 66(6):483–485. https://doi.org/10.1119/1.19042

Anderson LW, Krathwohl DR, Airasian PW, Cruikshank KA, Mayer RE, Pintrich PR, Raths R, Wittrock MC (2001) A taxonomy for learning, teaching, and assessing: a revision of Bloom's taxonomy of educational objectives. Allyn and Bacon, Pearson Education Group, Boston, MA

Barth M, Klappauf I, Friege G (2015) Das EU-projekt SAILS: Beispiele für den systematischen einsatz von begleitender lernerfolgsmessung (formative assessment) in forschendem physikunterricht. PhyDid B. https://ojs.dpg-physik.de/index.php/phydid-b/article/view/522

Bearden I, Dvorak L, Planinsic G (2022) Work group 2 position paper: Experiments and laboratory work in teacher education. J Phys Conf Ser 2297(1):012008. https://doi.org/10.1088/1742-6596/2297/1/012008

Brookes D, Etkina E, Planinsic G (2020) Implementing an epistemologically authentic approach to student-centered inquiry learning. Phys Rev Phys Educ Res 16(2):020148. https://doi.org/10.1103/PhysRevPhysEducRes.16.020148

Carlson EH (1986) Constructing laboratory courses. Am J Phys 54(11):972–976. https://doi.org/10.1119/1.14835

Chi MTH, Wylie R (2014) The ICAP framework: linking cognitive engagement to active learning outcomes. Educ Psychol 49(4):219–243. https://doi.org/10.1080/00461520.2014.965823

Dounas-Frazer DR, Lewandowski HJ (2018) The modelling framework for experimental physics: description, development, and applications. Eur J Phys 39(6):064005. https://doi.org/10.1088/1361-6404/aae3ce

Etkina E, van Heuvelen A, White-Brahmia S, Brookes DT, Gentile M, Murthy S, Rosengrant D, Warren A (2006) Scientific abilities and their assessment. Phys Rev Phys Educ Res 2(2):020103. https://doi.org/10.1103/PhysRevSTPER.2.020103

Hoehn JR, Lewandowski HJ (2020) Incorporating writing in advanced lab projects: a multiple case-study analysis. Phys Rev Phys Educ Res 16(2):020161. https://doi.org/10.1103/PhysRevPhysEducRes.16.020161

Holmes NG, Keep B, Wieman CE (2020) Developing scientific decision making by structuring and supporting student agency. Phys Rev Phys Educ Res 16(1):010109. https://doi.org/10.1103/PhysRevPhysEducRes.16.010109

Holmes NG, Wieman C (2018) Introductory physics labs: we can do better. Phys Today 71(1):38–45. https://doi.org/10.1063/PT.3.3816

May JM (2023) Historical perspectives on physics education research: A review. Phys Rev Phys Educ Res 19(2):020168. https://doi.org/10.1103/PhysRevPhysEducRes.19.020168

Meltzer DE, Thornton RK (2012) Resource letter alip-1: Active-learning instruction in physics. Am J Phys 80(6):478–496. https://doi.org/10.1119/1.3678299

Open Science Framework: Open science framework repository. https://osf.io/s6ry4/

Redish EF (2004) A theoretical framework for physics education research: modeling student thinking. In: Proceedings of the international school of physics "Enrico Fermi", research on physics education, vol 156, pp 1–63. https://doi.org/10.3254/978-1-61499-012-3-1

Stanley JT, Lewandowski HJ (2016) Lab notebooks as scientific communication: investigating development from undergraduate courses to graduate research. Phys Rev Phys Educ Res 12(2):020129. https://doi.org/10.1103/PhysRevPhysEducRes.12.020129

Stanley JT, Lewandowski HJ (2018) Recommendations for the use of notebooks in upper-division physics lab courses. Am J Phys 86(1):45–53. https://doi.org/10.1119/1.5001933

Teichmann E, Lewandowski HJ, Alemani M (2024) A scaffolded curriculum for introductory physics laboratories. In: Journal of physics: conference series. Accepted, 2024

Teichmann E, Lewandowski HJ, Alemani M (2022) Investigating students' views of experimental physics in German laboratory classes. Phys Rev Phys Educ Res 18(1):010135. https://doi.org/10.1103/PhysRevPhysEducRes.18.010135

Tufino E, Oss S, Alemani M (2024) Integrating python data analysis in an existing introductory laboratory course. Eur J Phys 45(4):045707. https://doi.org/10.1088/1361-6404/ad4fcc

Volkwyn TS, Allie S, Buffler A, Lubben F (2008) Impact of a conventional introductory laboratory course on the understanding of measurement. Phys Rev Phys Educ Res 4(1):010108. https://doi.org/10.1103/PhysRevSTPER.4.010108

Wieman C (2015) Comparative cognitive task analyses of experimental science and instructional laboratory courses. Phys Teach 53(7):349–351. https://doi.org/10.1119/1.4928349

Wieman C, Holmes NG (2015) Measuring the impact of an instructional laboratory on the learning of physics. Am J Phys 83(11):972–978. https://doi.org/10.1119/1.4931717

Wilcox BR, Lewandowski HJ (2018) A summary of research-based assessment of students' beliefs about the nature of experimental physics. Am J Phys 86(3):212–219. https://doi.org/10.1119/1.5009241

Zwickl BM, Finkelstein N, Lewandowski HJ (2013) The process of transforming an advanced lab course: goals, curriculum, and assessments. Am J Phys 81(1):63–70. https://doi.org/10.1119/1.4768890

Zwickl BM, Finkelstein N, Lewandowski HJ (2014) Incorporating learning goals about modeling into an upper-division physics laboratory experiment. Am J Phys 82(9):876–882. https://doi.org/10.1119/1.4875924

Zwickl BM, Hu D, Finkelstein N, Lewandowski HJ (2015) Model-based reasoning in the physics laboratory: framework and initial results. Phys Rev Phys Educ Res 11(2):020113. https://doi.org/10.1103/PhysRevSTPER.11.020113

Study of the Free Damped Oscillations of an Elastic Rubber Band Using a Smartphone

Aurelio Agliolo Gallitto, Claudio Fazio, and Onofrio Rosario Battaglia

Abstract The new interest in using smartphones as laboratory instruments for classroom physics experiments stems from their potential to conduct a diverse range of educational experiments using affordable equipment. This article presents a study on a damped oscillator, which includes an elastic rubber loop and a mass. We investigated the oscillations using a smartphone. The basic model was used to evaluate the experimental data, yielding insights into the viscoelastic and non-linear characteristics of the rubber material.

Keywords Smartphone-based experiments · Damped harmonic oscillator · Viscoelasticity · Modelling

1 Introduction

Smartphones may be utilised as laboratory instruments for conducting a wide range of educational experiments in an introductory physics lab, as demonstrated by various studies (Hammond and Assefa 2007; Kuhn and Vogt 2013; Oprea and Miron 2014; Agliolo Gallitto et al. 2022; kuhn and Vogt 2022). Smartphones are equipped with multiple sensors that, when controlled by suitable software (apps), enable users to investigate physical phenomena and conduct various measurements of physical quantities (Kuhn and Vogt 2012; Pili 2018; Listiaji et al. 2021; Organtini 2021). This is achieved by creating flexible and affordable set-ups that replace the need for expensive devices with smartphones.

Viscoelasticity is a term that describes the phenomenon in which elastic rubber materials display viscous and elastic properties simultaneously when they are subjected to deformation.

A. Agliolo Gallitto · C. Fazio · O. R. Battaglia (✉)
Dipartimento Di Fisica E Chimica - Emilio Segrè, University of Palermo, Palermo, Italy
e-mail: onofriorosario.battaglia@unipa.it

Z. Ješková and J. Hanč (eds.), *Advancing Physics Education*, Challenges in Physics Education, https://doi.org/10.1007/978-3-031-94207-5_15

In contrast to materials that are entirely elastic, viscoelastic substances have a viscous component that causes loss of energy when a force is applied and then removed. By observing the force–deformation (stress–strain) curve, hysteresis can be noticed. The area of the hysteresis curve is equal to the amount of energy that is lost during each cycle, as stated by Treloar (1975).

In this article, we show the usage of the smartphone accelerometer sensor (accelerometer), operated by the free phyphox app (Staacks et al. 2018; Pierratos and Polatoglou 2020; Agliolo Gallitto et al. 2021), to carry out measurements of vertical oscillations of a damped oscillator made of an elastic rubber-band loop and a mass. The experimental set-up enables the acquisition of data regarding the viscoelastic characteristics of the rubber material (Filipponi et al. 2010; Huang et al. 2020). We employ an elementary model to comprehend the dynamics of the system consisting of a rubber-band loop and a mass (Agliolo Gallitto et al. 2021; Agliolo Gallitto 2016; Deacon 1992).

2 Set-Up and Data

Determining the period of a vertical harmonic oscillator is a frequent activity in physics educational laboratories. Usually, this task is accomplished by manually measuring the time using a stopwatch. Nevertheless, the manual approach is unsuitable for investigating a damped harmonic oscillator due to the existence of damping, which hinders precise measurement of the oscillations in both time and amplitude. In order to surmount these challenges, we employ a smartphone as a means to collect empirical data.

The fluctuation of the damped oscillator, composed of a stretchable rubber band and a mass, is measured using the accelerometer embedded in the smartphone and operated through the phyphox app. Figure 1 depicts the experimental configuration. The smartphone is hung using a rubber-band loop with a cross-sectional area of 1 mm by 3 mm, enabling it to move up and down in a regular pattern. The smartphone, functioning as an oscillating mass, is firmly affixed to a custom-made expanded-PVC support. The smartphone's weight, including the support, is 265 grammes. The overall mass can be raised by attaching additional weights, as depicted in Fig. 1.

In order to record the acceleration of the device, the phyphox app makes use of the accelerometer that is built into the smartphone. This accelerometer is typically a micro-electro-mechanical system. According to Agliolo et al. (2015), the accelerometer is responsible for determining the gravitational acceleration, which is represented by the symbol g, whether the smartphone is either stationary or travelling at a constant speed.

When the smartphone is not being accelerated, the phyphox app provides a dedicated function known as "Acceleration without g" that enables users to identify a zero acceleration value. This option is available to users. This function makes use of a virtual sensor that, in most cases, makes use of data from other sensors in order to subtract the constant acceleration g. This application offers the capability to be

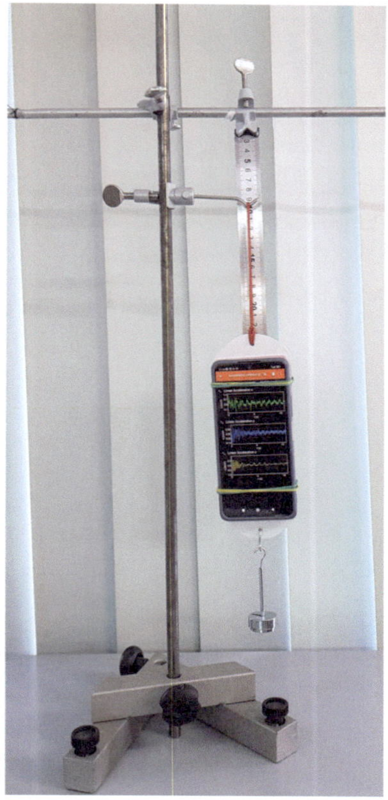

Fig. 1 Experimental set-up: smartphone is attached to a 1 × 3 mm rubber-band loop on a vertical shaft

remotely modified and monitored by any other device that is connected to a network (Staacks et al. 2018). Furthermore, it allows for the raw data that is generated to be stored in a file for further analysis in the future. As depicted in Fig. 1, smartphones come with three accelerometers that are built in. Each of these accelerometers is aligned with one of the three spatial axes that make up a standardised rectangular coordinate system.

After being moved a few centimetres away from its resting position and then released, the smartphone will oscillate up and down before returning to its original position. The phyphox app is responsible for recording the three components of the smartphone's acceleration at regular intervals of approximately 0.005 s.

An example of a screenshot taken by the phyphox app when it was in a vertical free motion is presented in Fig. 2. It is along the vertical y-axis that the oscillations are occurring. Furthermore, the values of the acceleration along the x-axis and the z-axis continue to be near zero. An estimation of the uncertainty of the measurements of acceleration and time is provided to us by the technical features of the smartphone that was used in the experiment. The values for the absolute errors for acceleration and time are computed to be 0.01 m/s^2 and 0.005 s, respectively. We

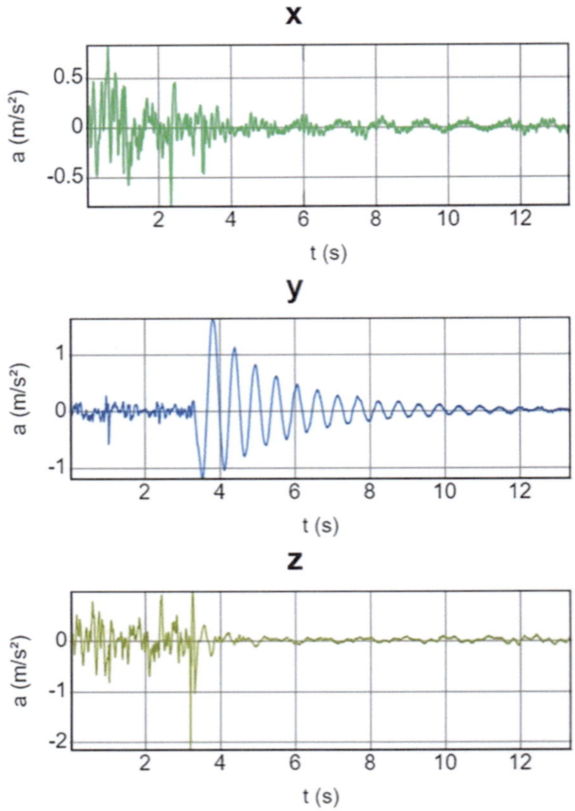

Fig. 2 A picture from the phyphox app that shows how the mass's acceleration changes over time during the damped waves. Since the cycles happen along the y-axis, the accelerations along the x- and z-axes don't change much during the experiment

are able to make an estimate of the elastic constant of the rubber-band loop using the static measurement approach, which is approximately $k = (43.5 \pm 0.6)$ N/m. It corresponds to an undamped angular frequency of $\omega_0 = \sqrt{k/m} = (9.94 \pm 0.07)$ rad/s when m is equal to 440 g. Figure 3 illustrates the values of acceleration in relation to time for a total mass equivalent to 440 g.

3 A Simple Damped Oscillations Model

The following differential equation is used to explain the free vertical motion of an ideal spring–mass oscillator with an elastic constant k and a mass m. This motion is governed by Newton's second law.

$$m \frac{d^2}{dt^2} y(t) + k\, y(t) - mg = 0 \tag{1}$$

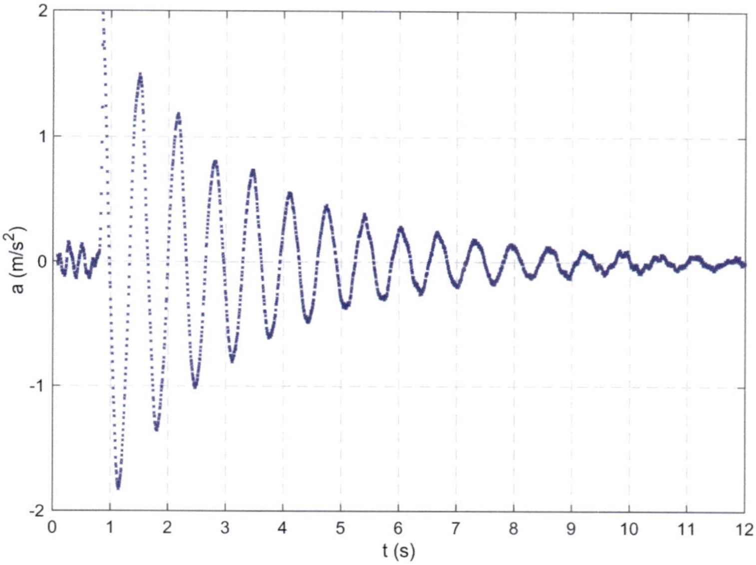

Fig. 3 Plot the experimental data of acceleration along the y-axis over time for a total mass of 440 g

which has a periodic solution as

$$y(t) = A\cos(\omega_0 t + \phi) + \frac{mg}{k} \tag{2}$$

where A is the amplitude of the oscillations, ϕ the phase constant and ω_0 the angular frequency that depends on k and m as

$$\omega_0 = \sqrt{k/m} \tag{3}$$

A sinusoidal function is used to describe the location of the mass as a function of time around the equilibrium point at $y_0 = mg/k = g/\omega_0^2$. This function has a constant amplitude and a constant frequency, and it does not depend on the oscillation amplitude Serway and Jewett (2018).

From (2), it is possible to obtain the mathematical expression of the acceleration as

$$a(t) = \frac{d^2}{dt^2}\, y(t) = -\omega_0^2 A\cos(\omega_0 t + \phi) \tag{4}$$

where $\omega_0^2 A$ is the acceleration amplitude.

In ideal systems, the rhythmic movement that we have described before occurs because of a linear restoring force that adheres to Hooke's law. These systems continue to oscillate indefinitely because of this force. In every genuine mechanical system, frictional forces cause motion to be impeded, which prevents endless oscillation from occurring. The mechanical energy of the system is steadily reduced over time as a result of friction, which ultimately leads to motion that is saturated. An oscillator being referred to as a damped harmonic oscillator is one in which the frictional force is directly proportional to the velocity of the oscillator.

The rubber-band oscillator is subject to elasticity and restoring force, in addition to significant drag and friction characteristics. In order to describe a rubber-band oscillator, it is possible to use a simplified mathematical model for Newton's second law. This model is obtained by connecting in series an ideal spring with an elastic constant of k and a damper with viscous damping of coefficient β. The damper exerts a frictional force that is proportional to the velocity, as described below.

$$F_v = -\beta v_y = -\beta \frac{\mathrm{d}}{\mathrm{d}t} y(t) \tag{5}$$

By taking into account the frictional force (1) becomes as reported below

$$m \frac{\mathrm{d}^2}{\mathrm{d}t^2} y(t) + \beta \frac{\mathrm{d}}{\mathrm{d}t} y(t) + k\, y(t) - mg = 0 \tag{6}$$

such that it is possible to rewrite it into the form

$$\frac{\mathrm{d}^2}{\mathrm{d}t^2} y(t) + 2\gamma \frac{\mathrm{d}}{\mathrm{d}t} y(t) + \omega_0^2 y - g = 0 \tag{7}$$

where $\omega_0 = \sqrt{k/m}$ is the undamped angular frequency of the oscillator, $\gamma = \dfrac{\beta}{2m}$ the damping coefficient and g the gravity acceleration.

When $\gamma < \omega_0$, the solution of (7) is

$$y(t) = Ae^{-\gamma t} \cos(\omega t + \phi) + \frac{mg}{k} \tag{8}$$

where $\omega = \sqrt{\omega_0^2 - \gamma^2}$.

Equation (8) is used to represent the phenomenon known as damped oscillations. This is due to the fact that the amplitude of $y(t)$ is no longer constant but rather declines exponentially as a function of time (Serway and Jewett 2018; Castro-Palacio et al. 2013). Feynman (1963) provides a comprehensive mathematical description of the damped harmonic oscillator. According to (8), it can be deduced that the acceleration likewise undergoes periodic variations throughout time.

$$a(t) = Ae^{-\gamma t}[(\gamma^2 - \omega^2)\cos(\omega t + \phi) + 2\gamma\omega \sin(\omega t + \phi)] \tag{9}$$

4 The Results

To establish the damping coefficient, we can use the experimental values of acceleration a vs. time from Fig. 3. By creating a semi-log graph with a logarithmic scale on the y-axis, we can plot the acceleration values at the crest (a_{max}) against time.

The semi-log graph in Fig. 4 displays the values of a_{max} plotted against time t.

By taking into account the parameter γ, which is responsible for determining the reduction in peak values, and fitting this data using a regression linear approach, we were able to acquire $\gamma = 0.352 \pm 0.016\ s^{-1}$, along with a correlation coefficient of $R = 0.995$.

A significant factor that plays a role in determining the behaviour of the system is the value of the damping coefficient γ.

The system is considered to be under-damped when $\gamma < \omega_0$ is proved, as it is in our example. This means that the system oscillates, but at a little different frequency compared to the scenario when it is not dampened, and the amplitude steadily decreases until it reaches zero. The oscillation motion is maintained when the viscosity is relatively low; nevertheless, the amplitude of vibration decreases over time, and the motion eventually stops.

Now, let's examine the model that was outlined in the preceding part. Figure 5 depicts the relationship between acceleration a and time for a mass of 440 g. The symbols represent the experimental data, whereas the continuous red line is derived from a non-linear fitting using (9). The optimal value for γ is determined to

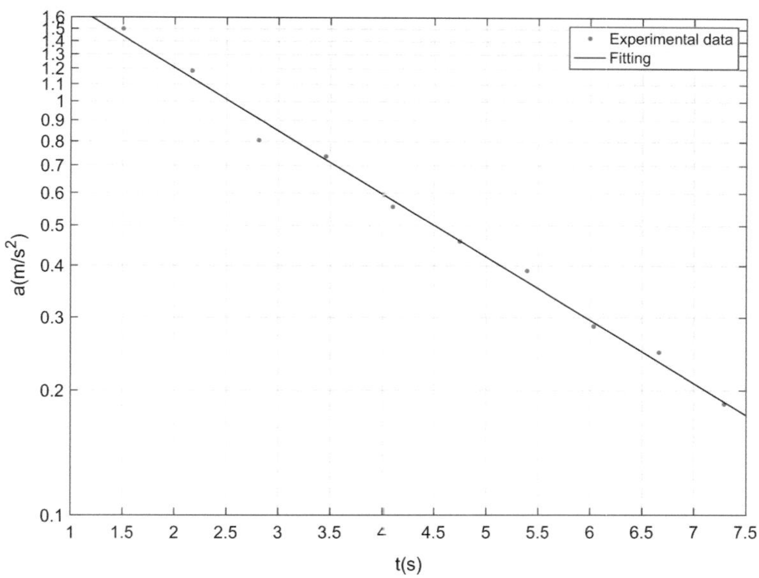

Fig. 4 Semi-log graph of the acceleration values measured at the crest, a_{max} versus time for a mass of 440 g

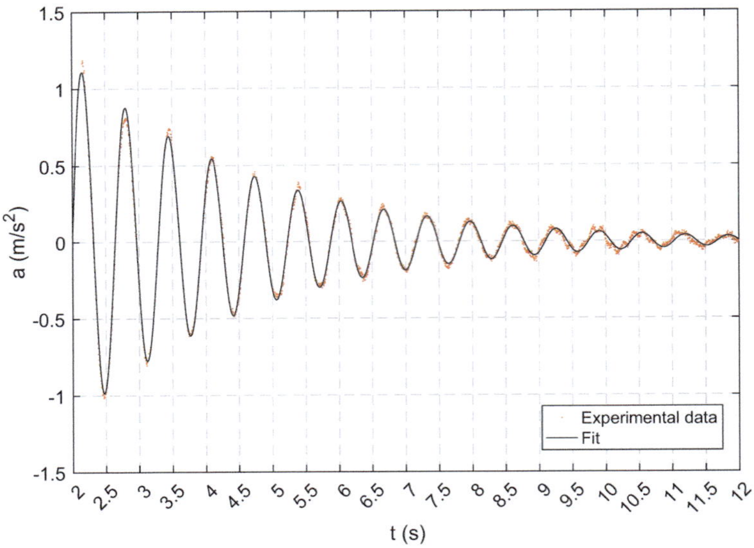

Fig. 5 Plot of a versus time. Symbols are experimental data; the continuous red line is a plot of (9), with $\gamma = 0.3679\ \mathrm{s}^{-1}$ and $\omega = 9.7249\ \mathrm{s}^{-1}$

be $0.3679\ \mathrm{s}^{-1}$ with a confidence interval of [0.3651; 0.3707]. Similarly, the optimal value for ω is found to be $9.7249\ \mathrm{s}^{-1}$ with a confidence interval of [9.7221; 9.7277]. It is noteworthy that the γ value acquired by the non-linear fit aligns with the value previously obtained by assuming simply an exponential drop in amplitude.

5 Discussion

Viscoelastic substances have a strain rate that is dependent on time because of their viscosity, which is expressed as a percentage. There is a connection between the chemical rearrangement and the viscoelastic qualities characteristic of elastic rubber. Components of the lengthy polymer chain shift their locations in response to the application of force. This particular movement or rearrangement is referred to as a "creep". Even though these sections of the chains of polymers are reorganising in order to follow the internal tension, polymers continue to be solid materials. This process, which causes the material to undergo back stress, is what causes the material to remain solid. After the removal of the external force, the polymer will return to its initial form due to the back stress, as stated in the research conducted by Ferry (1980).

Various viscoelastic materials, including amorphous polymers, semicrystalline polymers, biopolymers and even live tissue and cells, can be modelled as linear combinations of springs and dashpots (or dampers, which are mechanical devices that oppose motion by the use of viscous friction) (Huang et al. 2020). Particularly, Filipponi (2010) demonstrated that the dynamical behaviour of a mass–rubber band oscillator cannot be explained by a straightforward damped harmonic oscillator model. Instead, it necessitates a more complex model that takes into account viscoelastic aspects in an appropriate manner. Nevertheless, in a subsequent study, Huang (2020), demonstrated that the mechanical properties of rubber materials can be represented by the static elastic modulus (elastic coefficient), viscoelastic degree and viscoelastic coefficient.

It is important to acknowledge that the modelling presented here may have a didactic limitation on intrinsic mathematical complexity, particularly in the context of lower education levels. In this regard, it may be advantageous to consider the physical system under investigation as a dynamic system in order to employ modelling tools that are specifically designed for dynamic systems to study it. In general, the investigation of dynamic systems can be particularly challenging due to the necessity of resolving differential equations. Nevertheless, the resolution of differential equations can be readily surmounted by computer simulation environments like Coach (Heck and Ellermeijer 2009; Heck et al. 2009) or Netlogo (1999). These environments are, in fact, highly intuitive, and they enable the analysis of even the most intricate systems through a presentation that contains only a few elements.

In this section, we will not go into detail about the models that were constructed to take into consideration the viscoelastic features because doing so would go beyond the scope of the current paper. In spite of this, it is essential to point out that the elastic material that was used in the experiments that were detailed here might exhibit a behaviour that is not consistent with the linear force model's expected behaviour. This happens when the oscillation amplitude is large enough. Figure 6 illustrates this point exactly. To be more specific, when time values of around six seconds or more are taken into consideration, it is clear that the fitting curve displays a discernible disparity in comparison to the experimental data. Furthermore, the angular frequency of the under-damped harmonic oscillator is determined as expected by the linear model, which is $\omega = \sqrt{\omega_0^2 - \gamma^2} = 9.93 \pm 0.07$ rad/s. This is achieved by inserting the value that was previously obtained for γ. Because this value disagrees with the value that was acquired by the non-linear fitting, it demonstrates once more that the linear model is not appropriate for this oscillating system when it is operating in that domain.

It is evident that the material of which the elastic is made shows a non-linear behaviour that we could take into account by modifying in our model the mathematical expression of the elastic force. We found that if we insert this mathematical expression

$$F_e = k_0 \left(y(t) + y^2(t) + y^3(t) \right) \tag{10}$$

for the elastic force in (7) we obtain a better agreement with experimental data as shown in the Fig. 7.

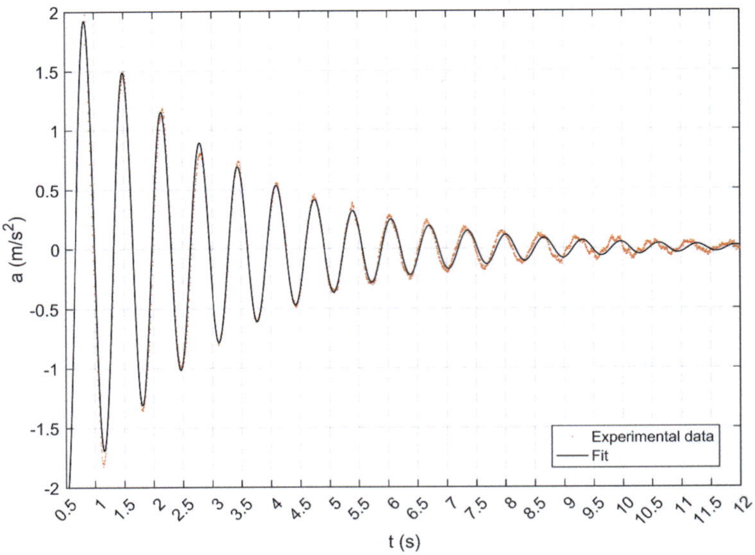

Fig. 6 Plot of a versus time for a mass of 440 g. Symbols are experimental data; the continuous red line is a plot of (9), with $\gamma = 0.39\ \mathrm{s}^{-1}$ and $\omega = 9.67\ \mathrm{s}^{-1}$

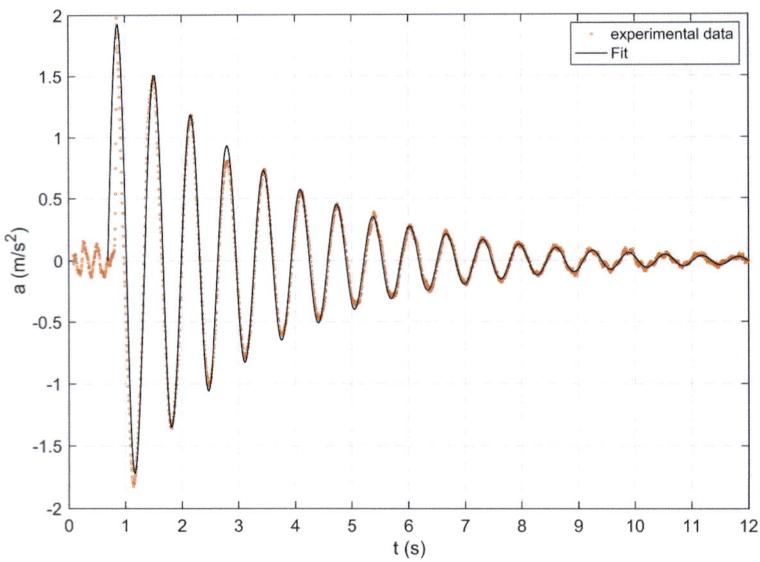

Fig. 7 Plot of a versus time for a mass of 440 g. Symbols are experimental data; the continuous red line is a plot of the numerical solution of (10), with $\gamma = 0.376\ \mathrm{s}^{-1}$ and $\omega = 8.64\ \mathrm{s}^{-1}$

6 Conclusion

A physics experiment that aims to measure the damping coefficient of an oscillator that is made up of a rubber loop band and a mass has been characterised by us as an educational activity that may be carried out by using smartphones to collect data. Our ability to investigate the damping oscillations was made much simpler by the use of the smartphone. Using smartphone applications, it offers a novel approach to teaching and learning that is intended to supplement the conventional method of education. This approach not only encourages students to develop an interest in physics (Gonzalez 2015; Lindquist 2007; Ayu et al. 2009; Hochberg et al. 2018) but also provides students with increased mobility for the purpose of distance learning (Vázquez-Cano 2014). The use of smartphones as measuring equipment in physics experiments, on the other hand, opened the door to the potential of creating and building multifunctional, low-cost laboratories. These laboratories might also be used for experiments that are conducted at a very low cost, such as remote learning or lecture experiments. Students at the undergraduate level can gain the skills to undertake some elementary error analysis.

To study the issue here presented, experiments must be conducted, models created and then compared with the results. Examining rubber's behaviour in undergraduate studies might be useful for bridging theoretical principles with real-world uses and for assessing fundamental models for a widely used material.

References

Agliolo Gallitto A (2016) Introduzione al Laboratorio di Fisica: gli errori nelle misure sperimentali. University of Palermo

Agliolo Gallitto A, Battaglia OR, Cavallaro G, Lazzara G, Lisuzzo L, Fazio C (2022) Exploring historical scientific instruments by using mobile media devices. Phys Teach 60:204–208. https://doi.org/10.1119/5.0032111

Agliolo Gallitto A, Battaglia OR, Fazio C (2021) Dynamic measurement of the elastic constant of a helicoidal spring by a smartphone. Phys Educ 56:035010. https://doi.org/10.1088/1361-6552/abdfac

Agliolo Gallitto A, Lupo L (2015) A mechanical model of the smartphone's accelerometer. Phys Educ 50:547–646. https://doi.org/10.1088/0031-9120/50/6/646

Ayu MA, Taylor K, Mantoro T (2009) Active learning: engaging students in the classroom using mobile phones. In: IEEE symposium on industrial electronics and applications (ISIEA), pp 711–715

Castro-Palacio JC, Velázquez-Abad L, Giménez MH, Monsoriu JA (2013) Using a mobile phone acceleration sensor in physics experiments on free and damped harmonic oscillations. Am J Phys 81:472–475. https://doi.org/10.1119/1.4793438

Deacon CG (1992) Error analysis in the introductory physics laboratory. Phys Teach 30:368–371. https://doi.org/10.1119/1.2343578

Ferry J (1980) Viscoelastic properties of polymers. Wiley, New York

Feynman RP, Leighton RB, Sands M (1963) The Feynman lectures on physics, vol 1, 1 edn., chap 24-2. Addison-Wesley, Reading, Massachusetts. https://www.feynmanlectures.caltech.edu/I_24.html

Filipponi A, Di Michele L, Ferrante C (2010) Viscoelastic behavior of a mass-rubber band oscillator. Am J Phys 78:437–444. https://doi.org/10.1119/1.3276052
Gonzalez MA et al (2015) Teaching and learning physics with smartphones. J Cases Inf Technol 17:31–50. https://doi.org/10.4018/JCIT.2015010103
Hammond EC, Assefa M (2007) Cell phones in the classroom. Phys Teach 45(5):312. https://doi.org/10.1119/1.2731283
Heck A, Ellermeijer T (2009) Giving students the run of sprinting models. Am J Phys 77:1028–1038. https://doi.org/10.1119/1.3187150
Heck A, Kedzierska E, Ellermeijer T (2009) Design and implementation of an integrated computer working environment. J Comput Math Sci Teach 28:147–161. https://www.learntechlib.org/p/30302/
Hochberg K, Kuhn J, Müller A (2018) Using smartphones as experimental tools—effects on interest, curiosity, and learning in physics education. J Sci Educ Technol 27:385–403. https://doi.org/10.1007/s10956-018-9731-7
Huang Y, He S, Yu A, Ren J (2020) Study on the free vibration of a rubber band oscillator. IOP Conf Ser Mater Sci Eng 782:022093. https://doi.org/10.1088/1757-899X/782/2/022093
Kuhn J, Vogt P (2012) Analyzing spring pendulum phenomena with a smartphone acceleration sensor. Phys Teach 50:504–505. https://doi.org/10.1119/1.4758162
Kuhn J, Vogt P (2013) Applications and examples of experiments with mobile phones and smartphones in physics lessons. Front Sens 1:67–73
Kuhn J, Vogt P (eds) (2022) Smartphones as mobile minilabs in physics. Springer Nature, Heidelberg
Lindquist D et al (2007) Exploring the potential of mobile phones for active learning in the classroom. ACM SIGCSE Bull 39:384–388
Listiaji P, Darmawan MS, Daeni F, Karmuji (2021) Comparison between the use of acceleration sensor and video tracker on smartphone for spring oscillation experiment. Phys Educ 56:013001. https://doi.org/10.1088/1361-6552/abb88b
Oprea M, Miron C (2014) Mobile phones in the modern teaching of physics. Rom Rep Phys 66:1236–1252
Organtini G (2021) Physics experiments with Arduino and smartphones. Springer Nature Switzerland AG
Pierratos T, Polatoglou HM (2020) Utilizing the phyphox app for measuring kinematics variables with a smartphone. Phys Educ 55:025019. https://doi.org/10.1088/1361-6552/ab6951
Pili U (2018) A dynamic-based measurement of a spring constant with a smartphone light sensor. Phys Educ 53:033002. https://doi.org/10.1088/1361-6552/aaa927
Serway RA, Jewett JW (2018) Physics for scientists and engineers with modern physics, 10th edn. Cengage Learning
Staacks S, Hütz S, Heinke H, Stampfer C (2018) Advanced tools for smartphone-based experiments: Phyphox. Phys Educ 53:045009. https://doi.org/10.1088/1361-6552/aac05e
Treloar LRG (1975) The physics of rubber elasticity. Clarendon Press, Oxford
Vázquez-Cano E (2014) Mobile distance learning with smartphones and apps in higher education. Educ Sci Theory Pract 14:1505–1520. https://doi.org/10.12738/estp.2014.4.2012
Wilensky U (1999) Netlogo. Online. Available at https://ccl.northwestern.edu/netlogo/

Teaching the Subtleties of Entanglement via the Delayed-Choice Two-Slit Experiment with Polarizers

Leanne Doughty and James K. Freericks

Abstract In instruction, most discussions about entanglement start with either the Einstein–Podolsky–Rosen experiment, or Bohm's variant of it with spin. This introduces an extra issue related to signaling, causality, and "spooky action at a distance." Often students focus primarily on these extraneous issues and miss the basic properties of entanglement, in particular the way a product-state superposition displays interference effects, but entangled superpositions do not. In this work, we discuss an alternative way to teach entanglement using a two-slit experiment with crossed polarizers at the slits that provide "which-way" information destroying interference and then illustrate how one can remove "which-way" information via a delayed-choice variant. This work complements recent experimental illustrations of these setups using single-photon light sources that can be used in tandem with the pedagogy.

Keywords Single-particle entanglement · Two-slit experiment · Quantum measurement · Quantum pedagogy · Retrocausality

1 Introduction

We are in the second quantum revolution, where we can manipulate and control individual quanta. The phenomenon of entanglement plays a critical role in many of the applications of quantum mechanics to quantum information science. It is critical to robustly teach the phenomenon of entanglement for students who will later enter the quantum workforce. Current instruction is weak in developing the ideas that underlie entanglement and the ideas that are needed to use entanglement in applications. In particular, while much instruction discusses complementarity in the context of having "which-way" information, far less instruction explains the role entanglement plays in "which-way" information, complementarity, and measurement.

L. Doughty · J. K. Freericks (✉)
Department of Physics, Georgetown University, Washington DC, USA
e-mail: james.freericks@georgetown.edu

Z. Ješková and J. Hanč (eds.), *Advancing Physics Education*, Challenges in Physics Education, https://doi.org/10.1007/978-3-031-94207-5_16

It is important to be clear on what we want students to learn. The high-level learning goal that we have is the following:

> *Develop the theoretical basis for describing measurements in systems that are in product-state superpositions which have interference and entangled superpositions which do not. Be able to independently analyze different experiments within this framework to determine how they behave.*

Our high-level learning goal is achieved by having the students successfully complete four more specific learning goals:

1. Recognize the difference between product-state and entangled superpositions and identify how systems can be prepared in entangled superpositions via irreversible measurement or reversible tagging.
2. Analyze experiments using the Copenhagen paradigm when the system is in a product-state superposition and when it is in an entangled superposition.
3. Describe how "which-way" information relates to product states and entangled states and identify the degrees of freedom used to create "which-way" information.
4. Explain how delayed-choice experiments work, and why they do not employ retrocausality but rather are determined by the state of the system at the moment of measurement.

Entanglement is usually discussed in terms of entanglement between two particles. The particles are often spatially separated and then, when measured, we discover there are correlations between the measurements of the two particles. The initial ideas for this stem from the Einstein–Podolsky–Rosen paper (Einstein et al. 1935). In that work, the two different particles' momentum (and effectively also their positions) were entangled. But, this example is not often taught in current instruction. Instead, we often teach the Bohm variant (Bohm 1951), where the spins of the two particles are entangled in a spin-singlet state. In either case, one has entangled particles that are separated, and can have measurements performed on either of them or both. This then requires the instructor to discuss "spooky action at a distance," relativity of simultaneity, signaling, and so forth, which takes away from discussing entanglement itself. We do acknowledge that many want to incorporate this extraneous discussion into the teaching of entanglement, but if one does do this, it is important to determine whether that discussion assists or detracts from the specific learning goals for the teaching of entanglement.

Because entanglement just requires two independent degrees of freedom, we propose that a simpler way to start discussing entanglement is with single-particle entanglement, such as a photon spatial position and its polarization, or the spatial position of an atom and its spin. The simplification we have in this case is that there is never any spatial separation of two particles, because the entanglement is just with the degrees of freedom of a single particle (note that the entanglement often entangles position as one of the degrees of freedom, but measurements always take place at a single location because there is only one particle and single quanta can be measured to be only in one location). By taking away all of the "spooky action at a distance" and other issues, one can focus on the differences in measurements between product states and entangled states.

Once entanglement is well understood by the students, one can subsequently examine the more traditional examples and then carefully explain the subtle issues that occur with spatially separated pairs of entangled particles.

The ideas discussed in this article were first presented at the GIREP conference in the summer of 2023. They have been discussed by others previously. A beautiful article on how to make such experiments appeared in *Scientific American* (Hillmer and Kwiat 2007), but without the formal developments presented here.

2 Single-Particle Entanglement

Often when we think of entanglement, we think in terms of two particles. But this is not a requirement. Entanglement is simply defined as a quantum superposition state that employs a tensor-product basis and cannot be factorized across the tensor product. Hence, in order to have an entangled state, we do require a tensor-product structure to the quantum state, because there is no concept of factorization of a state described by a single ket, as there is nothing to factorize in that case. But, a tensor-product structure can easily be formed for a single particle that has multiple degrees of freedom, such as a quantum state describing the wavepacket of a photon and its polarization, or similarly of an atom with its spin. Once one recognizes that we can have single-particle entanglement, it becomes clear we can discuss entanglement and its effects without needing to have spatially separated measurements. We believe this greatly aids in instruction because it removes some of the more confounding ideas from an initial exposure to the concept.

One of the most important aspects for understanding entanglement is to be sure to develop what happens to a system that is measured while in an entangled state. This requires a generalization of the standard Copenhagen interpretation to include measurement when the system has two or more degrees of freedom encoded in a superposition of tensor-product states. Of particular importance is developing the rules for measurement when one of the degrees of freedom is *not* measured.

The rule itself is simple. You determine what is being measured, form a tensor-product (measurement) state corresponding to the eigenstates for that measurement, construct the projection operator corresponding to that measurement state, operate it on the current state being measured and form the modulus squared of the final probability amplitude. If a degree of freedom is not being measured, such as polarization of light, for example, when we just measure the position the photon hits on a screen, then we create a projection operator that involves a sum over the complete set of states of the unmeasured degree of freedom, because that operator is represented by the identity operator. Next, when we square the probability amplitude, we will obtain the sum of the squares of the probability amplitudes corresponding to each unmeasured degree of freedom. This procedure will become clearer when we go through the details of calculations below. These rules for quantum measurement are standard for the Copenhagen interpretation.

The way we implement them follows Feynman's *QED* book (Feynman 1985). There, he emphasizes that we must identify the events for a given experiment and the alternative ways each event occurs. An event corresponds to a degree of freedom that is measured, or could be measured. Then, for each event, we find all alternative ways they can occur, add the amplitudes for each way, which is the superposition if there is more than one way, and then take the modulus squared of the sum. We repeat for each event and sum to find the total probability for all events. This is illustrated in equations below when we analyze Eq. (2).

3 Entanglement and the Two-Slit Experiment

In the *Character of the Physical Law* (Feynman 2017), Richard Feynman described how all of quantum mechanics boiled down to understanding "that experiment with the two little holes"—that is, the two-slit experiment. In this statement, he was referring to the property of superposition, along with wave-particle duality, which allows us to see interference patterns develop one photon at a time. Later, though, he admitted that quantum mechanics has more mystery to it, and one needs to add entanglement to the core ideas needed for mastery. Keeping to the spirit of Feynman, in this section, we show how one can use the two-slit experiment to also describe the phenomenon of entanglement.

In Fig. 1, we illustrate the two-slit experiment with a source to the left that can be a single-quanta source. Next, we have a screen with two slits in it. Then, far to the

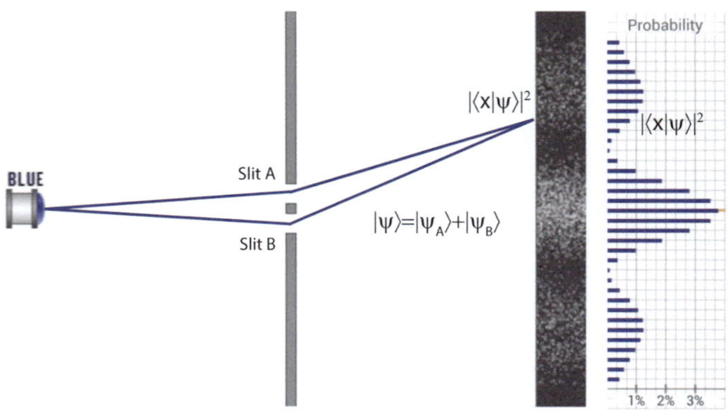

Fig. 1 Schematic of the two-slit experiment. The source is to the left, there is a screen with two slits (A and B) in the middle, and a detector screen on the right. The grainy pattern shows the result of a simulated single-photon experiment after it has collected data over a long period of time. The plot behind it shows the theoretical probability as a function of position

right, we have a detector screen, which detects the quanta that have passed through the device.

While it is common to discuss the qualitative features of a two-slit experiment, we want to combine the Feynman approach (Feynman 1985) that uses a simplified path-integral formalism (with arrows to describe alternative ways events occur) and Dirac notation, (with bra-ket notation rather than Feynman's arrows). This will then allow us to easily generalize the description to the case with entanglement.

When we start from the source, Feynman's method has us start with an arrow of length 1, pointing at 12 O'clock. As we move from source to slit, the arrow rotates at a rate given by the frequency for the blue light. When we reach the slits, the arrow is shrunk by a factor of 0.1, to signify that approximately 1% of the photons incident on the slit make it through when the other slit is blocked. Then the shrunken arrow continues to rotate until we reach the location x on the detector screen. We then combine the arrows corresponding to each alternative way that we leave the source, go through a slit and arrive at a specific location at the detector (two ways are drawn in the figure with the solid lines from source to final screen, which is a simplifying approximation used for these demonstrations). The combined arrow is then mod-squared, which gives us the probabilities. Since there are two arrows (one for each slit) and the relative rotation of the angles lies anywhere from 0 to 2π depending on x, we will find probabilities that range from 0 to 4% (recall each arrow is length 0.1).

Now, we want to re-express these results in terms of Dirac notation. We start from a state at the source $|\psi_S\rangle$. After passing through the slits, we have a superposition of states $|\psi\rangle = |\psi_A\rangle + |\psi_B\rangle$, where $|\langle\psi_A|\psi_A\rangle|^2 = 0.01$ and $|\langle\psi_B|\psi_B\rangle|^2 = 0.01$ to represent the 1% going through each slit, when the other slit is closed. So, we find $\psi(x) = \langle x|\psi_A\rangle + \langle x|\psi_B\rangle$ and the probability for detection at x is given by $|\langle x|\psi_A\rangle + \langle x|\psi_B\rangle|^2$. This result shows interference, as depicted in Fig. 1, because we add the amplitudes together before squaring them, and the amplitudes are not orthogonal to each other, in general (note that the figure shows what we expect from an actual experiment, not from our simplified approximation).

It is easy to see that both the Feynman approach and the approach that employs Dirac notation yield identical results as they must.

We now describe the experiment where we add detectors at each slit as shown in Fig. 2. This experiment gives us "which-way" information due to the intermediate detectors that tell us which slit the photon went through. Note that eventually we will show how one can detect photons at the slits in a very gentle way that does not affect the photon state at all, but these initial detectors we work with do localize the photons to a specific slit causing them to receive a large momentum kick from the uncertainty principle. There is nothing in quantum mechanics that forbids having a detector that measures the slit the photon goes through without changing its wavepacket. It just requires us to work with additional degrees of freedom of the photon as we explain later.

It is common to describe the detection in the detector as a collapse of the wave-function and we do this here for this first experimental setup with detectors. We will see below that one can have detectors that leave the system in an entangled state after detection, a concept often not discussed in instruction. In this first experiment, we

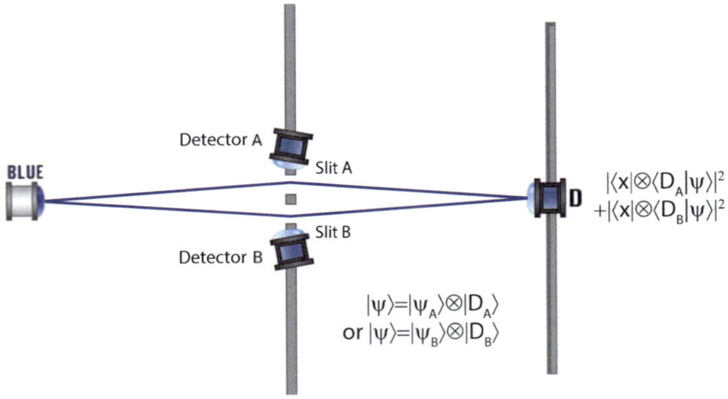

Fig. 2 Schematic of the two-slit experiment with detectors. Now, we have added detectors at each slit, which work 100% of the time. When we run an experiment with single photons, we always find either D_A registers a photon or D_B registers a photon

have three orthogonal states for the detector: (i) the state where the detector has not fired, which we call $|D_\emptyset\rangle$; (ii) the state where the A detector has fired ($|D_A\rangle$); and the state where the B detector has fired ($|D_B\rangle$). So, we start the system in the state $|\psi\rangle = |\psi_S\rangle \otimes |D_\emptyset\rangle$. After the photon passes through the slit, we go into the state $|\psi\rangle = |\psi_A\rangle \otimes |D_A\rangle$ or $|\psi_B\rangle \otimes |D_B\rangle$, depending on which detector has fired. Since the detectors work 100% of the time, these are the only possibilities. Now, when we do the measurement, we will project onto the state $|x\rangle \otimes |D_A\rangle$ or $|x\rangle \otimes |D_B\rangle$ depending on which detector fired. Interestingly, we can also project just onto the state $|x\rangle$ because we already collapsed the state in the first measurement at the slits, and the quantum state is no longer in a superposition. In either approach, we find a uniform probability of 1% for each detector, or 2% overall if we add up the results of both detectors. Note that this is an approximation to the single-slit diffraction pattern that we should have because we only used one alternative way through each slit. By adding more paths, and hence more alternative ways, we can obtain more and more accurate results for the single-slit diffraction patterns, but that is not needed for the material we cover here.

The "which-way" information here is encoded in the state right after the photon passes the slit. Depending on which detector went off, we know which slit the photon went through and we also know that there is just one way to go through each slit. As we often say when describing this experiment, when we have "which-way" information, we lose interference. That clearly occurs here, because the state after passing through the slits has just one state in the superposition, so there cannot be any interference.

The analysis of this experiment is quite cut and dry, and often has a description that goes along with it that in order to localize the photon to one slit or the other, the uncertainty principle requires the system to have its superposition broken. This perspective is often described as black-or-white—either we observe at the slits and destroy the interference, or we do not, with no middle ground. As we will see next, this is not the case. We can make an entangled state either via reversible tagging, or irreversible measurement, and then the procedure is more complex and subtle. Indeed, this is the main example we want to describe in this work. In addition, there are no realistic detectors to watch photons at the slits in the fashion that we described above, especially with 100% efficiency. Instead, a much more practical experiment is to work with the polarization degree of freedom of the photon and use it to create entangled states via measurement. A figure showing this setup is given in Fig. 3.

In this experiment, we start with a photon from the source that is diagonally polarized, or $|\psi_S\rangle \otimes \frac{1}{\sqrt{2}}(|V\rangle + |H\rangle)$. Next, it passes through the perpendicular polarizers at the slits—vertical at slit A and horizontal at slit B. This creates the entangled state given by

$$|\psi\rangle = \frac{1}{\sqrt{2}}(|\psi_A\rangle \otimes |V\rangle + |\psi_B\rangle \otimes |H\rangle), \tag{1}$$

because we project with $|V\rangle\langle V|$ at slit A and $|H\rangle\langle H|$ at slit B. Note that passing through the polarizers at the slits is a measurement. Half of the time the photon is absorbed, and half of the time it emerges in a definite polarization state. This is automatically taken into account by the application of the projection operators. In addition, we have a measurement here because passing through the slits and polarizers is an irreversible process—we cannot reconstruct the initial state after we have passed through the polarizers. Furthermore, we have created an entangled state, as the superposition cannot be factorized. In addition, we now have "which-way" information, in the sense that by subsequently measuring the polarization of the photon, we know immediately which slit it went through.

In this case, the measurement by the polarizer used an irreversible process to create the entangled state. If you instead work with atoms, Stern–Gerlach analyzers, and more general apparatus constructed from them, you can reversibly create entangled

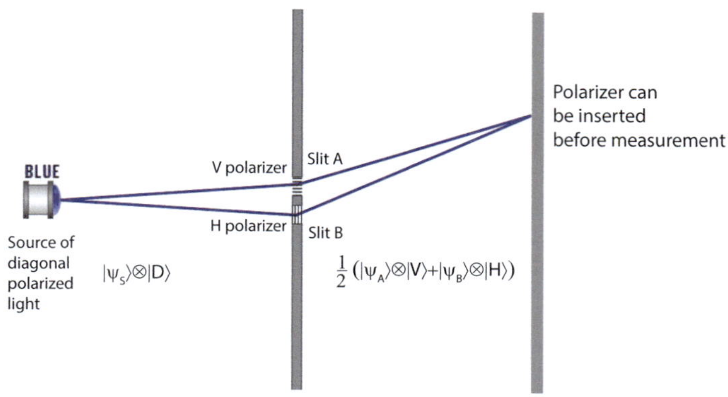

Fig. 3 Schematic of the two-slit experiment with polarizers at the slits. We have added perpendicular polarizers at each slit, respectively—V at slit A and H at slit B. When we run an experiment with single photons, the polarizers either absorb the incident photon (half of the time) or create an entangled superposition of states (half of the time). We transform the experiment into a delayed-choice experiment when we add polarizers just before the light hits the screen

states via a process we call tagging (Courtney et al. 2020). We will not discuss tagging further in this work.

Now, we know that if we have "which-way" information, we lose interference, and this is precisely what happens. Unlike the case of the two-slit experiment without polarizers, which we analyzed above, and we showed that it had interference, in this case, we find we do not have any interference when we measure at the detector screen. As we mentioned before, the measurement process requires us to project separately onto the states $|x\rangle \otimes |V\rangle$ and $|x\rangle \otimes |H\rangle$ and square each projection and then add them together. This is because the event is to measure the photon at position x, not measuring the polarization. In equations, we formulate this by applying the projection operator $|x\rangle\langle x| \otimes \mathbb{I} = |x\rangle\langle x| \otimes |V\rangle\langle V| + |x\rangle\langle x| \otimes |H\rangle\langle H|$. Then we use the fact that the measurement window for $|x\rangle$ is finite being determined by the physical size of the detector, so we can take the positions as discretely located at the detector "pixels" and then $\langle x|x\rangle = 1$ and the polarization states are orthonormal as well. This gives

$$
\begin{aligned}
P(x) &= \left| |x\rangle\langle x| \otimes |V\rangle\langle V| \tfrac{1}{\sqrt{2}}\big(|\psi_A\rangle \otimes |V\rangle + |\psi_B\rangle \otimes |H\rangle\big)\right|^2 \\
&\quad + \left| |x\rangle\langle x| \otimes |H\rangle\langle H| \tfrac{1}{\sqrt{2}}\big(|\psi_A\rangle \otimes |V\rangle + |\psi_B\rangle \otimes |H\rangle\big)\right|^2 \\
&= \tfrac{1}{2}(|\langle x|\psi_A\rangle|^2 + |\langle x|\psi_B\rangle|^2),
\end{aligned} \tag{2}
$$

because $\langle V|H\rangle = \langle H|V\rangle = 0$. Now we have the sum of the squares of the probability amplitudes, so the probability is $\frac{1}{2}(1\% + 1\%) = 1\%$. Note that it is one-half as large as the probability we obtained for the case with detectors at the slits. This is because of the absorption by the polarizers when we passed through the slits.

The first and last example in this section—the case with no polarizers and the case with crossed polarizers at the slits—illustrate precisely the scenario we want to describe. In the first case, we have a product-state superposition (if we include the polarization in the description of the state, which is the same for each term in the superposition) and in the second case, we have an entangled superposition. When the superposition is a product state, we have interference, but when we have an entangled state, the interference disappears. And this occurs whether we measure the polarization at the detector screen or not! In many respects, it is this last point that is perhaps the most confounding.

Note that in the case with polarizers, we have detectors at the slits, but we have no uncertainty argument that enters into the measurement at the slits. To show this is true, we will erase the "which-way" information and restore the interference and we only pay the price of having some more loss of photons that pass through polarizers. This is the subject of the next section.

4 Delayed-Choice Two-Slit Experiment

Now we describe how we can do a delayed-choice variant of the experiment. The experimental setup is the same as in Fig. 3. It is just that now we introduce a polarizer before we do the measurement at the screen. We will consider three cases: (i) a vertical polarizer; (ii) a horizontal polarizer; (iii) and a diagonal polarizer. In the vertical case, we project the state in Eq. (1) onto the state $|V\rangle\,\langle V|$ before we do the measurement. This yields a probability of $|\langle x|\psi_A\rangle|^2/2 = 0.5\%$ and corresponds to one-half the signal one would obtain for light going only through slit A. In the horizontal case, we have a probability of $|\langle x|\psi_B\rangle|^2/2 = 0.5\%$, which corresponds to one-half the signal one would obtain for light only going through slit B. But, in the diagonal case, the state becomes

$$\frac{1}{2}(|\psi_A\rangle + |\psi_B\rangle) \otimes \frac{1}{\sqrt{2}}(|V\rangle + |H\rangle), \tag{3}$$

just before the measurement. You can see that this is a product state (because we factorized it across the tensor-product symbol). When we perform the measurement at x, we find

$$P(x) = \frac{1}{4}|\langle x|\psi_A\rangle + \langle x|\psi_B\rangle|^2, \tag{4}$$

and the interference is restored. This is the probability for the two-slit experiment with no polarizers or detectors divided by four. The reduction happens because we lose half of the photons when we pass through the polarizers at the slits and we lose another half at the diagonal polarizer before measurement. So the probability range now runs from 0 to 1%.

There are many that try to discuss the delayed-choice experiment in terms of the concept of retrocausality. The way the argument goes is that the photon does not know whether to interfere with itself or not until the time of measurement. If it is to interfere, it must go back in time to interfere, because when it went through the original slit with polarizers it was measured and put into a state with no interference. The fallacy of this argument is that there is never any need to go backwards in time. The simple rule is that the measurement of the quantum state always corresponds to what the current state is. It was the manipulation we did which transformed the state from an entangled superposition to a product-state superposition that caused the interference to return. Whenever we are able to make such a transition, we can restore interference effects. But, this never requires us to go backwards in time.

In our opinion, the topic of retrocausality should be avoided, unless students are already aware of it, in which case the fallacy should be dispelled. Otherwise, just sticking to the notion that the result of a measurement is always dictated by what the quantum state is at the moment of the measurement will be sufficient for the student to understand how to analyze complex experiments like these.

5 Experimental Implementation

Unlike the thought experiments that use detectors at the slits (which we are unaware of any concrete realization with photons), the case where polarizers at the slits are used to create entangled states can be performed in an undergraduate laboratory as discussed by recent work from Kiko Galvez and his collaborators (Luo et al. 2024).

In recent years, a number of companies have made available affordable single-photon sources, detectors, and other equipment so that quantum optics demonstrations can be performed by undergraduate students. One of the most problematic issues in these experiments is often the alignment problem, but currently available hardware using optical fibers in a closed setup has solved this issue, making the experiments much more feasible to carry out.

So, it is now realistic to create lecture demonstrations, or even laboratory experiments that can illustrate the delayed-choice two-slit experiment with polarizers at the slits in a quantum class. The results from Luo et al. (2024) show that one obtains good interference patterns for the delayed-choice two-slit experiment with polarizers using single-photon sources and collections times of under a minute per channel.

The notion that the final polarizer acts simply as a filter is a common theme in describing the behavior of delayed-choice experiments in a number of different ways. The idea that the polarizer acts like a filter comes from the fact that if we add the H and V polarizer images, with probability given by $\frac{1}{2}(\langle x|\psi_A\rangle|^2 + |\langle x|\psi_B\rangle|^2)$, it is the same as we obtain if we add the images of the two diagonal polarizers $(\mathrm{V}\pm\mathrm{H})/\sqrt{2}$, which is the probability $\frac{1}{4}\left|\langle x|\psi_A\rangle + \langle x|\psi_B\rangle\right|^2 + \left|\langle x|\psi_A\rangle - \langle x|\psi_B\rangle\right|^2$. Hence, we can think of the application of the polarizers as different types of filters of the total signal—the V and H polarizers filter uniformly, so we see random images, while the D and D' polarizers filter nonuniformly, so we see interference patterns for both (that are complementary, so they add to a uniform pattern).

The filtering interpretation holds here because of the absorption by the polarizers—passing through a polarizer is a measurement. If we could implement an experiment using reversible tagging rather than irreversible measurement, we would see an enhancement of the interference images, and they would not be simply additive. While it is more complicated to do this, there are examples where it can be done, so it is better not to describe all delayed-choice experiments in terms of just filtering.

6 Discussion of Pedagogical Aspects

When developing a pedagogy for entanglement, it is important to create the learning goals first. Given all of the confusion surrounding delayed-choice experiments and retrocausality, we feel it is critical to understand the difference seen in quantum measurements for product states versus entangled states, especially if the entangled degrees of freedom are not measured. Once this learning goal has been mastered, it is easy to understand how decoherence occurs due to entanglement with a reservoir and subsequent measurement of the system, where the reservoir degrees of freedom are unmeasured. It also helps students properly understand why entanglement plays an important role throughout quantum information science and helps enable a number of different types of experiments. The notion that everything is encoded in the current quantum state, rather than having to worry about retrocausality, helps students understand the underlying framework that governs much of quantum computation as well. One should not underestimate the importance of having a strong conceptual and quantitative understanding of entanglement for modern quantum students.

In our opinion, there is an important need for students to practice working with the quantum states in the Dirac formalism, correlating the action on the state with what is happening in the experiment at a given moment in time, and properly implementing the Copenhagen interpretation for projective measurements in tensor-product states. This means one needs to scaffold this work with a sufficient number of exercises that cover these ideas. If one uses the two-slit experiment in instruction, then the application of these ideas within the context of the Stern–Gerlach experiment can provide significant practice (Courtney et al. 2020). It is also possible to extend these ideas to the Mach–Zehnder interferometer by using a polarizing beam splitter for the first beam splitter of the interferometer. We plan on discussing those examples in future work (note that this problem has been analyzed by Marshman et al. (2016) but in a different context from how we are discussing it). One can also have students go through a number of additional details in the two-slit experiment as well.

Once the concept of local entanglement is well understood by the students, it is then appropriate to discuss the spatially separated examples with two-particle entanglement. This would then include opportunities to describe many of the subtle phenomena related to "spooky action at a distance," relativity, and signaling. If students have a firm foundation for what entanglement is, they should better appreciate these examples. Of course, it is also recommended to introduce materials leading to Bell's inequality and how it can be tested for violations in quantum experiments.

Another realm that can be explored is entanglement in time. This is accomplished in the so-called Franson interferometer (Franson 1989), which examines the arrival times of two entangled photons, with interference occurring for photons that arrive at the same time (even though the travel time for the two photons is different for the two alternative ways the coincident arrival occurs). Extending the concepts from entanglement in space to entanglement in time is a very interesting new idea that we should expose students to as well.

As one determines the different activities for the students to pursue, it is useful to make sure that you map those exercises to the lower-level learning goals to ensure sufficient practice is available to the students that they can master these different concepts. It is also important to measure student success for each learning goal to inform changes to pedagogy. Here, we outline and discuss our analysis of students' responses to one assessment item from the course related to our learning goals around entanglement. Our analysis includes 53 student responses that come from four course implementations (2 online and 2 in-person).

The exam question presents students with a two-slit experiment setup similar to Fig. 3 where there is a horizontal polarizer over the lower B slit and the light from the source is emitted at 45° so that the initial state is $|S\rangle \otimes \frac{1}{\sqrt{2}}(|V\rangle + |H\rangle)$. The question reminds students that in the case of no polarizers at the slits the state will go to a superposition after passing through the slits ($|S\rangle \rightarrow |\psi_A\rangle + |\psi_B\rangle$). Part (a) is an open-response question that asks students to determine the quantum state after the photon passes through the slits, but before it hits the screen. Part (b) is a multiple-choice follow-up question that asks students what type of pattern will be on the screen after the photons have hit it. The choices are a partial or full interference pattern, a particle pattern from slit A, a particle pattern from slit B, or a particle pattern from both slit A and slit B. The question then introduces an additional polarizer in between the slits and the screen. Parts (c) and (d) ask students to again determine the quantum state after the photon passes the polarizer and the type of pattern on the screen in the same question format as before.

The results were similar across all implementations. 42% of students correctly determined the quantum state in part (a) as $\frac{1}{\sqrt{2}}|\psi_A\rangle \otimes |V\rangle + \frac{1}{\sqrt{2}}|\psi_B\rangle \otimes (|V\rangle + |H\rangle)$ or $\frac{1}{\sqrt{2}}|\psi_B\rangle \otimes |H\rangle + \frac{1}{\sqrt{2}}|\psi_A\rangle \otimes (|V\rangle + |H\rangle)$ or some other equivalent form. Another 26% of students correctly determined the state superposition but it appears that they had difficulty with the normalization, writing incorrect coefficients in their responses. Though the coefficient combinations were varied across responses, the most common response in this category was $\frac{1}{\sqrt{2}}|\psi_A\rangle \otimes (|V\rangle + |H\rangle) + |\psi_B\rangle \otimes |H\rangle$, where students are just missing the correct coefficient for the second term. The remaining 32% of responses for part (a) either did not represent all possible superposition's (e.g., $\frac{1}{\sqrt{2}}|\psi_A\rangle \otimes |V\rangle + \frac{1}{\sqrt{2}}|\psi_B\rangle \otimes |H\rangle$) or did not represent the entanglement (e.g., $\frac{1}{2}|\psi_A\rangle \otimes |V\rangle$). With the addition of another polarizer in part (c) the amount of students determining correct quantum states decreased to 19%. However, the same amount of students correctly determined the state superposition and only had difficulty finding the coefficients. This assessment item is part of the first midterm exam and so it is likely that students still need more practice working with these types of problems; it is viewed as one of the hardest questions on the exam. Another consideration is that this exam is the first place in the course where students do not receive immediate feedback on their submitted responses. In all other course components, including homework problem sets, students responses are graded immediately and students are allowed multiple attempts.

With regards to the pattern on the screen, the first setup will result in a partial interference pattern and the second will result in a full interference pattern. So, the correct response will be the same multiple-choice option, "a partial or full interference pattern," for both part (b) and part (d) of the question. 85% of students selected this option for part (d). However, only 36% selected this option for part (b), with 53% selecting "a particle pattern from both slit A and slit B." Based on student feedback following the exam, many students were unclear about the difference between a partial interference pattern and a particle pattern from both slit A and slit B, leading them to select the incorrect response in this case.

It is clear from this work that teaching entanglement, and having students grasp the concepts in full, is a difficult challenge. To ensure a better understanding, we should verify that students are able to facilely move between the concepts and the mathematical formalism, paying attention to normalization factors before they move onto other materials.

7 Summary

Many would say the second quantum revolution is all about exploiting entanglement for applications. Entanglement has a number of different aspects to it—two-particle entanglement has a quantum state where the properties of one particle are correlated with the properties of another. It also is a critical aspect of the mechanism by which quantum states decohere when in contact with an environment. But, one of the most important fundamental properties of entanglement is how the measurement in an entangled state differs from the measurement in a product state. It is this difference that the current work focuses on.

Our high-level learning goal is for students to be able to recognize whether a state is a product state or an entangled state and analyze measurements of the quantum state in such a way that they can reason what the response will be for different experiments depending upon what the state is at the moment of measurement.

Our approach is that these learning goals are best achieved when working first with single-particle entanglement rather than two-particle entanglement. In this case, we avoid having to discuss "spooky action at a distance," relativistic effects, including the lack of causality for space-like separated events, and we do not need to discuss anything related to signaling between the entangled quanta. This allows the students to focus on the primary ideas—what is the difference between a simple quantum state with no tensor-product structure, a tensor-product state that factorizes, and one that does not. Understanding these high-level concepts allows students to analyze complex experimental situations including how one can create entangled states, how they can be modified and manipulated, how one can convert from an entangled state to a product state, and how quantum information is manipulated for useful applications in the process.

Once the single-particle entanglement is understood, then it makes sense to discuss two-particle entanglement, which allows students to explore the Bell inequalities, the verification that quantum mechanics violates local realism, and so forth. Finally, students can also explore advanced ideas such as those exemplified by the Franson interferometer, where one forms quantum states entangled in time as opposed to in space.

It is time to modernize quantum instruction in the realm of quantum entanglement. We hope these suggestions of new quantum pedagogy will assist others in developing enriching experiences for their students. It is also clear that it is challenging to bring students to fully understand all of these ideas. We need to continually improve how we help students master them.

Acknowledgements This work was supported by the Air Force Office of Research under grant number FA9550-23-1-0378. J.K.F. was also supported by the McDevitt bequest at Georgetown University. J.K.F. planned the project. L.D. analyzed the student responses. All authors contributed to writing up the work.

References

Bohm D (1951) Quantum theory. Prentice-Hall, Englewood Cliffs, NJ

Courtney WF, Vieira LB, Julienne PS, Freericks JK (2020) Incorporating the Stern-Gerlach delayed-choice quantum eraser into the undergraduate quantum mechanics curriculum. Am J Phys 88(4):298–307. https://doi.org/10.1119/10.0000519

Einstein A, Podolsky B, Rosen N (1935) Can quantum-mechanical description of physical reality be considered complete? Phys Rev 47(10):777–780. https://doi.org/10.1103/PhysRev.47.777

Feynman RP (1985) QED: the strange theory of light and matter. Princeton University Press, Princeton, NJ

Feynman RP (2017) The character of physical law. MIT Press, Cambridge, MA

Franson JD (1989) Bell inequality for position and time. Phys Rev Lett 62(19):2205–2208. https://doi.org/10.1103/PhysRevLett.62.2205

Hillmer R, Kwiat P (2007) A do-it-yourself quantum eraser. Sci Am 296(5):90–95. https://doi.org/10.1038/scientificamerican0507-90

Luo BJ, Francis L, Rodríguez-Fajardo V, Galvez EJ, Khoshnoud F (2024) Young's double-slit interference demonstration with single photons. Am J Phys 92(4):308–316. https://doi.org/10.1119/5.0179131

Marshman E, Singh C (2016) Interactive tutorial to improve student understanding of single photon experiments involving a Mach-Zehnder interferometer. Eur J Phys 37(2):024001. https://doi.org/10.1088/0143-0807/37/2/024001

Developing Augmented Reality (AR) Teaching Tools for Undergraduate Physics: The Virtual Optical Table

Philippa L. Petts, Helen Cramman, and G. Peter Swift

Abstract Extended reality (XR) technology, blending the physical and virtual worlds, holds significant potential for transforming education, particularly in STEM. Augmented and virtual reality (AR and VR) offer interactive experiences that enhance students' ability to visualize and explore complex concepts. In physics, AR proves particularly valuable by making the invisible visible through immersive simulations. This proof-of-concept project developed AR-based simulations for a virtual optical table and basic electromagnetism. These were then evaluated by undergraduate and postgraduate students, as well as faculty, using a mixed-methods survey and thematic analysis of semi-structured interviews. Participants reported the simulations to be immersive, engaging, and offering significant educational potential. Participants felt the technology had potential to improve conceptual understanding of complex topics. While technical challenges like device accessibility and cost were identified, the AR simulations lay the groundwork for further integration into physics education.

Keywords Interactive learning technologies · Augmented reality simulations · Immersive teaching tools · Extended reality

1 Introduction

Extended Reality (XR), encompassing technologies such as augmented reality (AR) and virtual reality (VR), represents a continuum of immersive experiences that blend the physical and digital worlds, enabling interactive educational experiences. AR,

P. L. Petts (✉) · G. P. Swift
Department of Physics, Durham University, Durham, UK
e-mail: p.l.petts@durham.ac.uk

H. Cramman
School of Education, Durham University, Durham, UK

Durham Centre for Academic Development, Durham University, Durham, UK

Z. Ješková and J. Hanč (eds.), *Advancing Physics Education*, Challenges in Physics Education, https://doi.org/10.1007/978-3-031-94207-5_17

a key component of XR, gained broader recognition in the mid-1990s as part of "mixed reality" technologies within the reality-virtuality continuum (Milgram et al. 1995). AR supplements the real world with virtual objects that coexist with real-world objects, characterized by three-dimensional (3D) alignment and real-time interactivity (Azuma et al. 2001). Initially, AR applications focused on fields like medicine and entertainment (Azuma et al. 2001), but by the late 1990s, its educational potential began to be explored (Szalavári et al. 1998).

1.1 Current Use of XR in Physics Education

Today, AR is widely integrated into educational contexts, particularly within the natural sciences, including physics (Chang et al. 2022). The most common implementations of AR in education involve mobile devices such as smartphones and tablets, which are accessible and widely adopted (Zhang et al. 2022). Although AR glasses offer a more immersive experience, their use in educational settings is much less common, and empirical research on their effectiveness is scarce (Buchner et al. 2022). Despite challenges related to hardware costs, rapid technological advancements, and concerns about obsolescence, the potential for AR to transform undergraduate physics education remains compelling. AR provides immersive and interactive learning experiences, making abstract and complex physical phenomena visible and tangible to students, particularly valuable in STEM disciplines where internalizing such concepts is critical.

1.2 Benefits and Challenges of AR in Physics Education

Augmented Reality (AR) has significant potential to enhance physics education by increasing student engagement and facilitating the understanding of complex concepts. By enabling interactive visualization of abstract phenomena, AR environments improve memory retention, deepen conceptual understanding, and enhance problem-solving skills (Bacca et al. 2014; Ibáñez and Delgado-Kloos 2018). These benefits are especially notable in STEM disciplines, where AR's ability to visualize and interact with otherwise invisible phenomena is crucial for learning (Malone et al. 2023). AR also positively impacts motivation and attitudes, with students often favoring AR-enhanced environments over traditional methods (Chang et al. 2022).

However, the benefits of AR must be weighed against challenges, particularly concerning cognitive load. While some studies suggest AR can reduce cognitive load by simplifying complex concepts, findings in this area are inconsistent. In some cases, AR environments show no significant difference in cognitive load compared to traditional learning, highlighting the importance of careful design and implementation of AR tools (Altmeyer et al. 2020).

Additionally, technical challenges, such as high costs and rapid technological changes, complicate the adoption of AR in education (Buchner et al. 2022). Head-mounted displays (HMDs), while offering an immersive AR experience, are expensive and may not be accessible to all institutions. Lower cost alternatives, such as tablets and smartphones, provide a more affordable option but lack the same level of immersion as HMDs. This trade-off between cost and immersion is a key consideration when integrating AR technology into educational settings. These challenges underscore the need for well-designed AR tools thoughtfully integrated into curricula.

1.3 Gaps in the Literature on AR Glasses and Headsets

Despite the growing research on mobile AR in education, there is a significant gap in studies focusing on AR glasses/HMDs. These devices offer distinct advantages, such as hands-free interaction and enhanced spatial visualization, particularly beneficial in physics laboratory settings. However, empirical research on the effectiveness of AR glasses is limited, leaving their potential to improve cognitive load and learning outcomes speculative (Laumann et al. 2024). This gap is especially concerning given the inconsistent findings related to cognitive load in mobile AR applications. Investigating whether AR glasses can provide a more effective alternative is essential.

This study seeks to address this gap by providing empirical evidence on the use of AR glasses in physics education, aiming to determine participant opinions on their practical use, potential, and limitations. By focusing on AR glasses, this research offers a timely contribution to the field, advancing our understanding of how different AR technologies can be effectively integrated into STEM education. Moreover, addressing this gap will help educators make informed decisions about the best tools to enhance student learning in increasingly digital and immersive educational environments.

1.4 XR Unique Hooks

XR emerges as a powerful educational tool, offering distinct advantages that surpass traditional teaching methods (Petts 2023). The list below summarizes some of these key affordances:

- **Making the invisible visible**: XR technology's capacity to render abstract concepts tangible is pivotal for cultivating conceptual understanding in STEM subjects such as electromagnetic fields. This visual representation of intangible phenomena facilitates comprehension by providing concrete depictions of theoretical concepts (Wijdenes et al. 2018; Dunleavy and Dede 2014).

- **3D Interactions**: XR's provision of immersive 3D interactions is crucial for comprehending intricate scenarios like molecular structures and human biology (Fombona-Pascual et al. 2022). By presenting these concepts in their authentic 3D format, XR surpasses the constraints of traditional 2D representations, thereby enhancing student comprehension.
- **Prohibitive**: XR technology affords students the opportunity to engage with scenarios typically restricted due to safety concerns or equipment limitations (Kaufmann and Dünser 2007). This expanded access broadens learning opportunities and encourages hands-on exploration, enriching the educational experience.
- **Physicality**: The emphasis on physical interaction within XR simulations bridges the gap between theoretical concepts and practical applications (Tang et al. 2003). This tactile engagement fosters a deeper understanding of system behavior, thereby potentially improving learning outcomes.
- **Overlay of the real world**: Augmented reality (AR) within XR has the ability to integrate hidden scientific principles with real-world situations (Dunleavy and Dede 2014; Tang et al. 2003). This integration strengthens the correlation between theoretical knowledge and practical application, enriching the educational process.
- **Playfulness**: XR simulations introduce an element of exploration and playfulness, catering to users of all ages and fostering deep conceptual understanding. The freedom to explore beyond physical constraints enhances engagement and facilitates a nuanced comprehension of concepts especially when in collaboration with peers (Phon et al. 2014; Kumar et al. 2022).

In addition to these benefits, the significance of personalized learning and inclusivity in XR technology should be acknowledged. XR platforms can cater to individual learning needs and preferences, enhancing the educational experience (Xie et al. 2019; Li and Wong 2021). Furthermore, XR has the potential to improve accessibility for diverse learners, including those with disabilities, by offering alternative interaction modes and representations (Carreon et al. 2020; Jdaitawi and Kan'an 2021). In addition to the enumerated benefits, it is pertinent to acknowledge the significance of personalized learning and accessibility and inclusivity in XR technology. Personalized learning pathways, facilitated by XR platforms, can cater to individual student needs and preferences, thereby enhancing the educational experience (Xie et al. 2019; Li and Wong 2021). These aspects underscore the transformative potential of XR in education.

2 Development of the Virtual Optical Table and Electromagnetism Suite

The development team, including two first-year undergraduate computer science students employed as summer interns for 6 weeks, created two simulation suites using Unity 3D (Unity Technologies 2024) and the Vuforia AR SDK (PTC Inc. 2022). Assets were developed in Blender (Blender Foundation 2024), and the simulations were designed for the Microsoft Hololens (second generation) HMD, with a VR-compatible version for the Meta Quest 2. All coding, including hand tracking, was developed from scratch using an Agile product development approach, with careful attention to accurately replicating real-world counterparts, such as adding a safety warning label on the virtual laser.

The initial simulation, a basic electromagnetic setup, served as a preliminary exercise to familiarize the team with the software and development methods. It displayed dynamic 3D field lines generated by electric charges and bar magnets, allowing users to interact with and manipulate the setup in real time (see Fig. 1 left). This provided a flexible platform for experimenting with electromagnetic field interactions.

The primary focus of the project was the development of an interactive optical table with various lasers and lens components. This optical table, presented as a floating sheet, allows users to resize and move it spatially. Components, including lasers, convex and concave lenses, and reflecting plates, are stored on a virtual shelf and can be easily added to the table, snapping into place for customized configurations. Slider rails enable dynamic adjustment of the distance between components in real time. Figure 1 (right) shows an example setup with a laser and three lenses. The available components include both ideal lenses and more realistic ones that exhibit spherical and chromatic aberrations, modeled to behave in a physically realistic manner.

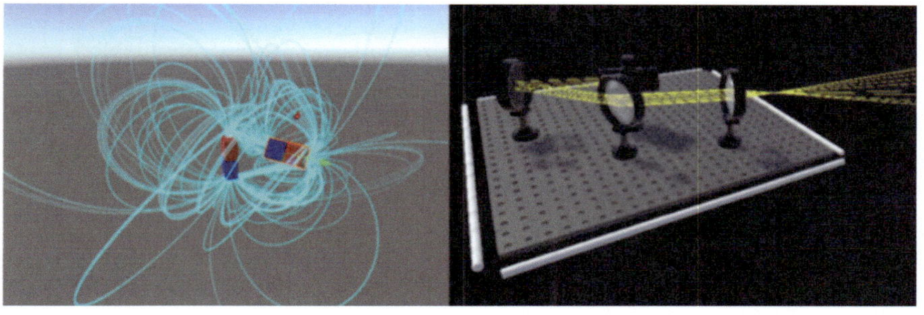

Fig. 1 (Left) An image of the 3D magnetic fields from two bar magnets from the initial electromagnetism simulation. (Right) An image of the interactive AR optical table showing a laser setup configured with three lenses

3 Evaluation

The potential efficacy of these interactive AR simulations as teaching tools was evaluated using a mixed-methods approach. A total of 28 participants were involved in the study, comprising undergraduate and postgraduate students and faculty. Participants were recruited from the undergraduate student body and the Physics department staff and postgraduate students through advertising. There were no specific inclusion or exclusion criteria.

The evaluation design was sequential, beginning with a survey followed by semi-structured interviews. This approach aligns with Creswell's (2017) framework (Creswell and Creswell 2017) for integrating quantitative and qualitative data to provide a comprehensive understanding of the research problem. The survey, tested on academic colleagues within the research section prior to deployment, consisted of Likert scale, multiple-choice, and open-ended questions designed to address the research questions. All participants who completed the survey were invited to participate in the interviews, and eight accepted the invitation. These interviews were instrumental in providing deeper insights into the user experience and perceptions of the AR tools.

The research questions (RQs) focused on the user experience from the students' perspective and the staff opinions on the potential of this technology as a teaching tool, including possible use cases. However, it should be noted that the students did not have the opportunity to perform the real experiments, which are neither particularly complicated nor expensive. Their responses were based solely on the AR "experiments," not on a comparison with traditional experiments. This limitation is acknowledged, and while the focus of this study was on the AR experience, comparing AR-based and traditional experiments is suggested as an area for future research. Additionally, all undergraduate participants will have used similar equipment during their laboratory modules.

The study addressed the following research questions: How do undergraduate physics students and faculty staff perceive the potential use cases of augmented reality virtual optical tables in laboratory settings (RQ1)? What are the perceived benefits of utilizing augmented reality virtual optical tables for teaching complex physics concepts (RQ2)? And what are the main disadvantages or challenges identified by students and faculty when using augmented reality virtual optical tables compared to traditional laboratory setups (RQ3)?

The survey results and thematic analysis of the interview data are presented in the subsequent sections, with the survey instrument included in the appendix for reference.

3.1 Survey Results

A total of 28 participants (20 male and 8 female) took part in the testing and completed the survey afterward. The participants included undergraduate students ($N = 21$), one postgraduate, and six faculty staff members, including academic, research, and support staff. The student participants were fairly evenly distributed across all four levels of study (Year 1 = 3, Year 2 = 6, Year 3 = 8, Year 4 = 5).

In terms of prior experience with XR technology, 46.4% of participants had used a VR device before, while 35.7% had not used either VR or AR before. Only two participants reported having previously used an AR headset, highlighting the novelty of the experience for most participants (Fig. 2, top left).

Participants were asked to rate their agreement with various statements using a standard five-point Likert scale. The results were overwhelmingly positive regarding the potential of AR technology in education. As shown in Fig. 2 (bottom-left), 100% of the participants agreed that "AR simulations could be a useful teaching tool," with 61.9% choosing "Strongly agree" and 38.1% choosing "Agree."

Furthermore, when asked whether "using the AR optics simulation would improve their conceptual understanding of the topic," 76.2% of participants responded positively, as illustrated in Fig. 2 (bottom-right). This indicates that the majority of participants believed that AR technology could enhance their understanding of complex physics concepts.

Interestingly, the data from Fig. 2 (top-right) revealed that 76.2% of participants also felt that "the AR optics simulation offered advantages over solely lab-based experiments." This suggests that participants not only saw the value of AR as a

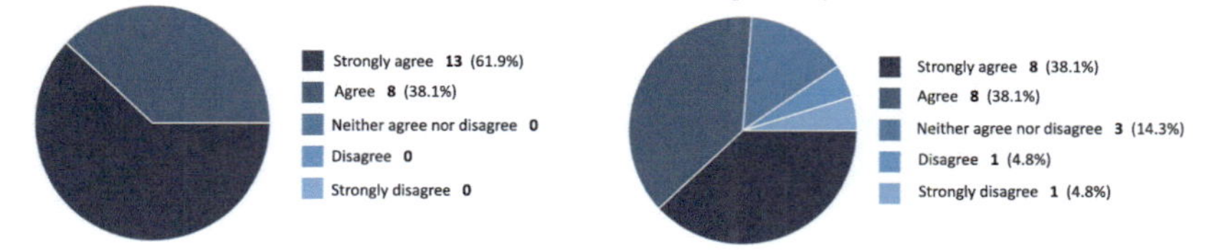

Fig. 2 Participants' experiences and perceptions of AR technology. (Top left) Proportion of participants having previously used AR/VR hardware. (Top right) Percentage of participants agreeing that AR simulations provide advantages over traditional lab-based experiments. (Bottom left) All participants agreed that AR is a valuable teaching tool. (Bottom right) A majority of participants (76.2%) felt that AR optics simulations improve their conceptual understanding of complex physics concepts

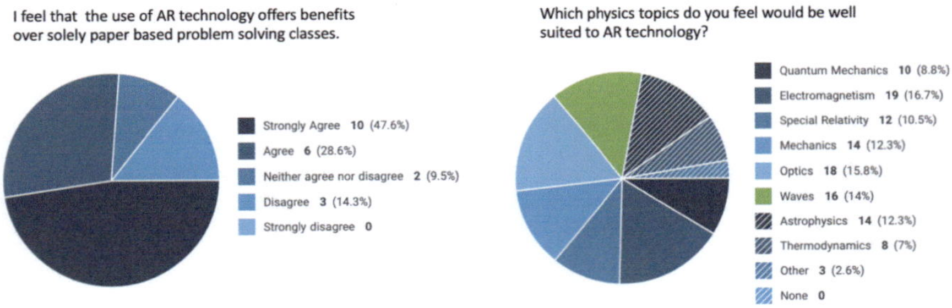

Fig. 3 (Left) Responses to "AR technology offers benefits over paper-based problem classes." (Right) Areas of physics participants believe are well suited to AR instruction, with responses evenly distributed across topics

supplementary tool but also recognized its potential to enhance or even replace certain aspects of traditional lab-based learning.

Additionally, the survey explored participants' views on the benefits of AR technology compared to paper-based problem-solving classes. As shown in Fig. 3 (left), 76.2% of participants agreed that "the use of AR technology offers benefits over solely paper-based problem-solving classes," with 47.6% selecting "Strongly agree" and 28.6% choosing "Agree." This reinforces the idea that AR can provide a more engaging and interactive learning experience compared to traditional methods.

The survey also inquired about which areas of physics participants felt AR technology would be most beneficial. The responses were wide-ranging, covering various topics across the physics curriculum. This indicates that participants believed AR technology had potential benefits across a broad spectrum of physics education, rather than being limited to specific areas (Fig. 3, right).

3.2 *Thematic Analysis*

The thematic analysis was conducted using Braun and Clarke's 2021 six-phase framework with the aid of NVivo software for organizing and coding the qualitative data. First, the data was familiarized by transcribing and repeatedly reading the interviews, while noting initial insights. Initial codes were then generated, tagging relevant segments without a pre-existing framework. These codes were then grouped into potential themes, facilitated by the use of visualization tools. These were then reviewed for coherence, refined for accuracy, and specific names were assigned. Finally, the analysis was compiled, with themes interpreted in relation to the research questions and existing literature. NVivo's use ensured systematic analysis, though as only one coder was involved, inter-rater reliability measures were not applicable. The results of this analysis are presented below.

3.2.1 Laboratories

Participants expressed enthusiasm for the integration of XR technology within laboratory settings. They emphasized its potential as a preparatory tool for future in-person sessions, noting its ability to allow for familiarization with experimental equipment in a controlled and risk-free environment. This aspect was particularly valued given the apprehension surrounding the handling of delicate and expensive equipment, as well as concerns related to health and safety protocols, including the use of lasers and other potentially hazardous materials.

Student 5 highlighted the safety benefits of using XR in laboratory settings, stating:

> "It would be much safer to kind of use this simulated laser in the simulation so people could still kind of learn the things that they need to learn and get an intuition for these things without maybe exposing themselves to risks that could be avoided."

Participants articulated how practicing setups within XR simulations would not only bolster their confidence but also optimize their efficiency during actual laboratory sessions, allowing for a more focused utilization of time and resources.

3.2.2 Exploration

The theme of exploration emerged as a prominent aspect of participants' experiences with XR technology. Beyond merely replicating real-world scenarios, it was suggested that virtual tools offered a liberating sense of freedom, enabling participants to transcend the confines of predetermined problems commonly encountered in traditional laboratory classes or physics problems. For instance, Student 3 emphasized:

> "There's a freedom that it brings because you don't have to worry about breaking anything or accidentally hurting anyone or accidentally hurting yourself."

This opportunity for unrestricted exploration resonated deeply with participants, evoking memories of their inherent curiosity and attraction to the field of physics from a young age.

Moreover, the immersive nature of XR environments facilitated experimentation beyond the boundaries of physical reality, with participants engaging in activities that were otherwise impractical or impossible. Student 4 illustrated this by stating:

> "It's just like where we can do things in the simulation that we can't do in real life. We can't change gravity. We can't do stuff like that... So going into that dangerous zone of things that you can't physically do. Exactly, I think that's where a lot of this technology could help."

Student 5 also highlighted the educational potential of this exploratory freedom, noting:

> "It's kind of through these little segues that you gain a deeper understanding of the material. And these types of Hololens simulations could really help with these sorts of theoretical situations."

The ability to explore and experiment in a risk-free, virtual environment not only enhances engagement but also deepens understanding, making XR technology a powerful tool for fostering curiosity and exploration in the learning process.

3.2.3 Intuition

The theme of intuition surfaced as participants reflected on the potential of interactive XR technology to enhance their conceptual understanding of complex physics topics, particularly those intertwined with mathematical principles. Participants expressed a desire for simulations that would enable them to explore physical situations governed by intricate equations, thereby bridging the gap between theoretical concepts and tangible visual representations.

Student 5 emphasized the value of visual representation in understanding complex mathematical concepts, stating:

> "When you try to decompose a wave with a Fourier decomposition... it would be a bit more, I guess, physically intuitive rather than just seeing expressions and numbers."

Similarly, Student 1 discussed the benefit of linking physical intuition with mathematical representations:

> "For example, having that equation in our field of view while we're messing around with it and seeing how the equation changes at different points... bridging the gap between the mathematics and the physical representation."

Moreover, Student 4 pointed out the advantage of XR technology in exploring scenarios that are difficult to visualize, noting:

> "So, for example, if you're wanting to show somebody how a wave function behaves in an infinite square well or something, you'd show that, but then they might be able to change the dimensions of the box, change the energy level, change the actual potential. So it would be a focused thing that they want to do, but they'd be able to look around it as well."

Visualizing changes in the physical environment linked to underlying equations was deemed invaluable in fostering a deeper comprehension and intuition for abstract mathematical concepts. The integration of XR technology was seen as a way to make these abstract concepts more tangible and accessible, thereby enhancing students' overall understanding and learning experience.

3.2.4 Collaboration

Collaborative opportunities within XR environments emerged as a recurring theme, with participants highlighting the value of interactive group problem-solving scenarios and shared exploration of key concepts alongside peers. The collaborative nature of XR technology was identified as facilitating peer-to-peer interaction but also engendering a sense of camaraderie and collective engagement in the learning process.

Student 1 highlighted how XR technology can foster group learning, stating:

> "As a group, you could work together to try and... Like some person might have an idea of, oh, this will work if we do this. Somebody else might not have a clue about that, but they'll know something else, and they can work together to build this... There's an element of peer learning in there as well."

Student 4 also envisioned XR technology as an effective tool for collaborative exploration in tutorials:

> "In a tutorial setting, it's going to work really well because a lot of the tutorial questions were like, what if this happens and if you change this little thing? And I think it's going to be really interesting to see how we can do the tutorial questions using something like that—something where we can change the parameters that we can't really do in an experimental setting."

Participants envisioned collaborative XR experiences as an effective means of fostering teamwork, communication, and mutual learning, thereby enriching their educational experiences.

3.2.5 Technological Concerns

Finally, participants expressed concerns regarding the maturity of current XR technology, particularly in terms of seamlessness and processing power limitations. The current cost of the equipment was also identified as a significant barrier, as well as logistical issues around using and maintaining the equipment.

Student 2 highlighted concerns about the current limitations of XR technology, particularly regarding field of view and user experience:

> "I feel again (a) bigger field of view I think is the main (issue)... It definitely has a future but it feels a few years off now. And I feel, I don't know, maybe ten years' time or something it would have a place but I think currently the university would be better to put the money elsewhere."

Similarly, Student 8 discussed the frustration and technical challenges that can arise with current XR technology:

> "I thought the field of view of the device was quite restricting. I can probably find that quite frustrating especially when the back plane of the content was also view... I think frustration and technical challenge are probably the biggest hurdles that are going to need to be jumped until this becomes kind of popular."

Additionally, Student 5 pointed out concerns related to the cost and the need for fine-tuning of XR technology:

> "There was definitely some things that could be fine-tuned... And I suppose the price aspect of getting each student a Hololens and setting up the rooms and all these stuff. But yeah, I think that's it."

Despite recognizing the transformative potential of XR technology in education, participants highlighted the need for continued refinement and advancement to address existing technological constraints and optimize the user experience. The prominence of this theme underscored the importance of ongoing research and development efforts in enhancing the usability and effectiveness of XR technology within educational contexts.

4 Conclusions

This pilot project has demonstrated an enthusiasm within the participant group for developing high-quality, immersive teaching tools for undergraduate physics students, highlighting the significant potential of XR technology to enhance physics education. The AR optical table was deemed effective for increasing student engagement, understanding, and enjoyment in learning complex physics concepts. While there were some concerns about the current cost and accessibility of AR devices, these challenges are typical for emerging technologies and are expected to diminish as the technology matures and becomes more widespread.

Participants were overwhelmingly positive about the potential of XR technology in physics education. They expressed enthusiasm for future developments that could further integrate XR into educational practices. Looking ahead, VR technology holds significant promise, enabling users to explore environments governed by different physical laws and overcome the limitations of traditional teaching methods. Similarly, AR offers unique opportunities to enhance learning by overlaying virtual information onto real-world objects, particularly in laboratory settings. This study highlights the importance of continued research and development to fully explore the potential of XR using HMDs in physics education.

Acknowledgements We gratefully acknowledge the Durham Centre for Academic Development (DCAD) Collaborative Innovation Grants, and especially Ross Parker, for funding and supporting this project. Special thanks to Benjamin J. Shattock and Christopher E. Teo for their coding expertise, and to Dr. Craig D. Stewart for providing the Hololens HMD, which enabled the timely completion of this work.

5 Survey Questions

5.1 Personal Information

1. Please enter your name. You will be assigned a pseudonym to keep your responses anonymous when analyzed.
2. How would you describe your gender? If you prefer to self-describe, please specify under Other.

- Male
- Female
- Prefer not to say
- Other: _______________

3. Are you a student or staff member?

- UG Student
- PG Student
- Staff member

4. What role do you currently hold?

- Academic Staff
- Research Staff
- Professional Services Staff
- Other: _______________

5. How long have you been involved with physics teaching?

- 0–5 years
- 5–10 years
- 10–15 years
- 15+ years
- Other: _______________
- N/A

6. What year of study are you currently in?

- First
- Second
- Third
- Fourth
- Other: _______________

7. Have you ever used an AR or VR device before?

- AR headset
- VR headset
- Neither
- Both

8. If you have used an AR or VR device before, please indicate how below:

- VR gaming: Frequent use/Occasional use/Used once or a few times/No use/N/A
- VR development: Frequent use/Occasional use/Used once or a few times/No use/N/A
- AR gaming: Frequent use/Occasional use/Used once or a few times/No use/N/A

- AR professional/development: Frequent use/Occasional use/Used once or a few times/No use/N/A

9. Which device did you test today?

- Microsoft Hololens
- Meta Quest 2
- Both

5.2 *Evaluation of AR Technology*

For each of the following statements, please indicate your level of agreement using the following scale: **Strongly Agree (SA)/Agree (A)/Neither Agree nor Disagree (N)/Disagree (D)/Strongly Disagree (SD)**.

- The AR simulation was enjoyable to use.
- The hand controls were easy to use, once I had gotten used to them.
- AR technology could be a valuable teaching tool.
- Using a simulation such as this would improve my conceptual understanding of basic optics.
- I feel that the use of AR simulations offers additional advantages over solely lab-based experiments.
- I feel that the use of AR simulations offers additional advantages over solely paper-based problem-solving classes.
- **(For staff and PG students only)** If a dedicated AR classroom and appropriate support existed, I would be interested in using AR materials in my own teaching in future.

5.3 *Final Thoughts*

1. Do you have any further comments to make about your experiences today and how this technology might be used in physics undergraduate teaching?
2. Do you think this is a technology worth exploring in relation to undergraduate physics teaching?

References

Altmeyer K, Kapp S, Thees M, Malone S, Kuhn J, Brünken R (2020) The use of augmented reality to foster conceptual knowledge acquisition in STEM laboratory courses—theoretical background and empirical results. Br J Educ Technol 51:611–628. https://doi.org/10.1111/bjet.12900

Azuma RT, Baillot Y, Behringer R, Feiner S, Julier S, MacIntyre B (2001) Recent advances in augmented reality. IEEE Comput Graph Appl 21(6):34–47. https://doi.org/10.1109/38.963459

Bacca J, Baldiris S, Fabregat R, Graf S (2014) Augmented reality trends in education: a systematic review of research and applications. Educ Technol Soc 17(4):133–149. https://www.jstor.org/stable/jeductechsoci.17.4.133

Blender Foundation: Blender (Version 3.2) (2024). https://www.blender.org. Computer software

Braun V, Clarke V (2021) Thematic analysis: a practical guide. SAGE Publications, London, UK. https://us.sagepub.com/en-us/nam/thematic-analysis/book248481

Buchner J, Buntins K, Kerres M (2022) The impact of augmented reality on cognitive load and performance: a systematic review. J Comput Assist Learn 38(2):285–303. https://doi.org/10.1111/jcal.12617

Carreon A, Smith SJ, Rowland A (2020) Augmented reality: creating and implementing digital classroom supports. J Spec Educ Technol 35:109–115. https://doi.org/10.1177/0162643419882423

Chang HY, Binali T, Liang JC, Chiou GL, Cheng KH, Lee SWY, Tsai CC (2022) Ten years of augmented reality in education: a meta-analysis of (quasi-) experimental studies to investigate the impact. Comput Educ 191. https://doi.org/10.1016/j.compedu.2022.104641

Creswell J, Creswell J (2017) Research design: qualitative, quantitative, and mixed methods approaches, 4th edn. SAGE Publications, Newbury Park

Dunleavy M, Dede C (2014) Augmented reality teaching and learning. In: Spector JM, Merrill MD, Elen J, Bishop MJ (eds) Handbook of research on educational communications and technology. Springer, New York, NY, pp 735–745. https://doi.org/10.1007/978-1-4614-3185-5_59

Fombona-Pascual A, Fombona J, Vicente R (2022) Augmented reality: a review of a way to represent and manipulate 3D chemical structures. J Chem Inf Model 62:1863–1872. https://doi.org/10.1021/acs.jcim.1c01255

Ibáñez MB, Delgado-Kloos C (2018) Augmented reality for STEM learning: a systematic review. Comput Educ 123:109–123. https://doi.org/10.1016/j.compedu.2018.05.002

Jdaitawi MT, Kan'an AF (2021) A decade of research on the effectiveness of augmented reality on students with special disability in higher education. Contemp Educ Technol 14:ep332. https://doi.org/10.30935/cedtech/11369

Kaufmann H, Dünser A (2007) Summary of usability evaluations of an educational augmented reality application. In: Shumaker R (ed) Virtual reality, vol 4563. Springer, Berlin, Heidelberg, pp 660–669. https://doi.org/10.1007/978-3-540-73335-5_71

Kumar A, Mantri A, Singh G, Kaur DP (2022) Impact of AR-based collaborative learning approach on knowledge gain of engineering students in embedded system course. Educ Inf Technol 27:6015–6036. https://doi.org/10.1007/s10639-021-10858-9

Laumann D, Schlummer P, Abazi A, Borkamp R, Lauströer J, Pernice W, Schuck C, Schulz-Schaeffer R, Heusler S (2024) Analyzing the effective use of augmented reality glasses in university physics laboratory courses for the example topic of optical polarization. J Sci Educ Technol 33:668–685. https://doi.org/10.1007/s10956-024-10112-0

Li KC, Wong BTM (2021) Features and trends of personalised learning: a review of journal publications from 2001 to 2018. Interact Learn Environ 29:182–195. https://doi.org/10.1080/10494820.2020.1811735

Malone S, Garzón J, Kuhn J (2023) Three decades of augmented reality in education: a second-order meta-analysis and research synthesis. https://doi.org/10.31219/osf.io/amw4t. OSF Preprints

Milgram P, Takemura H, Utsumi A, Kishino F (1995) Augmented reality: a class of displays on the reality-virtuality continuum. In: Proceedings of the SPIE telemanipulator and telepresence technology, vol 2351. SPIE, Boston, MA, pp 282–292. https://doi.org/10.1117/12.197321

Petts PL (2023) Extended reality (XR) unique hooks. https://doi.org/10.25416/NTR.24707163.v2

Phon DNE, Ali MB, Halim NDA (2014) Collaborative augmented reality in education: a review. In: International Conference on Teaching and Learning in Computing and Engineering. IEEE, Kuching, Malaysia, pp 78–83. https://doi.org/10.1109/LaTiCE.2014.23
PTC Inc. (2022) Vuforia SDK (Version 10.7). Software Development Kithttps://developer.vuforia.com/
Szalavári Z, Schmalstieg D, Fuhrmann A, Gervautz M (1998) Studierstube: an environment for collaboration in augmented reality. Virtual Reality 3(1):37–48. https://doi.org/10.1007/BF01409796
Tang A, Owen C, Biocca F, Mou W (2003) Comparative effectiveness of augmented reality in object assembly. In: Proceedings of the SIGCHI conference on human factors in computing systems, pp 73–80. https://doi.org/10.1145/642611.642626
Unity Technologies (2024) Unity game engine (version 2024.1.0f1). https://unity.com/
Wijdenes P, Borkenhagen D, Babione J, Ma I, Hallihan G (2018) Leveraging augmented reality training tool for medical education: a case study in central venous catheterization. In: Extended abstracts of the 2018 CHI conference on human factors in computing systems. ACM, Montreal, QC, Canada, pp 1–7. https://doi.org/10.1145/3170427.3174361
Xie H, Chu HC, Hwang GJ, Wang CC (2019) Trends and development in technology-enhanced adaptive/personalized learning: a systematic review of journal publications from 2007 to 2017. Comput Educ 140:103599. https://doi.org/10.1016/j.compedu.2019.103599
Zhang J, Li G, Huang Q, Feng Q, Luo H (2022) Augmented reality in K–12 education: a systematic review and meta-analysis of the literature from 2000 to 2020. Sustainability 14(15):9725. https://doi.org/10.3390/su14159725

MIX
Papier aus verantwortungsvollen Quellen
Paper from responsible sources
FSC® C105338

If you have any concerns about our products,
you can contact us on
ProductSafety@springernature.com

In case Publisher is established outside the EU,
the EU authorized representative is:
Springer Nature Customer Service Center GmbH
Europaplatz 3, 69115 Heidelberg, Germany

Printed by Libri Plureos GmbH
in Hamburg, Germany